Teacher, Student, and Parent One-Stop Internet Resources

Log on to booka.msscience.com

ONLINE STUDY TOOLS

- Section Self-Check Quizzes
- Interactive Tutor
- Chapter Review Tests
- Standardized Test Practice
- Vocabulary PuzzleMaker

ONLINE RESEARCH

- WebQuest Projects
- Prescreened Web Links
- Career Links
- Internet Labs

INTERACTIVE ONLINE STUDENT EDITION

- Complete Interactive Student Edition available at mhln.com

FOR TEACHERS

- Teacher Bulletin Board
- Teaching Today—Professional Development

SAFETY SYMBOLS

SAFETY SYMBOLS	HAZARD	EXAMPLES	PRECAUTION	REMEDY
DISPOSAL	Special disposal procedures need to be followed.	certain chemicals, living organisms	Do not dispose of these materials in the sink or trash can.	Dispose of wastes as directed by your teacher.
BIOLOGICAL	Organisms or other biological materials that might be harmful to humans	bacteria, fungi, blood, unpreserved tissues, plant materials	Avoid skin contact with these materials. Wear mask or gloves.	Notify your teacher if you suspect contact with material. Wash hands thoroughly.
EXTREME TEMPERATURE	Objects that can burn skin by being too cold or too hot	boiling liquids, hot plates, dry ice, liquid nitrogen	Use proper protection when handling.	Go to your teacher for first aid.
SHARP OBJECT	Use of tools or glassware that can easily puncture or slice skin	razor blades, pins, scalpels, pointed tools, dissecting probes, broken glass	Practice common-sense behavior and follow guidelines for use of the tool.	Go to your teacher for first aid.
FUME	Possible danger to respiratory tract from fumes	ammonia, acetone, nail polish remover, heated sulfur, moth balls	Make sure there is good ventilation. Never smell fumes directly. Wear a mask.	Leave foul area and notify your teacher immediately.
ELECTRICAL	Possible danger from electrical shock or burn	improper grounding, liquid spills, short circuits, exposed wires	Double-check setup with teacher. Check condition of wires and apparatus.	Do not attempt to fix electrical problems. Notify your teacher immediately.
IRRITANT	Substances that can irritate the skin or mucous membranes of the respiratory tract	pollen, moth balls, steel wool, fiberglass, potassium permanganate	Wear dust mask and gloves. Practice extra care when handling these materials.	Go to your teacher for first aid.
CHEMICAL	Chemicals can react with and destroy tissue and other materials	bleaches such as hydrogen peroxide; acids such as sulfuric acid, hydrochloric acid; bases such as ammonia, sodium hydroxide	Wear goggles, gloves, and an apron.	Immediately flush the affected area with water and notify your teacher.
TOXIC	Substance may be poisonous if touched, inhaled, or swallowed.	mercury, many metal compounds, iodine, poinsettia plant parts	Follow your teacher's instructions.	Always wash hands thoroughly after use. Go to your teacher for first aid.
FLAMMABLE	Flammable chemicals may be ignited by open flame, spark, or exposed heat.	alcohol, kerosene, potassium permanganate	Avoid open flames and heat when using flammable chemicals.	Notify your teacher immediately. Use fire safety equipment if applicable.
OPEN FLAME	Open flame in use, may cause fire.	hair, clothing, paper, synthetic materials	Tie back hair and loose clothing. Follow teacher's instruction on lighting and extinguishing flames.	Notify your teacher immediately. Use fire safety equipment if applicable.

Eye Safety
Proper eye protection should be worn at all times by anyone performing or observing science activities.

Clothing Protection
This symbol appears when substances could stain or burn clothing.

Animal Safety
This symbol appears when safety of animals and students must be ensured.

Handwashing
After the lab, wash hands with soap and water before removing goggles.

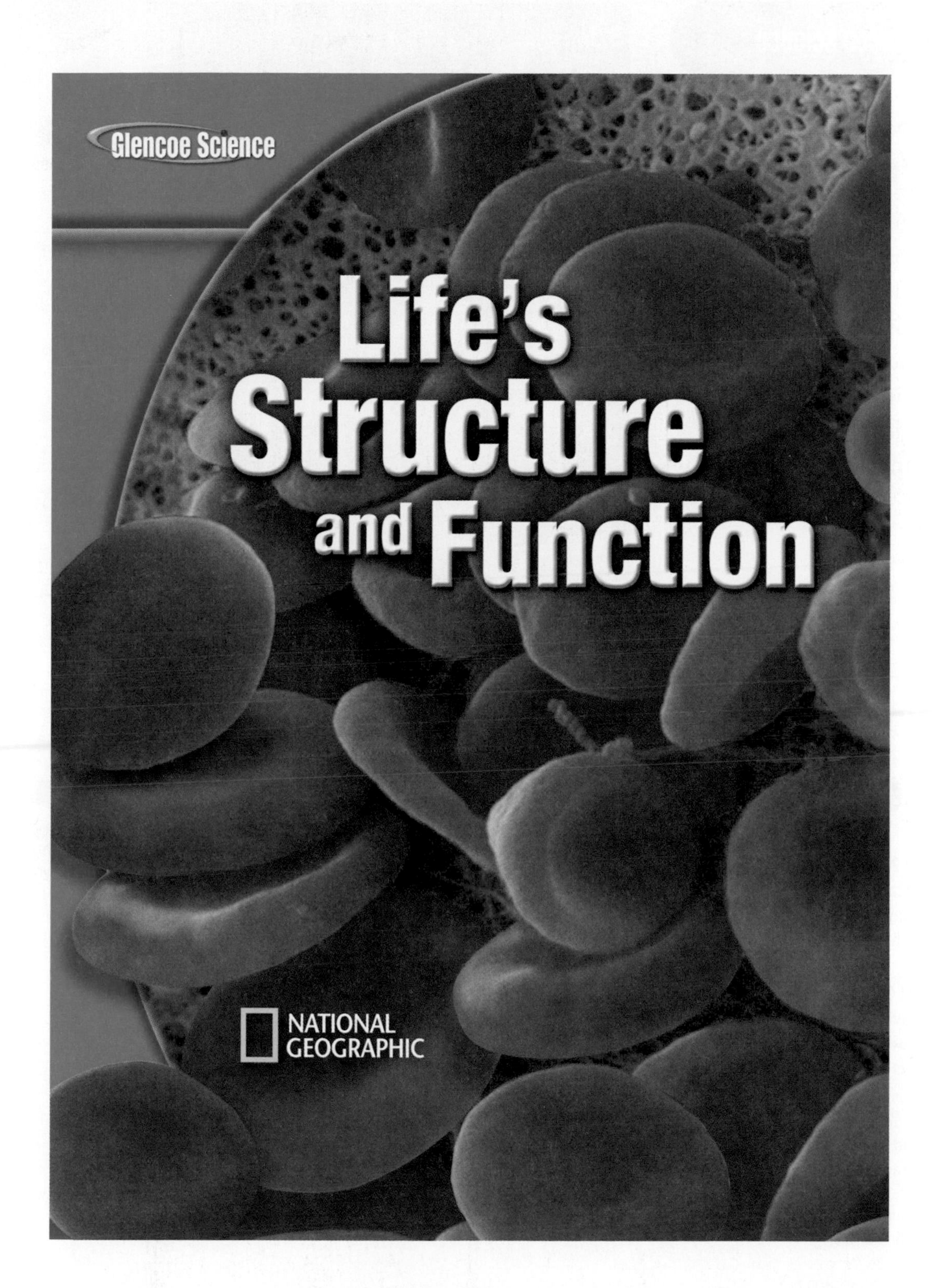

New York, New York Columbus, Ohio Chicago, Illinois Woodland Hills, California

Glencoe Science

Life's Structure and Function

These human red blood cells are part of a liquid tissue—blood. They deliver oxygen and remove wastes. The protein hemoglobin gives them their red color, and contains iron to transport oxygen and carbon dioxide.

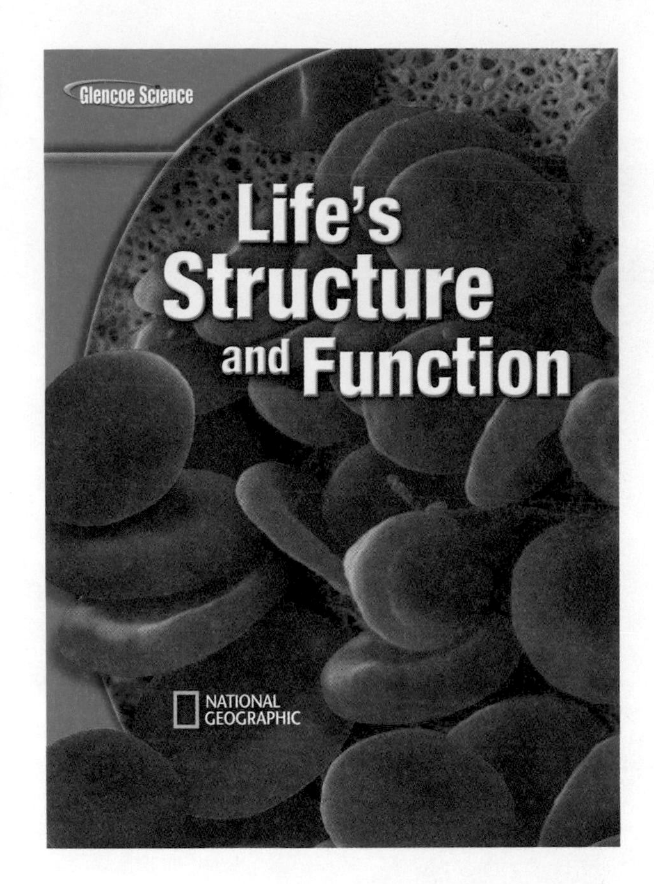

The McGraw·Hill Companies

Send all inquiries to:
Glencoe/McGraw-Hill
8787 Orion Place
Columbus, OH 43240-4027

ISBN: 978-0-07-877812-4
MHID: 0-07-877812-3

Printed in the United States of America.

6 7 8 9 10 DOW 12 11

Authors

Education Division
Washington, D.C.

Lucy Daniel, PhD
Teacher/Consultant
Rutherford County Schools
Rutherfordton, NC

Alton Biggs
Biology Teacher
Allen High School
Allen, TX

Dinah Zike
Educational Consultant
Dinah-Might Activities, Inc.
San Antonio, TX

Series Consultants

CONTENT

Connie Rizzo, MD, PhD
Department of Science/Math
Marymount Manhattan College
New York, NY

Dominic Salinas, PhD
Middle School Science Supervisor
Caddo Parish Schools
Shreveport, LA

MATH

Teri Willard, EdD
Mathematics Curriculum Writer
Belgrade, MT

READING

Elizabeth Babich
Special Education Teacher
Mashpee Public Schools
Mashpee, MA

Carol A. Senf, PhD
School of Literature,
Communication, and Culture
Georgia Institute of Technology
Atlanta, GA

SAFETY

Sandra West, PhD
Department of Biology
Texas State University-San Marcos
San Marcos, TX

ACTIVITY TESTERS

Nerma Coats Henderson
Pickerington Lakeview Jr. High School
Pickerington, OH

Mary Helen Mariscal-Cholka
William D. Slider Middle School
El Paso, TX

Science Kit and Boreal Laboratories
Tonawanda, NY

Series Reviewers

Maureen Barrett
Thomas E. Harrington Middle School
Mt. Laurel, NJ

Robin Dillon
Hanover Central High School
Cedar Lake, IN

Carolyn Elliott
South Iredell High School
Statesville, NC

Sueanne Esposito
Tipton High School
Tipton, IN

Cory Fish
Burkholder Middle School
Henderson, NV

Linda V. Forsyth
Retired Teacher
Merrill Middle School
Denver, CO

Michelle Mazeika
Whiting Middle School
Whiting, IN

Joe McConnell
Speedway Jr. High School
Indianapolis, IN

Amy Morgan
Berry Middle School
Hoover, AL

Mark Sailer
Pioneer Jr.-Sr. High School
Royal Center, IN

Dee Stout
Penn State University
University Park, PA

HOW TO...

Use Your Science Book

Why do I need my science book?

Have you ever been in class and not understood all of what was presented? Or, you understood everything in class, but at home, got stuck on how to answer a question? Maybe you just wondered when you were ever going to use this stuff?

These next few pages are designed to help you understand everything your science book can be used for . . . besides a paperweight!

Before You Read

- **Chapter Opener** Science is occurring all around you, and the opening photo of each chapter will preview the science you will be learning about. The **Chapter Preview** will give you an idea of what you will be learning about, and you can try the **Launch Lab** to help get your brain headed in the right direction. The **Foldables** exercise is a fun way to keep you organized.
- **Section Opener** Chapters are divided into two to four sections. The **As You Read** in the margin of the first page of each section will let you know what is most important in the section. It is divided into four parts. **What You'll Learn** will tell you the major topics you will be covering. **Why It's Important** will remind you why you are studying this in the first place! The **Review Vocabulary** word is a word you already know, either from your science studies or your prior knowledge. The **New Vocabulary** words are words that you need to learn to understand this section. These words will be in **boldfaced** print and highlighted in the section. Make a note to yourself to recognize these words as you are reading the section.

Glencoe Science
Life's Structure and Function
NATIONAL GEOGRAPHIC

As You Read

- **Headings** Each section has a title in large red letters, and is further divided into blue titles and small red titles at the beginnings of some paragraphs. To help you study, make an outline of the headings and subheadings.
- **Margins** In the margins of your text, you will find many helpful resources. The **Science Online** exercises and **Integrate** activities help you explore the topics you are studying. **MiniLabs** reinforce the science concepts you have learned.
- **Building Skills** You also will find an **Applying Math** or **Applying Science** activity in each chapter. This gives you extra practice using your new knowledge, and helps prepare you for standardized tests.
- **Student Resources** At the end of the book you will find **Student Resources** to help you throughout your studies. These include **Science, Technology,** and **Math Skill Handbooks,** an **English/Spanish Glossary,** and an **Index.** Also, use your **Foldables** as a resource. It will help you organize information, and review before a test.
- **In Class** Remember, you can always ask your teacher to explain anything you don't understand.

FOLDABLES™ Study Organizer

Science Vocabulary Make the following Foldable to help you understand the vocabulary terms in this chapter.

STEP 1 **Fold** a vertical sheet of notebook paper from side to side.

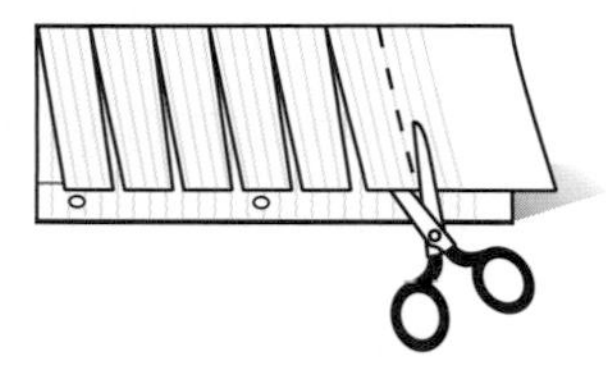

STEP 2 **Cut** along every third line of only the top layer to form tabs.

STEP 3 **Label** each tab with a vocabulary word from the chapter.

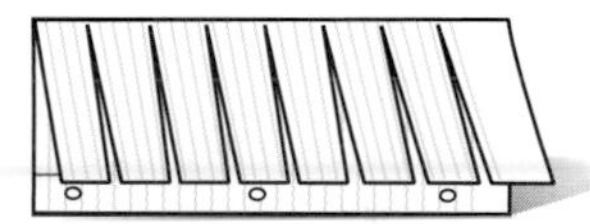

Build Vocabulary As you read the chapter, list the vocabulary words on the tabs. As you learn the definitions, write them under the tab for each vocabulary word.

In Lab

Working in the laboratory is one of the best ways to understand the concepts you are studying. Your book will be your guide through your laboratory experiences, and help you begin to think like a scientist. In it, you not only will find the steps necessary to follow the investigations, but you also will find helpful tips to make the most of your time.

- Each lab provides you with a **Real-World Question** to remind you that science is something you use every day, not just in class. This may lead to many more questions about how things happen in your world.
- Remember, experiments do not always produce the result you expect. Scientists have made many discoveries based on investigations with unexpected results. You can try the experiment again to make sure your results were accurate, or perhaps form a new hypothesis to test.
- Keeping a **Science Journal** is how scientists keep accurate records of observations and data. In your journal, you also can write any questions that may arise during your investigation. This is a great method of reminding yourself to find the answers later.

Before a Test

Admit it! You don't like to take tests! However, there *are* ways to review that make them less painful. Your book will help you be more successful taking tests if you use the resources provided to you.

- Review all of the **New Vocabulary** words and be sure you understand their definitions.
- Review the notes you've taken on your **Foldables,** in class, and in lab. Write down any question that you still need answered.
- Review the **Summaries** and **Self Check questions** at the end of each section.
- Study the concepts presented in the chapter by reading the **Study Guide** and answering the questions in the **Chapter Review.**

a or b?

?

T or F?

Look For...

- **Reading Checks** and **caption questions** throughout the text.
- the **Summaries** and **Self Check questions** at the end of each section.
- the **Study Guide** and **Review** at the end of each chapter.
- the **Standardized Test Practice** after each chapter.

Let's Get Started

To help you find the information you need quickly, use the Scavenger Hunt below to learn where things are located in Chapter 1.

1. What is the title of this chapter?
2. What will you learn in Section 1?
3. Sometimes you may ask, "Why am I learning this?" State a reason why the concepts from Section 2 are important.
4. What is the main topic presented in Section 2?
5. How many reading checks are in Section 1?
6. What is the Web address where you can find extra information?
7. What is the main heading above the sixth paragraph in Section 2?
8. There is an integration with another subject mentioned in one of the margins of the chapter. What subject is it?
9. List the new vocabulary words presented in Section 2.
10. List the safety symbols presented in the first Lab.
11. Where would you find a Self Check to be sure you understand the section?
12. Suppose you're doing the Self Check and you have a question about concept mapping. Where could you find help?
13. On what pages are the Chapter Study Guide and Chapter Review?
14. Look in the Table of Contents to find out on which page Section 2 of the chapter begins.
15. You complete the Chapter Review to study for your Chapter test. Where could you find another quiz for more practice?

Teacher Advisory Board

The Teacher Advisory Board gave the editorial staff and design team feedback on the content and design of the Student Edition. They provided valuable input in the development of the 2008 edition of ***Glencoe Science.***

John Gonzales
Challenger Middle School
Tucson, AZ

Rachel Shively
Aptakisic Jr. High School
Buffalo Grove, IL

Roger Pratt
Manistique High School
Manistique, MI

Kirtina Hile
Northmor Jr. High/High School
Galion, OH

Marie Renner
Diley Middle School
Pickerington, OH

Nelson Farrier
Hamlin Middle School
Springfield, OR

Jeff Remington
Palmyra Middle School
Palmyra, PA

Erin Peters
Williamsburg Middle School
Arlington, VA

Rubidel Peoples
Meacham Middle School
Fort Worth, TX

Kristi Ramsey
Navasota Jr. High School
Navasota, TX

Student Advisory Board

The Student Advisory Board gave the editorial staff and design team feedback on the design of the Student Edition. We thank these students for their hard work and creative suggestions in making the 2008 edition of ***Glencoe Science*** student friendly.

Jack Andrews
Reynoldsburg Jr. High School
Reynoldsburg, OH

Peter Arnold
Hastings Middle School
Upper Arlington, OH

Emily Barbe
Perry Middle School
Worthington, OH

Kirsty Bateman
Hilliard Heritage Middle School
Hilliard, OH

Andre Brown
Spanish Emersion Academy
Columbus, OH

Chris Dundon
Heritage Middle School
Westerville, OH

Ryan Manafee
Monroe Middle School
Columbus, OH

Addison Owen
Davis Middle School
Dublin, OH

Teriana Patrick
Eastmoor Middle School
Columbus, OH

Ashley Ruz
Karrer Middle School
Dublin, OH

The Glencoe middle school science Student Advisory Board taking a timeout at COSI, a science museum in Columbus, Ohio.

Contents

Contents

Nature of Science: Land Use in Floodplains—2

In each chapter, look for these opportunities for review and assessment:
- Reading Checks
- Caption Questions
- Section Review
- Chapter Study Guide
- Chapter Review
- Standardized Test Practice
- Online practice at booka.msscience.com

Get Ready to Read Strategies

Student Resources

Cross-Curricular Readings/Labs

DVD available as a video lab

NATIONAL GEOGRAPHIC **VISUALIZING**

TIME SCIENCE AND Society

TIME SCIENCE AND HISTORY

Oops! Accidents in SCIENCE

Science and Language Arts

SCIENCE Stats

Launch LAB

Mini LAB

Mini LAB Try at Home

One-Page Labs

Two Page Labs

Labs/Activities

Design Your Own Labs

Use the Internet Labs

Applying Math

Applying Science

INTEGRATE

Standardized Test Practice

The NATURE OF SCIENCE

Science Today

Genome Sequencing

Your genome determines your traits—everything from your eye color and blood type to the likelihood that you might get certain diseases. A genome is all of the DNA in a one-celled or many-celled organism. Each gene plays a part in the expression of a specific trait. Organisms like protists, fungi, plants, and animals have their genes on chromosomes in the nucleus of cells. In the human genome, there are about 30,000 genes on 23 pairs of chromosomes.

Sequencing the genome of any organism—bacteria, protist, fungus, plant, or animal—is complex. Each gene's message involves four chemicals called bases—adenine (A), cytosine (C), guanine (G), and thymine (T). The bases are linked in pairs—adenine with thymine and cytosine with guanine. Each gene is a unique chain of paired bases. The average size of a human gene is about 3,000 paired bases. The sequence carries instructions for making a specific protein. Depending on the string of bases in a gene, the protein might control the formation of a certain type of tissue or it might be an enzyme that drives a biochemical reaction. Many human disorders and diseases, including Huntington's Disease and sickle-cell disease, are the result of a person's genetic makeup.

Figure 1 The DNA in your cells makes up your genetic material.

Figure 2 99.99% of all human genes are the same from individual to individual. It takes only 0.01% of your genes for your unique combination of traits.

Decoding the Genome

To decode a genome, scientists first have to identify the bases and their correct sequence. Then they must determine which parts of the sequence are genes. Only a small percentage of the human genome are useful genes.

Powerful supercomputers and inventive software make it possible to collect, sequence, and analyze genetic data faster than ever before. In one method, scientists mark chromosomes and then cut them into manageable fragments. Through chemical processes, each of the bases is dyed a different color then displayed in a pattern read by super-fast machines. The machines convert the base sequences into digital data. A supercomputer then puts the fragments in the proper sequence, using markers from the first stage of the process.

By March 2001, the complete genome of some organisms in every kingdom—eubacteria, archaebacteria, protists, fungi, plants, and animals—was known, including the human genome. These accomplishments would not have been possible without modern computer technology and the research of many scientists worldwide.

Figure 3 The genetic material passed from parents to offspring determines individual characteristics.

Science

Scientists often collect and analyze data to find answers to questions or solve problems. When you collect data and analyze it to answer questions or solve problems, you are doing science.

Genetics is a part of life science, the study of all Earth's living organisms. It includes zoology and botany. In this book, you'll learn about the structures that make up organisms, including genes and their functions.

Science Today

Through science, scientists now have a better understanding than ever before of the world and its inhabitants. New scientific discoveries build on many that came before them. Future scientists might work to understand the long lines of genetic code that today's scientists have uncovered. Each step in decoding genomes will add to the understanding of what each of the genes does. Understanding the human genome is an important key to solving many medical problems. Some day, it might be possible to genetically identify health problems in advance and treat them before they develop.

Improved technology is another important part of science. Many advances in science would not be possible without new equipment to perform experiments and collect data. Sequencing the human genome would be impossible without the technology that allows scientists to see, manipulate, and record the genetic material. Sequencing machines and supercomputers have allowed scientists to map the human genome more quickly and accurately than was previously possible. The transfer of computerized data over the Internet also has made it possible for the Human Genome Project's scientists to share the details of their results instantly.

Figure 4 These machines, called 3700s, ran nonstop to sequence the human genome. They filter the DNA and digitally record it.

Benefits for Society

New scientific discoveries often benefit society. Science has made work easier, has helped to keep people safer, and led to medical advances that allow people to live longer and healthier lives. Scientists working on the Human Genome Project hope that their work will help provide a better explanation of how living organisms are constructed and how they function. A complete understanding of the genome will tell us more about the physical makeup of humans than we have ever known before.

Where Do Scientists Work?

Scientists work in a variety of places for a variety of reasons. Most of the American scientists who sequenced the human genome worked for either the United States government as part of the federally funded Human Genome Project, or for private companies. Both groups have the same goal even though their methods vary. These scientists worked with sensitive equipment in university laboratories and research centers. Scientists, who apply genetic research to medicine, work with the data provided by the new genetic maps. They might test hypotheses about genetic therapy through clinical experimentation.

Figure 5 Some results from the Human Genome Project are available on the Internet.

A Geneticist

Dr. John Carpten is a molecular geneticist working on the Human Genome Project. His focus is on the gene that causes some men to be at higher risk for developing prostate cancer. Geneticists like Dr. Carpten may do research in laboratories, analyze data in lab settings, or do diagnostic work with patients.

Dr. John Carpten

Figure 6 Some day, a physician might consult a patient's genome before prescribing medicine or other treatment for a disease.

The human genome is sequenced and the locations of genes for human traits and illnesses are known. If you had the power to choose a project that uses the genome map to create something new or solve a problem, what would you choose to do and why? How would the sequenced genome help you?

chapter 1

Exploring and Classifying Life

The BIG Idea

Life science includes the study of living and once-living things.

SECTION 1
What is science?
Main Idea Science is an organized way of studying things and finding answers to questions.

SECTION 2
Living Things
Main Idea Living things have certain characteristics in common.

SECTION 3
Where does life come from?
Main Idea There are many hypotheses about the origins of life.

SECTION 4
How are living things classified?
Main Idea Classification systems show relationships among living things.

Life Under the Sea

This picture contains many living things—including living coral. These living things have both common characteristics and differences. Scientists classify life according to similarities.

Science Journal List three characteristics that you would use to classify underwater life.

Start-Up Activities

Classify Organisms

Life scientists discover, describe, and name hundreds of organisms every year. How do they decide if a certain plant belongs to the iris or orchid family of flowering plants, or if an insect is more like a grasshopper or a beetle?

1. Observe the organisms on the opposite page or in an insect collection in your class.
2. Decide which feature could be used to separate the organisms into two groups, then sort the organisms into the two groups.
3. Continue to make new groups using different features until each organism is in a category by itself.
4. **Think Critically** How do you think scientists classify living things? List your ideas in your Science Journal.

Preview this chapter's content and activities at booka.msscience.com

Vocabulary Make the following Foldable to help you understand the vocabulary terms in this chapter.

STEP 1 **Fold** a vertical sheet of notebook paper from side to side.

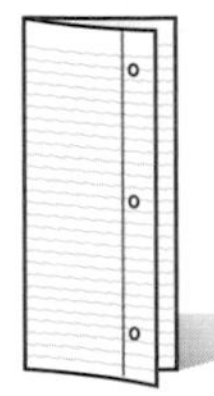

STEP 2 **Cut** along every third line of only the top layer to form tabs.

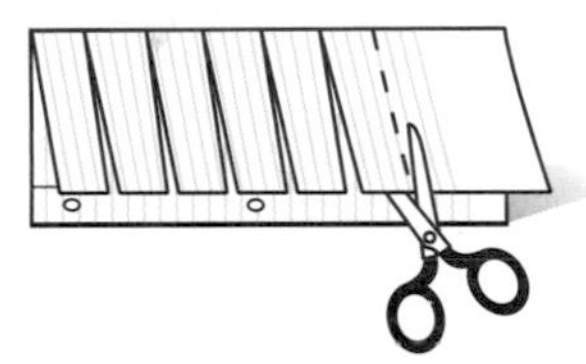

STEP 3 **Label** each tab.

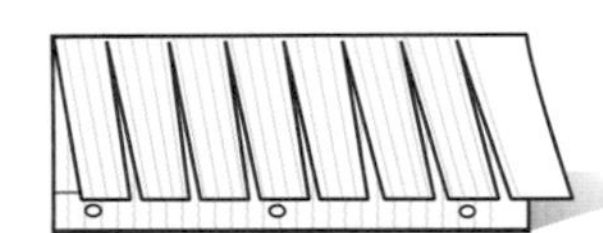

Build Vocabulary As you read the chapter, write the vocabulary words on the tabs. As you learn the definitions, write them under the tab for each vocabulary word.

Get Ready to Read

Preview

① Learn It! If you know what to expect before reading, it will be easier to understand ideas and relationships presented in the text. Follow these steps to preview your reading assignments.

1. Look at the title and any illustrations that are included.
2. Read the headings, subheadings, and anything in bold letters.
3. Skim over the passage to see how it is organized. Is it divided into many parts?
4. Look at the graphics—pictures, maps, or diagrams. Read their titles, labels, and captions.
5. Set a purpose for your reading. Are you reading to learn something new? Are you reading to find specific information?

② Practice It! Take some time to preview this chapter. Skim all the main headings and subheadings. With a partner, discuss your answers to these questions.

- Which part of this chapter looks most interesting to you?
- Are there any words in the headings that are unfamiliar to you?
- Choose one of the lesson review questions to discuss with a partner.

③ Apply It! Now that you have skimmed the chapter, write a short paragraph describing one thing you want to learn from this chapter.

As you preview this chapter, be sure to scan the illustrations, tables, and graphs. Skim the captions.

Target Your Reading

Use this to focus on the main ideas as you read the chapter.

1. **Before you read** the chapter, respond to the statements below on your worksheet or on a numbered sheet of paper.
   - Write an **A** if you **agree** with the statement.
   - Write a **D** if you **disagree** with the statement.

2. **After you read** the chapter, look back to this page to see if you've changed your mind about any of the statements.
   - If any of your answers changed, explain why.
   - Change any false statements into true statements.
   - Use your revised statements as a study guide.

Before You Read A or D	Statement	After You Read A or D
	1 If not supported by evidence collected over time, scientists reject a theory.	
	2 Some living things do not require water to survive.	
	3 There is just one way to approach a scientific problem.	
	4 Some organisms grow by enlarging cells.	
	5 Following safety rules in lab not only protects you but your classmates as well.	
	6 All living things use energy.	
	7 Living things can grow spontaneously from nonliving things.	
	8 Scientists rarely repeat experiments.	
	9 An organism's classification can change with the discovery of new information.	
	10 All living things reproduce.	

Science Online
Print out a worksheet of this page at booka.msscience.com

section 1

What is science?

as you read

What You'll Learn

- **Apply** scientific methods to problem solving.
- **Demonstrate** how to measure using scientific units.

Why It's Important

Learning to use scientific methods will help you solve ordinary problems in your life.

Review Vocabulary
experiment: using controlled conditions to test a hypothesis

New Vocabulary
- scientific methods
- hypothesis
- control
- variable
- theory
- law

The Work of Science

Movies and popcorn seem to go together. So before you and your friends watch a movie, sometimes you pop some corn in a microwave oven. When the popping stops, you take out the bag and open it carefully. You smell the mouthwatering, freshly popped corn and avoid hot steam that escapes from the bag. What makes the popcorn pop? How do microwaves work and make things hot? By the way, what are microwaves anyway?

Asking questions like these is one way scientists find out about anything in the world and the universe. Science is often described as an organized way of studying things and finding answers to questions.

Types of Science Many types of science exist. Each is given a name to describe what is being studied. For example, energy and matter have a relationship. That's a topic for physics. A physicist could answer most questions about microwaves.

On the other hand, a life scientist might study any of the millions of different animals, plants, and other living things on Earth. Look at the objects in **Figure 1.** What do they look like to you? A life scientist could tell you that some of the objects are living plants and some are just rocks. Life scientists who study plants are botanists, and those who study animals are zoologists. What do you suppose a bacteriologist studies?

Figure 1 Examine the picture carefully. Some of these objects are actually *Lithops* plants. They commonly are called stone plants and are native to deserts in South Africa.

Critical Thinking

Whether or not you become a trained scientist, you are going to solve problems all your life. You probably solve many problems every day when you sort out ideas about what will or won't work. Suppose your CD player stops playing music. To figure out what happened, you have to think about it. That's called critical thinking, and it's the way you use skills to solve problems.

If you know that the CD player does not run on batteries and must be plugged in to work, that's the first thing you check to solve the problem. You check and the player is plugged in so you eliminate that possible solution. You separate important information from unimportant information—that's a skill. Could there be something wrong with the first outlet? You plug the player into a different outlet, and your CD starts playing. You now know that it's the first outlet that doesn't work. Identifying the problem is another skill you have.

Solving Problems

Scientists use the same types of skills that you do to solve problems and answer questions. Although scientists don't always find the answers to their questions, they always use critical thinking in their search. Besides critical thinking, solving a problem requires organization. In science, this organization often takes the form of a series of procedures called **scientific methods. Figure 2** shows one way that scientific methods might be used to solve a problem.

Figure 2 The series of procedures shown below is one way to use scientific methods to solve a problem.

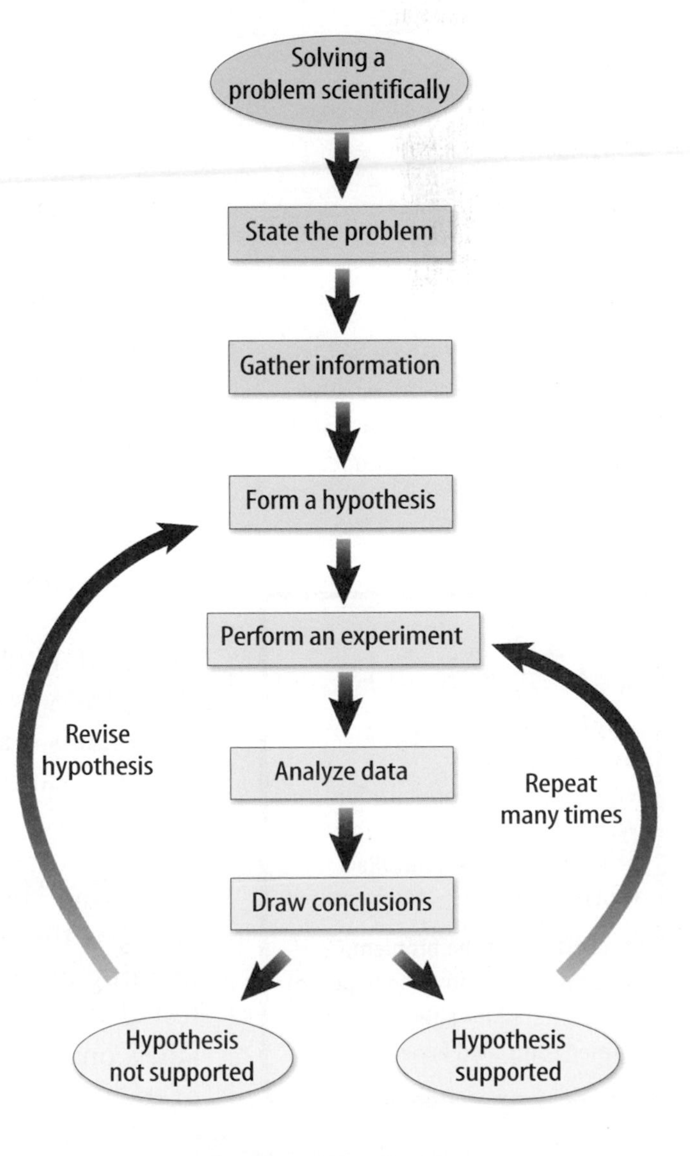

State the Problem Suppose a veterinary technician wanted to find out whether different types of cat litter cause irritation to cats' skin. What would she do first? The technician begins by observing something she cannot explain. A pet owner brings his four cats to the clinic to be boarded while he travels. He leaves his cell phone number so he can be contacted if any problems arise. When they first arrive, the four cats seem healthy. The next day however, the technician notices that two of the cats are scratching and chewing at their skin. By the third day, these same two cats have bare patches of skin with red sores. The technician decides that something in the cats' surroundings or their food might be irritating their skin.

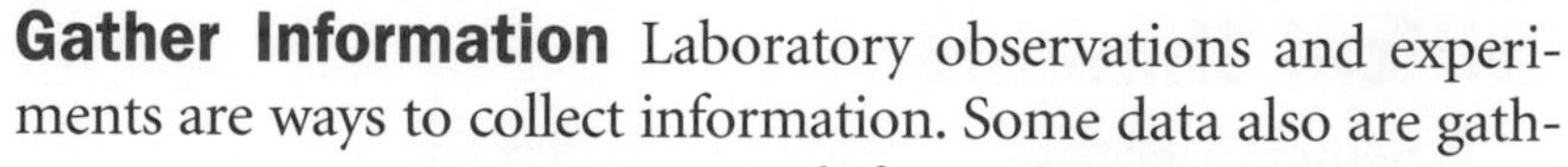

Gather Information Laboratory observations and experiments are ways to collect information. Some data also are gathered from fieldwork. Fieldwork includes observations or experiments that are done outside of the laboratory. For example, the best way to find out how a bird builds a nest is to go outside and watch it. **Figure 3** shows some ways data can be gathered.

Laboratory investigations

Computer models

Fieldwork

Figure 3 Observations can be made in many different settings. **List** *three other places where scientific observations can be made.*

The technician gathers information about the problem by watching the cats closely for the next two days. She knows that cats sometimes change their behavior when they are in a new place. She wants to see if the behavior of the cats with the skin sores seems different from that of the other two cats. Other than the scratching and chewing at their skin, all four cats' behavior seems to be the same.

The technician calls the owner and tells him about the problem. She asks him what brand of cat food he feeds his cats. Because his brand is the same one used at the clinic, she decides that food is not the cause of the skin irritation. She decides that the cats probably are reacting to something in their surroundings. There are many things in the clinic that the cats might react to. How does she decide what it is?

During her observations she notices that the cats seem to scratch and chew themselves most after using their litter boxes. The cat litter used by the clinic contains a deodorant. The technician calls the owner and finds out that the cat litter he buys does not contain a deodorant.

Topic: Controlled Experiments
Visit booka.msscience.com for Web links to information about how scientists use controlled experiments.

Activity List the problem, hypothesis, and how the hypothesis was tested for a recently performed controlled experiment.

Form a Hypothesis Based on this information, the next thing the veterinary technician does is form a hypothesis. A **hypothesis** is an explanation that can be tested. After discussing her observations with the clinic veterinarian, she hypothesizes that something in the cat litter is irritating the cats' skin.

Test the Hypothesis with an Experiment The technician gets the owner's permission to test her hypothesis by performing an experiment. In an experiment, the hypothesis is tested using controlled conditions. The technician reads the labels on two brands of cat litter and finds that the ingredients of each are the same except that one contains a deodorant.

Controls The technician separates the cats with sores from the other two cats. She puts each of the cats with sores in a cage by itself. One cat is called the experimental cat. This cat is given a litter box containing the cat litter without deodorant. The other cat is given a litter box that contains cat litter with deodorant. The cat with deodorant cat litter is the control.

A **control** is the standard to which the outcome of a test is compared. At the end of the experiment, the control cat will be compared with the experimental cat. Whether or not the cat litter contains deodorant is the variable. A **variable** is something in an experiment that can change. An experiment should have only one variable. Other than the difference in the cat litter, the technician treats both cats the same.

Reading Check *How many variables should an experiment have?*

Analyze Data The veterinary technician observes both cats for one week. During this time, she collects data on how often and when the cats scratch or chew, as shown in **Figure 4.** These data are recorded in a journal. The data show that the control cat scratches and chews more often than the experimental cat does. The sores on the skin of the experimental cat begin to heal, but those on the control cat do not.

Draw Conclusions The technician then draws the conclusion—a logical answer to a question based on data and observation—that the deodorant in the cat litter probably irritated the skin of the two cats. To accept or reject the hypothesis is the next step. In this case, the technician accepts the hypothesis. If she had rejected it, new experiments would have been necessary.

Although the technician decides to accept her hypothesis, she realizes that to be surer of her results she should continue her experiment. She should switch the experimental cat with the control cat to see what the results are a second time. If she did this, the healed cat might develop new sores. She makes an ethical decision and chooses not to continue the experiment. Ethical decisions, like this one, are important in deciding what science should be done.

Analyzing Data

Procedure

1. Obtain a **pan balance.** Follow your teacher's instructions for using it.
2. Record all data in your **Science Journal.**
3. Measure and record the mass of a **dry sponge.**
4. Soak this sponge in **water.** Measure and record its mass.
5. Calculate how much water your sponge absorbed.
6. Combine the class data and calculate the average amount of water absorbed.

Analysis

What other information about the sponges might be important when analyzing the data from the entire class?

Figure 4 Collecting and analyzing data is part of scientific methods.

Report Results When using scientific methods, it is important to share information. The veterinary technician calls the cats' owner and tells him the results of her experiment. She tells him she has stopped using the deodorant cat litter.

The technician also writes a story for the clinic's newsletter that describes her experiment and shares her conclusions. She reports the limits of her experiment and explains that her results are not final. In science it is important to explain how an experiment can be made better if it is done again.

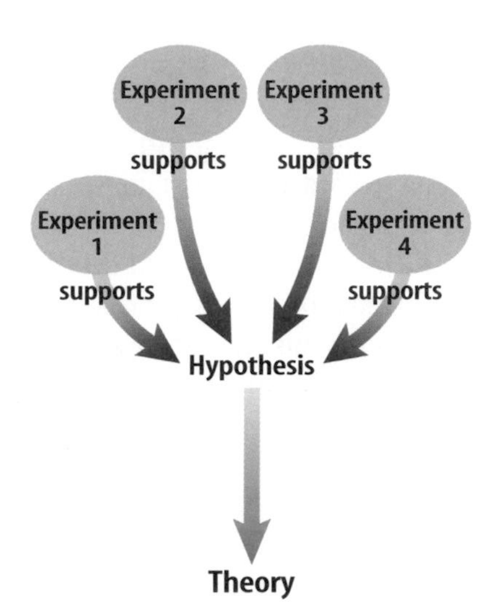

Figure 5 If data collected from several experiments over a period of time all support the hypothesis, it finally can be called a theory.

Developing Theories

After scientists report the results of experiments supporting their hypotheses, the results can be used to propose a scientific theory. When you watch a magician do a trick you might decide you have an idea or "theory" about how the trick works. Is your idea just a hunch or a scientific theory? A scientific **theory** is an explanation of things or events based on scientific knowledge that is the result of many observations and experiments. It is not a guess or someone's opinion. Many scientists repeat the experiment. If the results always support the hypothesis, the hypothesis can be called a theory, as shown in **Figure 5.**

Reading Check *What is a theory based on?*

A theory usually explains many hypotheses. For example, an important theory in life sciences is the cell theory. Scientists made observations of cells and experimented for more than 100 years before enough information was collected to propose a theory. Hypotheses about cells in plants and animals are combined in the cell theory.

A valid theory raises many new questions. Data or information from new experiments might change conclusions and theories can change. Later in this chapter you will read about the theory of spontaneous generation and how this theory changed as scientists used experiments to study new hypotheses.

Laws A scientific **law** is a statement about how things work in nature that seems to be true all the time. Although laws can be modified as more information becomes known, they are less likely to change than theories. Laws tell you what will happen under certain conditions but do not necessarily explain why it happened. For example, in life science you might learn about laws of heredity. These laws explain how genes are inherited but do not explain how genes work. Due to the great variety of living things, laws that describe them are few. It is unlikely that a law about how all cells work will ever be developed.

Scientific Methods Help Answer Questions You can use scientific methods to answer all sorts of questions. Your questions may be as simple as "Where did I leave my house key?" or as complex as "Will global warming cause the polar ice caps to melt?" You probably have had to find the answer to the first question. Someday you might try to find the answer to the second question. Using these scientific methods does not guarantee that you will get an answer. Often scientific methods just lead to more questions and more experiments. That's what science is about—continuing to look for the best answers to your questions.

Applying Science

Does temperature affect the rate of bacterial reproduction?

Some bacteria make you sick. Other bacteria, however, are used to produce foods like cheese and yogurt. Understanding how quickly bacteria reproduce can help you avoid harmful bacteria and use helpful bacteria. It's important to know things that affect how quickly bacteria reproduce. How do you think temperature will affect the rate of bacterial reproduction? A student makes the hypothesis that bacteria will reproduce more quickly as the temperature increases.

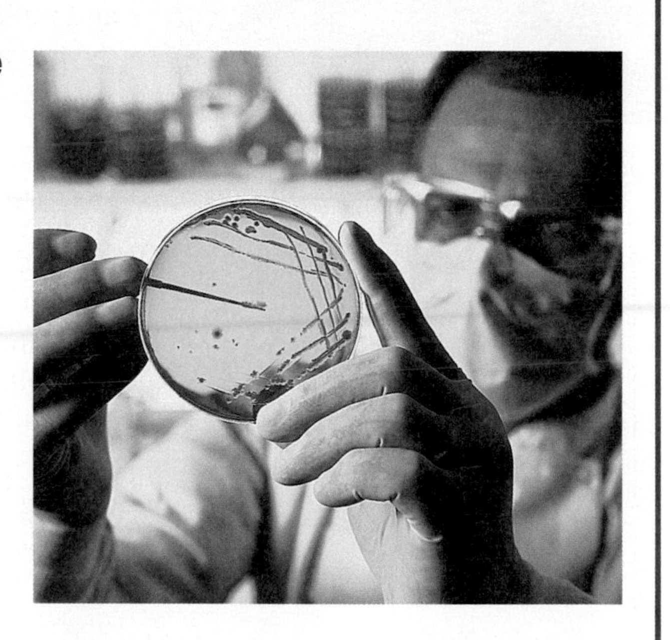

Identifying the Problem

The table below lists the reproduction-doubling rates at specific temperatures for one type of bacteria. A rate of 2.0 means that the number of bacteria doubled two times that hour (e.g., 100 to 200 to 400).

Bacterial Reproductive Rates

Temperature (°C)	Doubling Rate per Hour
20.5	2.0
30.5	3.0
36.0	2.5
39.2	1.2

Look at the table. What conclusions can you draw from the data?

Solving the Problem

1. Do the data in the table support the student's hypothesis?
2. How would you write a hypothesis about the relationship between bacterial reproduction and temperature?
3. Make a list of other factors that might have influenced the results in the table.
4. Are you satisfied with these data? List other things that you wish you knew.
5. Describe an experiment that would help you test these other ideas.

The label of this juice bottle shows you that it contains 473 mL of juice.

Figure 6 Your food often is measured in metric units. Nutritional information on the label is listed in grams or milligrams.

Measuring with Scientific Units

An important part of most scientific investigations is making accurate measurements. Think about things you use every day that are measured. Ingredients in your hamburger, hot dog, potato chips, or soft drink are measured in units such as grams and milliliters, as shown in **Figure 6.** The water you drink, the gas you use, and the electricity needed for a CD player are measured, too.

Why is it important to make accurate measurements?

In your classroom or laboratory this year, you will use the same standard system of measurement scientists use to communicate and understand each other's research and results. This system is called the International System of Units, or SI. For example, you may need to calculate the distance a bird flies in kilometers. Perhaps you will be asked to measure the amount of air your lungs can hold in liters or the mass of an automobile in kilograms. Some of the SI units are shown in **Table 1.**

Table 1 Common SI Measurements

Measurement	Unit	Symbol	Fractions and Multiples
Distance	Meter	m	1/1,000 m = 1 millimeter (mm) 1/100 m = 1 centimeter (cm) 1000 m = 1 kilometer (km)
Mass	Kilogram	kg	1/1,000 g = 1 milligram (mg) 1/1,000 kg = 1 gram (g) 1000 kg = 1 tonne (t) (metric ton)
Time	Second	s	60 s = 1 minute (min) 60 min = 1 hour (h)

Safety First

Doing science is usually much more interesting than just reading about it. Some of the scientific equipment that you will use in your classroom or laboratory is the same as what scientists use. Laboratory safety is important. In many states, a student can participate in a laboratory class only when wearing proper eye protection. Don't forget to wash your hands after handling materials. Following safety rules, as shown in **Figure 7,** will protect you and others from injury during your lab experiences. Symbols used throughout your text will alert you to situations that require special attention. Some of these symbols are shown below. A description of each symbol is in the Safety Symbols chart at the front of this book.

Figure 7 Proper eye protection should be worn whenever you see this safety symbol.
Predict *what might happen if you do not wear eye protection in the lab.*

section 1 review

Summary

The Work of Science

- Science is an organized way of studying things and finding answers to questions.

Solving Problems and Developing Theories

- Scientific methods are procedures used to solve problems and answer questions.
- A theory is an explanation based on many scientific observations.

Measuring with Scientific Units

- Scientists use the SI system for measurements.

Safety First

- Follow safety rules in the lab.

Self Check

1. **Describe** scientific methods.
2. **Infer** why it is important to test only one variable at a time during an experiment.
3. **Identify** the SI unit you would use to measure the width of your classroom.
4. **Compare and contrast** a theory with a hypothesis.
5. **Think Critically** Can the veterinary technician in this section be sure that deodorant caused the cats' skin problems? How could she improve her experiment?

Applying Skills

6. **Write a paper** that explains what the veterinary technician discovered from her experiment.

section 2

Living Things

as you read

What You'll Learn

- **Distinguish** between living and nonliving things.
- **Identify** what living things need to survive.

Why It's Important

All living things, including you, have many of the same traits.

Review Vocabulary
raw materials: substances needed by organisms to make other necessary substances

New Vocabulary
- organism
- cell
- homeostasis

What are living things like?

What does it mean to be alive? If you walked down your street after a thunderstorm, you'd probably see earthworms on the sidewalk, birds flying, clouds moving across the sky, and puddles of water. You'd see living and nonliving things that are alike in some ways. For example, birds and clouds move. Earthworms and water feel wet when they are touched. Yet, clouds and water are nonliving things, and birds and earthworms are living things. Any living thing is called an **organism.**

Organisms vary in size from the microscopic bacteria in mud puddles to gigantic oak trees and are found just about everywhere. They have different behaviors and food needs. In spite of these differences, all organisms have similar traits. These traits determine what it means to be alive.

Living Things Are Organized If you were to look at almost any part of an organism, like a plant leaf or your skin, under a microscope, you would see that it is made up of small units called cells. A **cell** is the smallest unit of an organism that carries on the functions of life. Some organisms are composed of just one cell while others are composed of many cells. Cells take in materials from their surroundings and use them in complex ways. Each cell has an orderly structure and contains hereditary material. The hereditary material contains instructions for cellular organization and function. **Figure 8** shows some organisms that are made of many cells. All the things that these organisms can do are possible because of what their cells can do.

Muscle cells

Color-enhanced LM Magnification: 106×

Nerve cells

Color-enhanced SEM Magnification: 2500×

Figure 8 Your body is organized into many different types of cells. Two types are shown here.

Living Things Respond Living things interact with their surroundings. Watch your cat when you use your electric can opener. Does your cat come running to find out what's happening even when you're not opening a can of cat food? The cat in **Figure 9** ran in response to a stimulus—the sound of the can opener. Anything that causes some change in an organism is a stimulus (plural, *stimuli*). The reaction to a stimulus is a response. Often that response results in movement, such as when the cat runs toward the sound of the can opener. To carry on its daily activity and to survive, an organism must respond to stimuli.

Living things also respond to stimuli that occur inside them. For example, water or food levels in organisms' cells can increase or decrease. The organisms then make internal changes to keep the right amounts of water and food in their cells. Their temperature also must be within a certain range. An organism's ability to keep the proper conditions inside no matter what is going on outside the organism is called **homeostasis.** Homeostasis is a trait of all living things.

Figure 9 Some cats respond to a food stimulus even when they are not hungry.
Infer *why a cat comes running when it hears a can opener.*

What are some internal stimuli living things respond to?

Living Things Use Energy Staying organized and carrying on activities like homeostasis require energy. The energy used by most organisms comes either directly or indirectly from the Sun. Plants and some other organisms use the energy in sunlight and the raw materials carbon dioxide and water to make food. You and most other organisms can't use the energy in sunlight directly. Instead, you take in and use food as a source of energy. You get food by eating plants or other organisms that ate plants. Most organisms, including plants, also must take in oxygen in order to release the energy of foods.

Some bacteria live at the bottom of the oceans and in other areas where sunlight cannot reach. They can't use the energy in sunlight to produce food. Instead, the bacteria use energy stored in some chemical compounds and the raw material carbon dioxide to make food. Unlike most other organisms, many of these bacteria do not need oxygen to release the energy that is found in their food.

Topic: Homeostasis
Visit booka.msscience.com for Web links to information about homeostasis.

Activity Describe the external stimuli and the corresponding internal changes for three different situations.

Living Things Grow and Develop When a puppy is born, it might be small enough to hold in one hand. After the same dog is fully grown, you might not be able to hold it at all. How does this happen? The puppy grows by taking in raw materials, like milk from its female parent, and making more cells. Growth of many-celled organisms, such as the puppy, is mostly due to an increase in the number of cells. In one-celled organisms, growth is due to an increase in the size of the cell.

Organisms change as they grow. Puppies can't see or walk when they are born. In eight or nine days, their eyes open, and their legs become strong enough to hold them up. All of the changes that take place during the life of an organism are called development. **Figure 10** shows how four different organisms changed as they grew.

The length of time an organism is expected to live is its life span. A dog can live for 20 years and a cat for 25 years. Some organisms have a short life span. Mayflies live only one day, but land tortoises can live for more than 180 years. Some bristlecone pine trees have been alive for more than 4,600 years. Your life span is about 80 years.

Figure 10 Complete development of an organism can take a few days or several years. The pictures below show the development of a dog, a human, a pea plant, and a butterfly.

Figure 11 Living things reproduce themselves in many different ways. A *Paramecium* reproduces by dividing into two. Beetles, like most insects, reproduce by laying eggs. Every spore released by the puffballs can grow into a new fungus.

Paramecium dividing

Living Things Reproduce Cats, dogs, alligators, fish, birds, bees, and trees eventually reproduce. They make more of their own kind. Some bacteria reproduce every 20 minutes while it might take a pine tree two years to produce seeds. **Figure 11** shows some ways organisms reproduce.

Without reproduction, living things would not exist to replace those individuals that die. An individual cat can live its entire life without reproducing. However, if cats never reproduced, all cats soon would disappear.

Reading Check *Why is reproduction important?*

What do living things need?

What do you need to live? Do you have any needs that are different from those of other living things? To survive, all living things need a place to live and raw materials. The raw materials that they require and the exact place where they live can vary.

A Place to Live The environment limits where organisms can live. Not many kinds of organisms can live in extremely hot or extremely cold environments. Most cannot live at the bottom of the ocean or on the tops of mountains. All organisms also need living space in their surroundings. For example, thousands of penguins build their nests on an island. When the island becomes too crowded, the penguins fight for space and some may not find space to build nests. An organism's surroundings must provide for all of its needs.

Social Development Human infants quickly develop their first year of life. Research to find out how infants interact socially at different stages of development. Make a chart that shows changes from birth to one year old.

Figure 12 You and a corn plant each take in and give off about 2 L of water in a day. Most of the water you take in is from water you drink or from foods you eat. **Infer** *where plants get water to transport materials.*

Raw Materials Water is important for all living things. Plants and animals take in and give off large amounts of water each day, as shown in **Figure 12.** Organisms use homeostasis to balance the amounts of water lost with the amounts taken in. Most organisms are composed of more than 50 percent water. You are made of 60 to 70 percent water. Organisms use water for many things. For example, blood, which is about 50 percent water, transports digested food and wastes in animals. Plants have a watery sap that transports materials between roots and leaves.

Living things are made up of substances such as proteins, fats, and sugars. Animals take in most of these substances from the foods they eat. Plants and some bacteria make them using raw materials from their surroundings. These important substances are used over and over again. When organisms die, substances in their bodies are broken down and released into the soil or air. The substances can then be used again by other living organisms. Some of the substances in your body might once have been part of a butterfly or an apple tree.

At the beginning of this section, you learned that things such as clouds, sidewalks, and puddles of water are not living things. Now do you understand why? Clouds, sidewalks, and water do not reproduce, use energy, or have other traits of living things.

section 2 review

Summary

What are living things like?

- A cell is the smallest unit of an organism that carries on the functions of life.
- Anything that causes some change in an organism is a stimulus.
- Organisms use energy to stay organized and perform activities like homeostasis.
- All of the changes that take place during an organism's life are called development.

What do living things need?

- Living things need a place to live, water, and food.

Self Check

1. **Identify** the source of energy for most organisms.
2. **List** five traits that most organisms have.
3. **Infer** why you would expect to see cells if you looked at a section of a mushroom cap under a microscope.
4. **Determine** what most organisms need to survive.
5. **Think Critically** Why is homeostasis important to organisms?

Applying Skills

6. **Use a Database** Use references to find the life span of ten animals. Use your computer to make a database. Then, graph the life spans from shortest to longest.

section 3

Where does life come from?

Life Comes from Life

You've probably seen a fish tank, like the one in **Figure 13,** that is full of algae. How did the algae get there? Before the seventeenth century, some people thought that insects and fish came from mud, that earthworms fell from the sky when it rained, and that mice came from grain. These were logical conclusions at that time, based on repeated personal experiences. The idea that living things come from nonliving things is known as **spontaneous generation.** This idea became a theory that was accepted for several hundred years. When scientists began to use controlled experiments to test this theory, the theory changed.

✓ Reading Check *Why did the theory of spontaneous generation change?*

Spontaneous Generation and Biogenesis From the late seventeenth century through the middle of the eighteenth century, experiments were done to test the theory of spontaneous generation. Although these experiments showed that spontaneous generation did not occur in most cases, they did not disprove it entirely.

It was not until the mid-1800s that the work of Louis Pasteur, a French chemist, provided enough evidence to disprove the theory of spontaneous generation. It was replaced with **biogenesis** (bi oh JE nuh suss), which is the theory that living things come only from other living things.

as you read

What You'll Learn

- **Describe** experiments about spontaneous generation.
- **Explain** how scientific methods led to the idea of biogenesis.

Why It's Important

You can use scientific methods to try to find out about events that happened long ago or just last week. You can even use them to predict how something will behave in the future.

Review Vocabulary
contaminate: to make impure by coming into contact with an unwanted substance

New Vocabulary
- spontaneous generation
- biogenesis

Figure 13 The sides of this tank were clean and the water was clear when the aquarium was set up. Algal cells, which were not visible on plants and fish, reproduced in the tank. So many algal cells are present now that the water is cloudy.

NATIONAL GEOGRAPHIC

VISUALIZING THE ORIGINS OF LIFE

Figure 14

For centuries scientists have theorized about the origins of life. As shown on this timeline, some examined spontaneous generation—the idea that nonliving material can produce life. More recently, scientists have proposed theories about the origins of life on Earth by testing hypotheses about conditions on early Earth.

1668 Francesco Redi put decaying meat in some jars, then covered half of them. When fly maggots appeared only on the uncovered meat (see below, left), Redi concluded that they had hatched from fly eggs and had not come from the meat.

John Needham heated broth in sealed flasks. When the broth became cloudy with microorganisms, he mistakenly concluded that they developed spontaneously from the broth. **1745**

Lazzaro Spallanzani boiled broth in sealed flasks for a longer time than Needham did. Only the ones he opened became cloudy with contamination. **1768**

Not contaminated

Contaminated

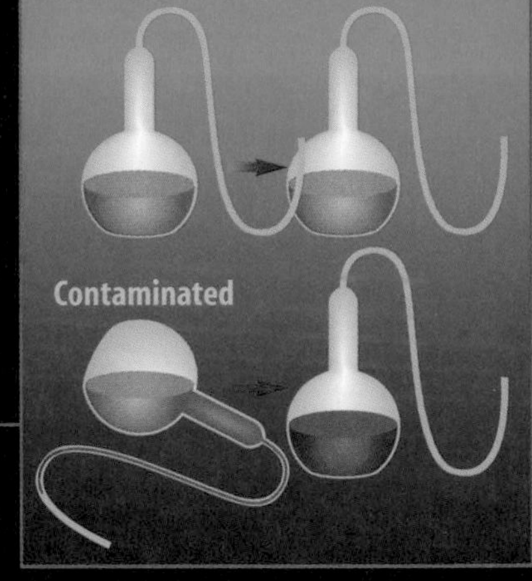

1859 Louis Pasteur disproved spontaneous generation by boiling broth in S-necked flasks that were open to the air. The broth became cloudy (see above, bottom right) only when a flask was tilted and the broth was exposed to dust in the S-neck.

1924 Alexander Oparin hypothesized that energy from the Sun, lightning, and Earth's heat triggered chemical reactions early in Earth's history. The newly-formed molecules washed into Earth's ancient oceans and became a part of what is often called the primordial soup.

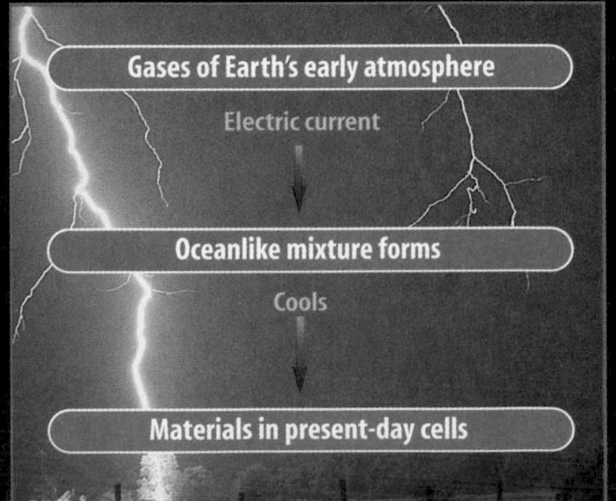

Stanley Miller and Harold Urey sent electric currents through a mixture of gases like those thought to be in Earth's early atmosphere. When the gases cooled, they condensed to form an oceanlike liquid that contained materials such as amino acids, found in present-day cells. **1953**

Life's Origins

INTEGRATE Astronomy

If living things can come only from other living things, how did life on Earth begin? Some scientists hypothesize that about 5 billion years ago, Earth's solar system was a whirling mass of gas and dust. They hypothesize that the Sun and planets were formed from this mass. It is estimated that Earth is about 4.6 billion years old. Rocks found in Australia that are more than 3.5 billion years old contain fossils of once-living organisms. Where did these living organisms come from?

Oparin's Hypothesis In 1924, a Russian scientist named Alexander I. Oparin suggested that Earth's early atmosphere had no oxygen but was made up of the gases ammonia, hydrogen, methane, and water vapor. Oparin hypothesized that these gases could have combined to form the more complex compounds found in living things.

Using gases and conditions that Oparin described, American scientists Stanley L. Miller and Harold Urey set up an experiment to test Oparin's hypothesis in 1953. Although the Miller-Urey experiment showed that chemicals found in living things could be produced, it did not prove that life began in this way.

For many centuries, scientists have tried to find the origins of life, as shown in **Figure 14.** Although questions about spontaneous generation have been answered, some scientists still are investigating ideas about life's origins.

Oceans Scientists hypothesize that Earth's oceans originally formed when water vapor was released into the atmosphere from many volcanic eruptions. Once it cooled, rain fell and filled Earth's lowland areas. Identify five lowland areas on Earth that are now filled with water. Record your answer in your Science Journal.

section 3 review

Summary

Life Comes from Life

- Spontaneous generation is the idea that living things come from nonliving things.
- The work of Louis Pasteur in 1859 disproved the theory of spontaneous generation.
- Biogenesis is the theory that living things come only from other living things.

Life's Origins

- Alexander I. Oparin hypothesized about the origin of life.
- The Miller-Urey experiment did not prove that Oparin's hypothesis was correct.

Self Check

1. **Compare and contrast** spontaneous generation with biogenesis.
2. **Describe** three controlled experiments that helped disprove the theory of spontaneous generation and led to the theory of biogenesis.
3. **Summarize** the results of the Miller-Urey experiment.
4. **Think Critically** How do you think life on Earth began?

Applying Skills

5. **Draw Conclusions** Where could the organisms have come from in the 1768 broth experiment described in **Figure 14?**

section 4

How are living things classified?

as you read

What You'll Learn

- **Describe** how early scientists classified living things.
- **Explain** how similarities are used to classify organisms.
- **Explain** the system of binomial nomenclature.
- **Demonstrate** how to use a dichotomous key.

Why It's Important

Knowing how living things are classified will help you understand the relationships that exist among all living things.

Review Vocabulary
common name: a nonscientific term that may vary from region to region

New Vocabulary
- phylogeny
- kingdom
- binomial nomenclature
- genus

Classification

If you go to a library to find a book about the life of Louis Pasteur, where do you look? Do you look for it among the mystery or sports books? You expect to find a book about Pasteur's life with other biography books. Libraries group similar types of books together. When you place similar items together, you classify them. Organisms also are classified into groups.

History of Classification When did people begin to group similar organisms together? Early classifications included grouping plants that were used in medicines. Animals were often classified by human traits such as courageous—for lions—or wise—for owls.

More than 2,000 years ago, a Greek named Aristotle observed living things. He decided that any organism could be classified as either a plant or an animal. Then he broke these two groups into smaller groups. For example, animal categories included hair or no hair, four legs or fewer legs, and blood or no blood. **Figure 15** shows some of the organisms Aristotle would have grouped together. For hundreds of years after Aristotle, no one way of classifying was accepted by everyone.

Figure 15 Using Aristotle's classification system, all animals without hair would be grouped together.
List *other animals without hair that Aristotle would have put in this group.*

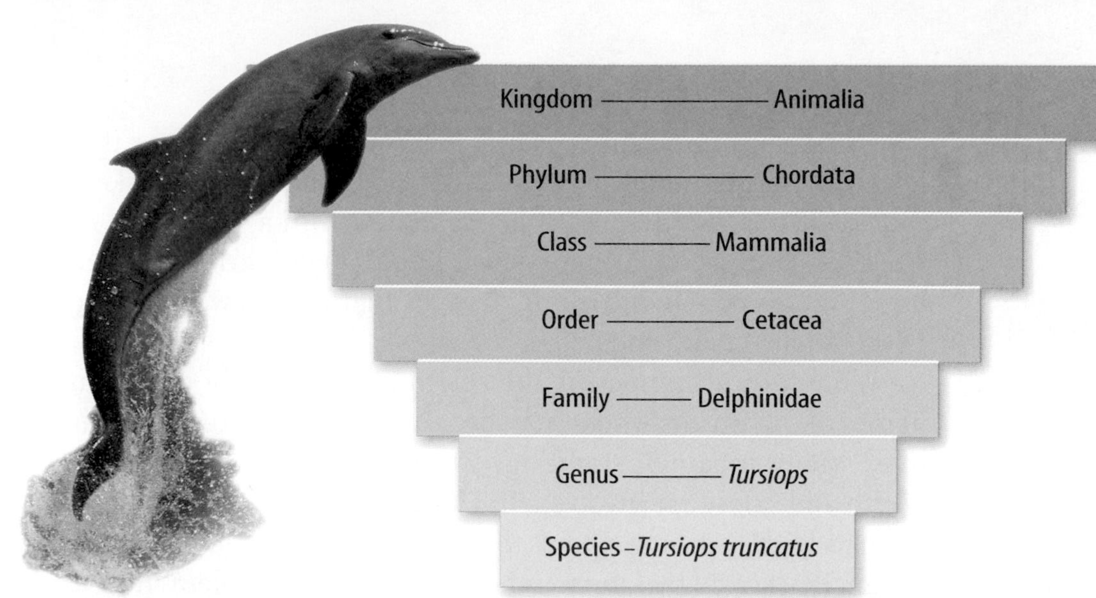

Figure 16 The classification of the bottle-nosed dolphin shows that it is in the order Cetacea. This order includes whales and porpoises.

Linnaeus In the late eighteenth century, Carolus Linnaeus, a Swedish naturalist, developed a new system of grouping organisms. His classification system was based on looking for organisms with similar structures. For example, plants that had similar flower structure were grouped together. Linnaeus's system eventually was accepted and used by most other scientists.

Modern Classification Like Linnaeus, modern scientists use similarities in structure to classify organisms. They also use similiarities in both external and internal features. Specific characteristics at the cellular level, such as the number of chromosomes, can be used to infer the degree of relatedness among organisms. In addition, scientists study fossils, hereditary information, and early stages of development. They use all of this information to determine an organism's phylogeny. **Phylogeny** (fi LAH juh nee) is the evolutionary history of an organism, or how it has changed over time. Today, it is the basis for the classification of many organisms.

What information would a scientist use to determine an organism's phylogeny?

Six Kingdoms A classification system commonly used today groups organisms into six kingdoms. A **kingdom** is the first and largest category. Organisms are placed into kingdoms based on various characteristics. Kingdoms can be divided into smaller groups. The smallest classification category is a species. Organisms that belong to the same species can mate and produce fertile offspring. To understand how an organism is classified, look at the classification of the bottle-nosed dolphin in **Figure 16.** Some scientists propose that before organisms are grouped into kingdoms, they should be placed in larger groups called domains. One proposed system groups all organisms into three domains.

Topic: Domains

Visit booka.msscience.com for Web links to information about domains.

Activity List all the domains and give examples of organisms that are grouped in each domain.

Scientific Names

Using common names can cause confusion. Suppose that Diego is visiting Jamaal. Jamaal asks Diego if he would like a soda. Diego is confused until Jamaal hands him a soft drink. At Diego's house, a soft drink is called pop. Jamaal's grandmother, listening from the living room, thought that Jamaal was offering Diego an ice-cream soda.

What would happen if life scientists used only common names of organisms when they communicated with other scientists? Many misunderstandings would occur, and sometimes health and safety are involved. In **Figure 17,** you see examples of animals with common names that can be misleading. A naming system developed by Linnaeus helped solve this problem. It gave each species a unique, two-word scientific name.

Figure 17 Common names can be misleading.

Binomial Nomenclature The two-word naming system that Linnaeus used to name the various species is called **binomial nomenclature** (bi NOH mee ul • NOH mun klay chur). It is the system used by modern scientists to name organisms. The first word of the two-word name identifies the genus of the organism. A **genus** is a group of similar species. The second word of the name might tell you something about the organism—what it looks like, where it is found, or who discovered it.

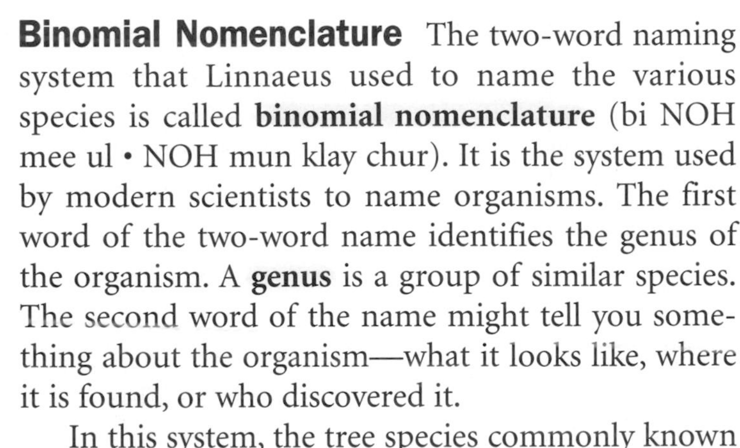

In this system, the tree species commonly known as red maple has been given the name *Acer rubrum.* The maple genus is *Acer.* The word *rubrum* is Latin for red, which is the color of a red maple's leaves in the fall. The scientific name of another maple is *Acer saccharum.* The Latin word for sugar is *saccharum.* In the spring, the sap of this tree is sweet.

Sea lions are more closely related to seals than to lions. **Identify** *another misleading common name.*

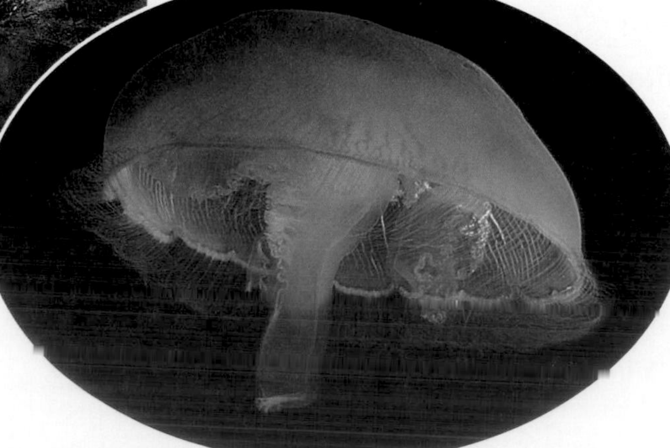

Jellyfish are neither fish nor jelly.

Figure 18 These two lizards have the same common name, iguana, but are two different species.

Uses of Scientific Names Two-word scientific names are used for four reasons. First, they help avoid mistakes. Both of the lizards shown in **Figure 18** have the name *iguana.* Using binomial nomenclature, the green iguana is named *Iguana iguana.* Someone who studied this *iguana,* shown in the left photo, would not be confused by information he or she read about *Dispsosaurus dorsalis,* the desert iguana, shown in the right photo. Second, organisms with similar evolutionary histories are classified together. Because of this, you know that organisms in the same genus are related. Third, scientific names give descriptive information about the species, like the maples mentioned earlier. Fourth, scientific names allow information about organisms to be organized easily and efficiently. Such information may be found in a book or a pamphlet that lists related organisms and gives their scientific names.

Reading Check *What are four functions of scientific names?*

Tools for Identifying Organisms

Tools used to identify organisms include field guides and dichotomous (di KAH tuh mus) keys. Using these tools is one way you and scientists solve problems scientifically.

Many different field guides are available. Most have illustrations or photographs of organisms, information about where each organism lives, and general descriptions of each species. You can identify species from around the world using the appropriate field guide.

Communicating Ideas

Procedure

1. Find a **magazine picture of a piece of furniture** that can be used as a place to sit and to lie down.
2. Show the picture to ten people and ask them to tell you what word they use for this piece of furniture.
3. Keep a record of the answers in your **Science Journal.**

Analysis

1. In your Science Journal, infer how using common names can be confusing.
2. How do scientific names make communication among scientists easier?

Try at Home

Dichotomous Keys A dichotomous key is a detailed list of identifying characteristics that includes scientific names. Dichotomous keys are arranged in steps with two descriptive statements at each step. If you learn how to use a dichotomous key, you can identify and name a species.

Did you know many types of mice exist? You can use **Table 2** to find out what type of mouse is pictured to the left. Start by choosing between the first pair of descriptions. The mouse has hair on its tail, so you go to 2. The ears of the mouse are small, so you go on to 3. The tail of the mouse is less that 25 mm. What is the name of this mouse according to the key?

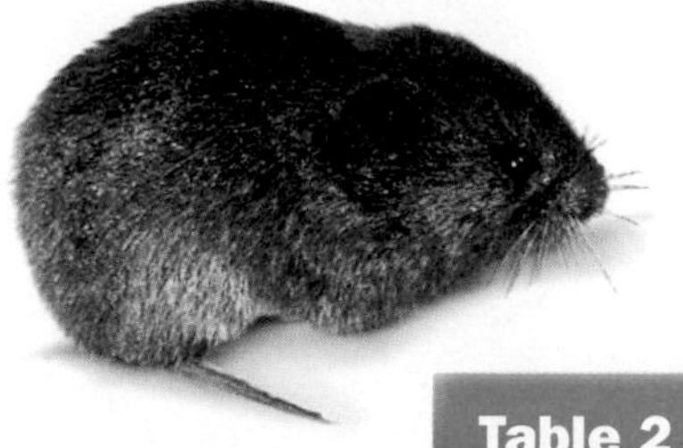

Table 2 Key to Some Mice of North America

1. Tail hair	**a.** no hair on tail; scales show plainly; house mouse, *Mus musculus* **b.** hair on tail, go to 2
2. Ear size	**a.** ears small and nearly hidden in fur, go to 3 **b.** ears large and not hidden in fur, go to 4
3. Tail length	**a.** less than 25 mm; woodland vole, *Microtus pinetorum* **b.** more than 25 mm; prairie vole, *Microtus ochrogaster*
4. Tail coloration	**a.** sharply bicolor, white beneath and dark above; deer mouse, *Peromyscus maniculatus* **b.** darker above than below but not sharply bicolor; white-footed mouse, *Peromyscus leucopus*

section 4 review

Summary

Classification

- Organisms are classified into groups based on their similarities.
- Scientists today classify organisms into six kingdoms.
- Species is the smallest classification category.

Scientific Names

- Binomial nomenclature is the two-word naming system that gives organisms their scientific names.

Tools for Identifying Organisms

- Field guides and dichotomous keys are used to identify organisms.

Self Check

1. **State** Aristotle's and Linnaeus' contributions to classifying living things.
2. **Identify** a specific characteristic used to classify organisms.
3. **Describe** what each of the two words identifies in binomial nomenclature.
4. **Think Critically** Would you expect a field guide to have common names as well as scientific names? Why or why not?

Applying Skills

5. **Classify** Create a dichotomous key that identifies types of cars.

Classifying Seeds

Scientists use classification systems to show how organisms are related. How do they determine which features to use to classify organisms? In this lab, you will observe seeds and use their features to classify them.

Real-World Question

How can the features of seeds be used to develop a dichotomous key to identify the seed?

Goals

- **Observe** the seeds and notice their features.
- **Classify** seeds using these features.

Materials

packets of seeds (10 different kinds)
magnifying lens
metric ruler

Safety Precautions

WARNING: *Some seeds may have been treated with chemicals. Do not put them in your mouth.*

Procedure

1. Copy the following data table in your Science Journal and record the features of each seed. Your table will have a column for each different type of seed you observe.

Seed Data

Feature	Type of Seed		
Color			
Length (mm)	Do not write in this book.		
Shape			
Texture			

2. Use the features to develop a dichotomous key.
3. Exchange keys with another group. Can you use their key to identify seeds?

Conclude and Apply

1. **Determine** how different seeds can be classified.
2. **Explain** how you would classify a seed you had not seen before using your data table.
3. **Explain** why it is an advantage for scientists to use a standardized system to classify organisms. What observations did you make to support your answer?

Communicating Your Data

Compare your conclusions with those of other students in your class. **For more help, refer to the Science Skill Handbook.**

Design Your Own

Using Scientific Methods

Goals

- **Design** and carry out an experiment using scientific methods to infer why brine shrimp live in the ocean.
- **Observe** the jars for one week and notice whether the brine shrimp eggs hatch.

Possible Materials

500-mL, widemouthed containers (3)
brine shrimp eggs
small, plastic spoon
distilled water (500 mL)
weak salt solution (500 mL)
strong salt solution (500 mL)
labels (3)
magnifying lens

Safety Precautions

WARNING: *Protect eyes and clothing. Be careful when working with live organisms.*

Real-World Question

Brine shrimp are relatives of lobsters, crabs, crayfish, and the shrimp eaten by humans. They are often raised as a live food source in aquariums. In nature, they live in the oceans where fish feed on them. They can hatch from eggs that have been stored in a dry condition for many years. How can you use scientific methods to determine whether salt affects the hatching and growth of brine shrimp?

Brine shrimp

Form a Hypothesis

Based on your observations, form a hypothesis to explain how salt affects the hatching and growth of brine shrimp.

Test Your Hypothesis

Make a Plan

1. As a group, agree upon the hypothesis and decide how you will test it. Identify what results will confirm the hypothesis.

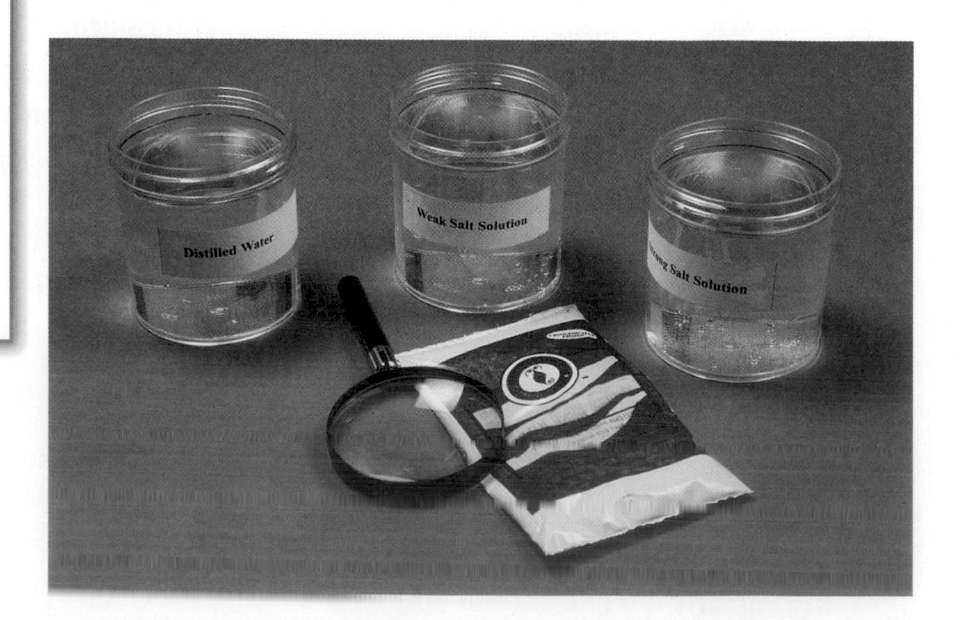

2. **List** steps that you need to test your hypothesis. Be specific. Describe exactly what you will do at each step.
3. **List** your materials.
4. **Prepare** a data table in your Science Journal to record your data.
5. Read over your entire experiment to make sure that all planned steps are in logical order.
6. **Identify** any constants, variables, and controls of the experiment.

Follow Your Plan

1. Make sure your teacher approves your plan before you start.
2. Carry out the experiment as planned by your group.
3. While doing the experiment, record any observations and complete the data table in your Science Journal.
4. Use a bar graph to plot your results.

Analyze Your Data

1. **Describe** the contents of each jar after one week.
2. **Identify** your control in this experiment.
3. **Identify** your variable in this experiment.

Conclude and Apply

1. **Explain** whether or not the results support your hypothesis.
2. **Predict** the effect that increasing the amount of salt in the water would have on the brine shrimp eggs.
3. **Compare** your results with those of other groups.

Communicating Your Data

Prepare a set of instructions on how to hatch brine shrimp to use to feed fish. Include diagrams and a step-by-step procedure.

Acari marmoset

Monkey BUSINESS

Manicore marmoset

In 2000, a scientist from Brazil's Amazon National Research Institute came across two squirrel-sized monkeys in a remote and isolated corner of the rain forest, about 2,575 km from Rio de Janeiro.

It turns out that the monkeys had never been seen before, or even known to exist.

Acari marmoset

The new species were spotted by a scientist who named them after two nearby rivers the Manicore and the Acari, where the animals were discovered. Both animals are marmosets, which is a type of monkey found only in Central and South America. Marmosets have claws instead of nails, live in trees, and use their extraordinarily long tail like an extra arm or leg. Small and light, both marmosets measure about 23 cm in length with a 38 cm tail, and weigh no more than 0.4 kg.

The Manicore marmoset has a silvery-white upper body, a light-gray cap on its head, a yellow-orange underbody, and a black tail.

The Acari marmoset's upper body is snowy white, its gray back sports a stripe running to the knee, and its black tail flashes a bright-orange tip.

Amazin' Amazon

The Amazon Basin is a treasure trove of unique species. The Amazon River is Earth's largest body of freshwater, with 1,100 smaller tributaries. And more than half of the world's plant and animal species live in its rain forest ecosystems.

Research and Report Working in small groups, find out more about the Amazon rain forest. Which plants and animals live there? What products come from the rain forest? How does what happens in the Amazon rain forest affect you? Prepare a multimedia presentation.

For more information, visit booka.msscience.com/time

Reviewing Main Ideas

Section 1 What is science?

1. Scientists use problem-solving methods to investigate observations about living and nonliving things.
2. Scientists use SI measurements to gather measurable data.
3. Safe laboratory practices help you learn more about science.

Section 2 Living Things

1. Organisms are made of cells, use energy, reproduce, respond, grow, and develop.
2. Organisms need energy, water, food, and a place to live.

Section 3 Where does life come from?

1. Controlled experiments finally disproved the theory of spontaneous generation.
2. Pasteur's experiment proved biogenesis.

Section 4 How are living things classified?

1. Classification is the grouping of ideas, information, or objects based on their similar characteristics.
2. Scientists today use phylogeny to group organisms into six kingdoms.
3. All organisms are given a two-word scientific name using binomial nomenclature.

Visualizing Main Ideas

Copy and complete this events-chain concept map that shows the order in which you might use a scientific method. Use these terms: analyze data, perform an experiment, *and* form a hypothesis.

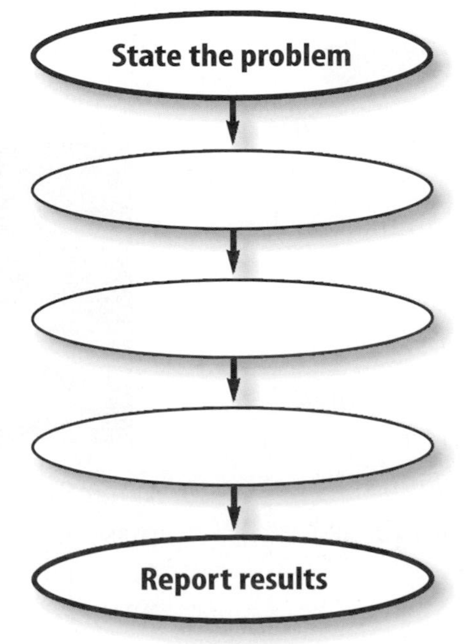

chapter 1 Review

Using Vocabulary

binomial nomenclature p. 26
biogenesis p. 21
cell p. 16
control p. 11
genus p. 26
homeostasis p. 17
hypothesis p. 10
kingdom p. 25
law p. 12
organism p. 16
phylogeny p. 25
scientific methods p. 9
spontaneous generation p. 21
theory p. 12
variable p. 11

Explain the differences in the vocabulary words in each pair below. Then explain how they are related.

1. control—variable
2. law—theory
3. biogenesis—spontaneous generation
4. binomial nomenclature—phylogeny
5. organism—cell
6. kingdom—phylogeny
7. hypothesis—scientific methods
8. organism—homeostasis
9. kingdom—genus
10. theory—hypothesis

Checking Concepts

Choose the word or phrase that best answers the question.

11. What category of organisms can mate and produce fertile offspring?
A) family **C)** genus
B) class **D)** species

12. What is the closest relative of *Canis lupus*?
A) *Quercus alba* **C)** *Felis tigris*
B) *Equus zebra* **D)** *Canis familiaris*

13. What is the source of energy for plants?
A) the Sun **C)** water
B) carbon dioxide **D)** oxygen

14. What makes up more than 50 percent of all living things?
A) oxygen **C)** minerals
B) carbon dioxide **D)** water

15. Who finally disproved the theory of spontaneous generation?
A) Oparin **C)** Pasteur
B) Aristotle **D)** Miller

16. What gas do some scientists think was missing from Earth's early atmosphere?
A) ammonia **C)** methane
B) hydrogen **D)** oxygen

17. What is the length of time called that an organism is expected to live?
A) life span **C)** homeostasis
B) stimulus **D)** theory

18. What is the part of an experiment that can be changed called?
A) conclusion **C)** control
B) variable **D)** data

19. What does the first word in a two-word name of an organism identify?
A) kingdom **C)** phylum
B) species **D)** genus

Use the photo below to answer question 20.

20. What SI unit could you use to measure the mass of the fish shown above?
A) meter **C)** gram
B) liter **D)** degree

Science Online booka.msscience.com/vocabulary_puzzlemaker

Thinking Critically

21. **Predict** what *Lathyrus odoratus,* the scientific name for a sweet pea plant, tells you about one of its characteristics.

Use the photo below to answer question 22.

22. **Determine** what problem-solving techniques this scientist would use to find how dolphins learn.

Use the graph below to answer question 23.

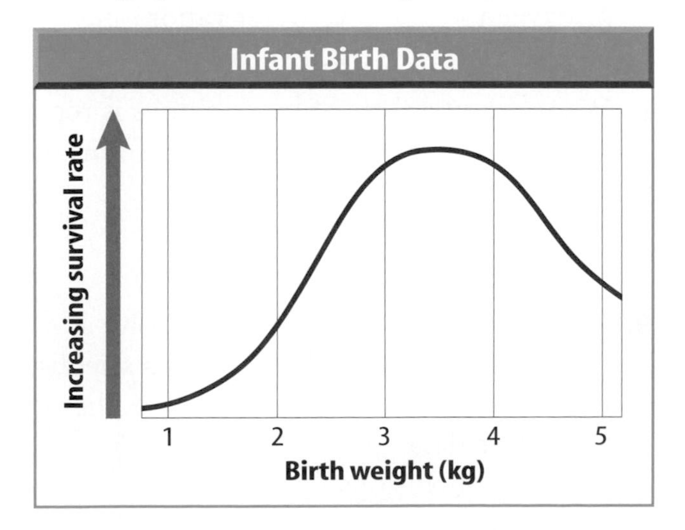

23. **Interpret Data** Do the data in the graph above support the hypothesis that babies with a birth weight of 2.5 kg have the best chance of survival? Explain.

24. **List** advantages of using SI units.

25. **Form a Hypothesis** A lima bean plant is placed under green light, another is placed under red light, and a third under blue light. Their growth is measured for four weeks to determine which light is best for plant growth. What are the variables in this experiment? State a hypothesis for this experiment.

Performance Activities

26. **Bulletin Board** Interview people in your community whose jobs require a knowledge of life science. Make a Life Science Careers bulletin board. Summarize each person's job and what he or she had to study to prepare for that job.

Applying Math

27. **Body Temperature** Normal human body temperature is 98.6°F. What is this temperature in degrees Celsius? Use the following expression, 5/9(°F−32), to find degrees Celsius.

Use the graph below to answer question 28.

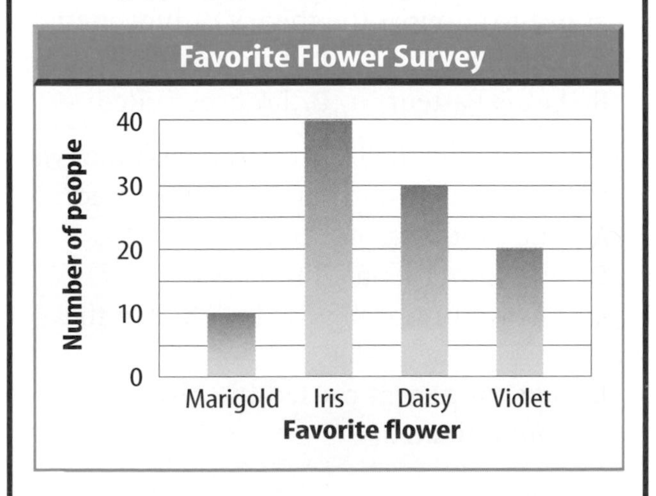

28. **Favorite Flower** The graph above shows how many people selected a certain type of flower as their favorite. According to the graph, what percentage of the people picked daisy as their favorite?

chapter 1 Standardized Test Practice

Part 1 Multiple Choice

Record your answers on the answer sheet provided by your teacher or on a sheet of paper.

1. A prediction that can be tested is a
   A. conclusion. **C.** hypothesis.
   B. variable. **D.** theory.

2. Which of the following units would a scientist likely use when measuring the length of a mouse's tail?
   A. kilometers **C.** grams
   B. millimeters **D.** milliliters

Use the illustrations below to answer questions 3 and 4.

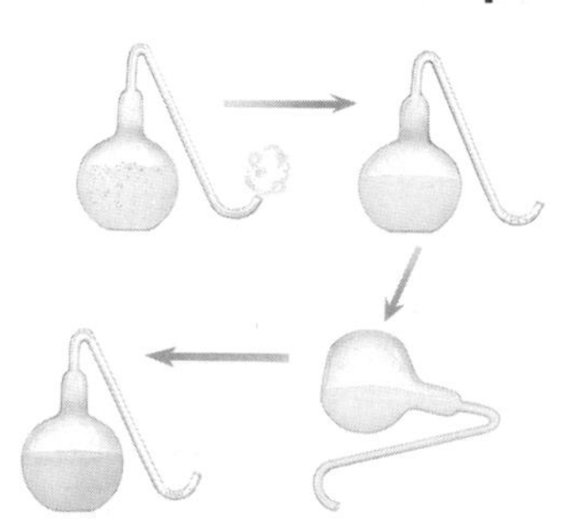

3. What scientist used the flasks pictured above to support the theory of biogenesis?
   A. John Needham **C.** Lazzaro Spallanzani
   B. Louis Pasteur **D.** Francesco Redi

4. Why did only the broth in the flask that was tilted become cloudy and contaminated?
   A. The broth was not boiled.
   B. Flies contaminated the broth.
   C. The broth was exposed to dust in the neck of the flask.
   D. Decaying meat caused the broth to be contaminated.

Test-Taking Tip

Practice Skills Remember that test-taking skills can improve with practice. If possible, take at least one practice test and familiarize yourself with the test format and instructions.

Use the photos below to answer questions 5 and 6.

5. The dog pictured above has increased in size. How did most of this size increase take place?
   A. an increase in cell size
   B. an increase in the number of cells
   C. an increase in cell water
   D. an increase in cell energy

6. What characteristic of life is illustrated by the change in the dog?
   A. reproduction
   B. homeostasis
   C. growth and development
   D. response to stimulus

7. What gas must most organisms take in to release the energy of foods?
   A. oxygen **C.** water vapor
   B. carbon dioxide **D.** hydrogen

8. What characteristic of living things is represented by a puffball releasing millions of spores?
   A. reproduction **C.** organization
   B. development **D.** use of energy

9. When using scientific methods to solve a problem, which of the following is a scientist most likely to do after forming a hypothesis?
   A. analyze data
   B. draw conclusions
   C. state a problem
   D. perform an experiment

10. What are the smallest units that make up your body called?
    A. cells **C.** muscles
    B. organisms **D.** fibers

Part 2 Short Response/Grid In

Record your answers on the answer sheet provided by your teacher or on a sheet of paper.

11. From where do bacteria that live in areas where there is no sunlight obtain energy?

12. Organisms take in and give off large amounts of water each day. What process do they use to balance the amount of water lost with the amount taken in?

13. After a rain storm, earthworms may be seen crawling on the sidewalk or road. How would the theory of spontaneous generation explain the origin of the worms?

14. List three things modern scientists study when they classify organisms.

Use the illustrations below to answer questions 15 and 16.

15. A science class set up the experiment above to study the response of plants to the stimulus of light. What hypothesis is likely being tested by this experiment?

16. After day 4, Fatima wanted to find out how plant 2 and plant 3 would grow in normal light. What did she have to do to find out?

Part 3 Open Ended

Record your answers on a sheet of paper.

17. Describe several different ways scientists gather information. Which of these ways would likely be used to collect data about which foods wild alligators eat in Florida?

18. Some scientists think that lightning may have caused chemicals in the Earth's early atmosphere to combine to begin the origin of life. Explain how the experiment of Miller and Urey does not prove this hypothesis.

Use the photo below to answer questions 19 and 20.

19. Both of these living things use energy. Describe the difference between the source of energy for each. In what similar ways would each of these organisms use energy?

20. How are the needs of the two organisms alike? Explain why the plant is raw material for the beetle. When the beetle dies, how could it be raw material for the plant?

21. Explain stimulus and response. How is response to a stimulus related to homeostasis?

22. What information would you need to write a field guide used to identify garden plants? What other information would you need if the guide included a dichotomous key?

23. Explain the difference between a kingdom and a species in the classification system commonly used.

chapter

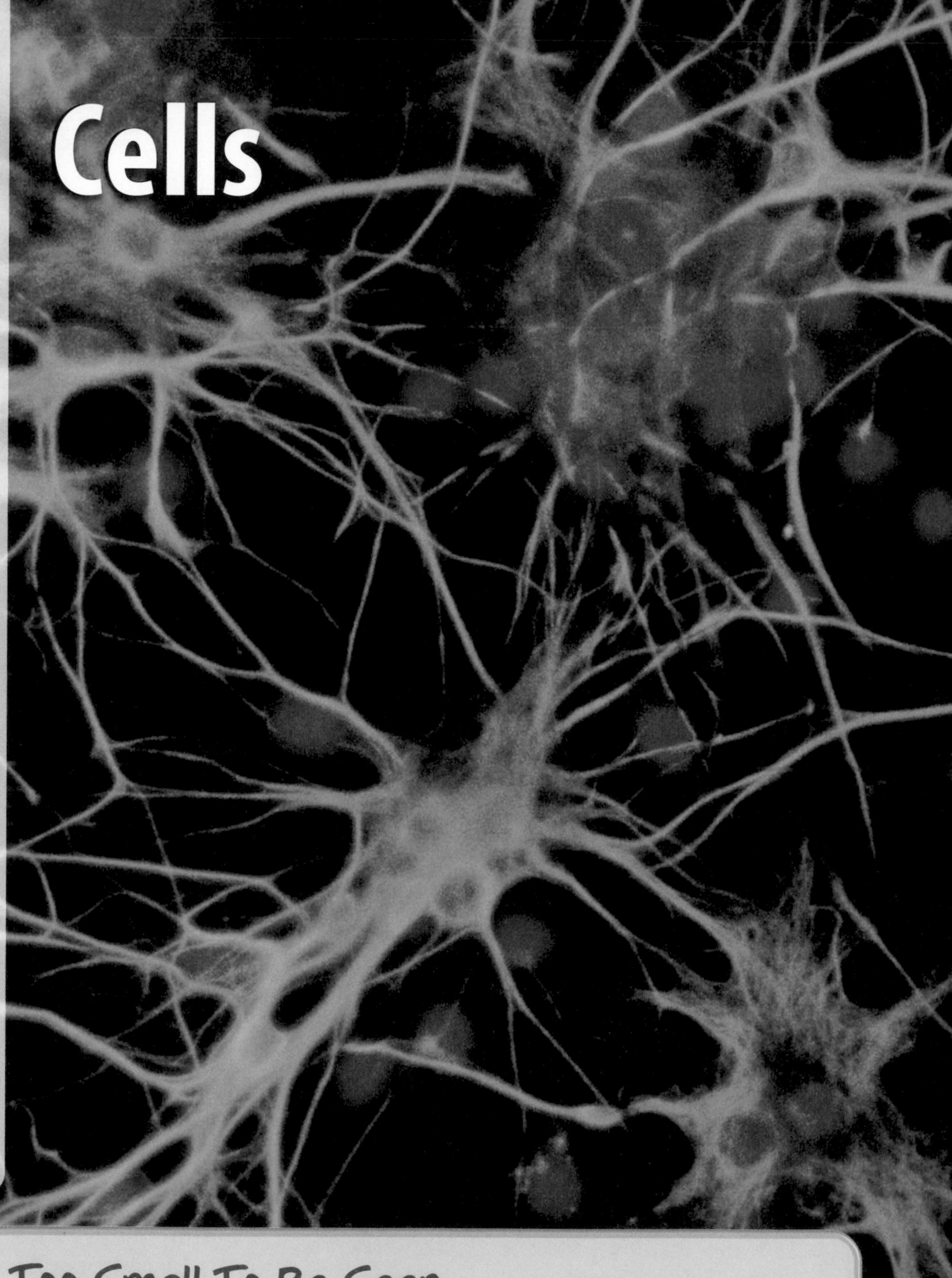

Cells

The BIG Idea

The different structures in a cell work together to ensure the cell's survival.

SECTION 1
Cell Structure

Main Idea Different cell types can have different structures, but some cell structures are common to all cells.

SECTION 2
Viewing Cells

Main Idea Scientists can study living things too small to be seen with only the human eye by using microscopes.

SECTION 3
Viruses

Main Idea Although viruses are not considered living things, they can affect all living things.

Too Small To Be Seen

The world around you is filled with organisms that you could overlook, or even be unable to see. Some of these organisms are one-celled and some are many-celled. You can study these organisms and the cells of other organisms by using microscopes.

Science Journal Write three questions that you would ask a scientist researching cancer cells.

Start-Up Activities

Magnifying Cells

If you look around your classroom, you can see many things of all sizes. Using a magnifying lens, you can see more details. You might examine a speck of dust and discover that it is a living or dead insect. In the following lab, use a magnifying lens to search for the smallest thing you can find in the classroom.

1. Obtain a magnifying lens from your teacher. Note its power (the number followed by ×, shown somewhere on the lens frame or handle).
2. Using the magnifying lens, look around the room for the smallest object that you can find.
3. Measure the size of the image as you see it with the magnifying lens. To estimate the real size of the object, divide that number by the power. For example, if it looks 2 cm long and the power is 10×, the real length is about 0.2 cm.
4. **Think Critically** Write a paragraph that describes what you observed. Did the details become clearer? Explain.

FOLDABLES™ Study Organizer

Cells Make the following Foldable to help you illustrate the main parts of cells.

STEP 1 **Fold** a vertical sheet of paper in half from top to bottom.

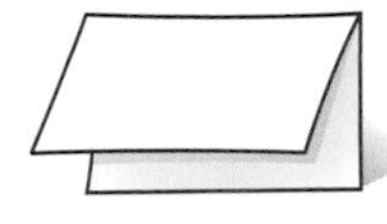

STEP 2 **Fold** in half from side to side with the fold at the top.

STEP 3 **Unfold** the paper once. **Cut** only the fold of the top flap to make two tabs.

STEP 4 **Turn** the paper vertically and **write** on the front tabs as shown.

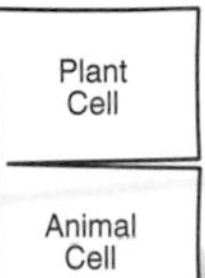

Illustrate and Label As you read the chapter, draw and identify the parts of plant and animal cells under the appropriate tab.

Preview this chapter's content and activities at booka.msscience.com

Get Ready to Read

Identify the Main Idea

1 Learn It! Main ideas are the most important ideas in a paragraph, lesson, or chapter. Supporting details are facts or examples that explain the main idea. Understanding the main idea allows you to grasp the whole picture.

2 Picture It! Read the following paragraph. Draw a graphic organizer like the one below to show the main idea and supporting details.

> Things that are too small to be seen with other microscopes can be viewed with an electron microscope. Instead of using lenses to direct beams of light, an electron microscope uses a magnetic field in a vacuum to direct beams of electrons. Some electron microscopes can magnify images up to one million times. To see electron microscope images, they must be photographed or electronically produced.
>
> *—from page 52*

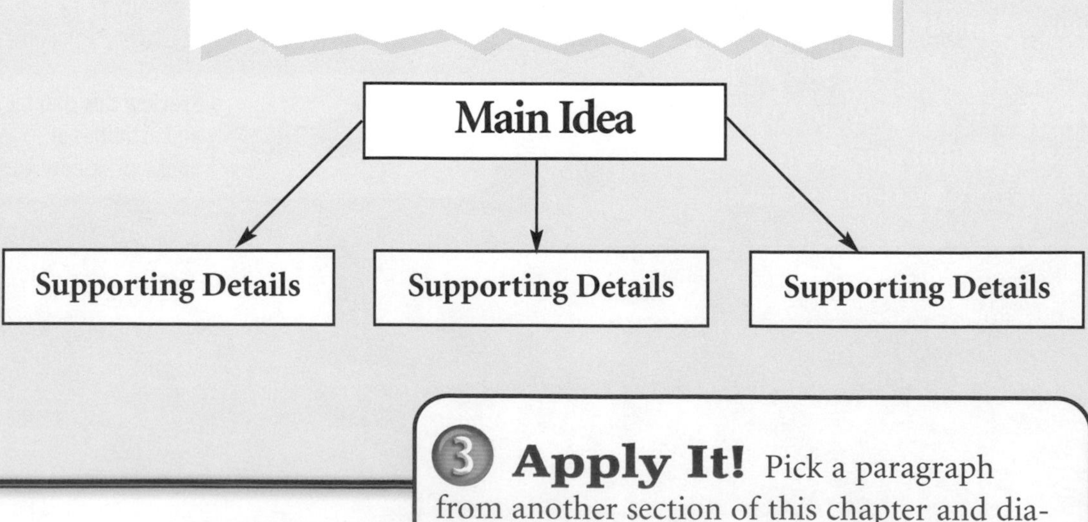

3 Apply It! Pick a paragraph from another section of this chapter and diagram the main ideas as you did above.

Reading Tip

The main idea is often the first sentence in a paragraph but not always.

Target Your Reading

Use this to focus on the main ideas as you read the chapter.

1. **Before you read** the chapter, respond to the statements below on your worksheet or on a numbered sheet of paper.
   - Write an **A** if you **agree** with the statement.
   - Write a **D** if you **disagree** with the statement.

2. **After you read** the chapter, look back to this page to see if you've changed your mind about any of the statements.
   - If any of your answers changed, explain why.
   - Change any false statements into true statements.
   - Use your revised statements as a study guide.

Before You Read A or D	Statement	After You Read A or D
	1 All new cells come from preexisting cells.	
	2 You must use a microscope to see most cells.	
	3 A flexible cell membrane surrounds every cell.	
	4 Chromosomes are in the nucleus of every cell.	
	5 A bacterium is larger than an animal cell.	
	6 A cell wall and cytoplasm control the shape of each cell.	
	7 Tissues are groups of similar types of cells that work together to perform a function.	
	8 The most powerful microscopes create images by focusing light through two or more lenses.	
	9 A cell's mitochondria transform light energy into chemical energy.	
	10 Viruses are harmful and never beneficial.	

Science Online
Print out a worksheet of this page at booka.msscience.com

section 1

Cell Structure

as you read

What You'll Learn

- **Identify** names and functions of each part of a cell.
- **Explain** how important a nucleus is in a cell.
- **Compare** tissues, organs, and organ systems.

Why It's Important

If you know how organelles function, it's easier to understand how cells survive.

Review Vocabulary

photosynthesis: process by which most plants, some protists, and many types of bacteria make their own food

New Vocabulary

- cell membrane
- cytoplasm
- cell wall
- organelle
- nucleus
- chloroplast
- mitochondrion
- ribosome
- endoplasmic reticulum
- Golgi body
- tissue
- organ

Common Cell Traits

Living cells are dynamic and have several things in common. A cell is the smallest unit that is capable of performing life functions. All cells have an outer covering called a **cell membrane.** A living membrane is made of one or more layers of linked molecules. Inside every cell is a gelatinlike material called **cytoplasm** (SI tuh pla zum). In the cytoplasm of every cell is hereditary material that controls the life of the cell.

Comparing Cells Cells come in many sizes. A nerve cell in your leg could be a meter long. A human egg cell is no bigger than the dot on this *i*. A human red blood cell is about one-tenth the size of a human egg cell. A bacterium is even smaller—8,000 of the smallest bacteria can fit inside one of your red blood cells.

A cell's shape might tell you something about its function. The nerve cell in **Figure 1** has extensions that send impulses to or receive impulses from other cells. A nerve cell cannot change shape but muscle cells and some blood cells can. Some plant stems have long, hollow cells with openings at their ends. These cells carry food and water throughout the plant.

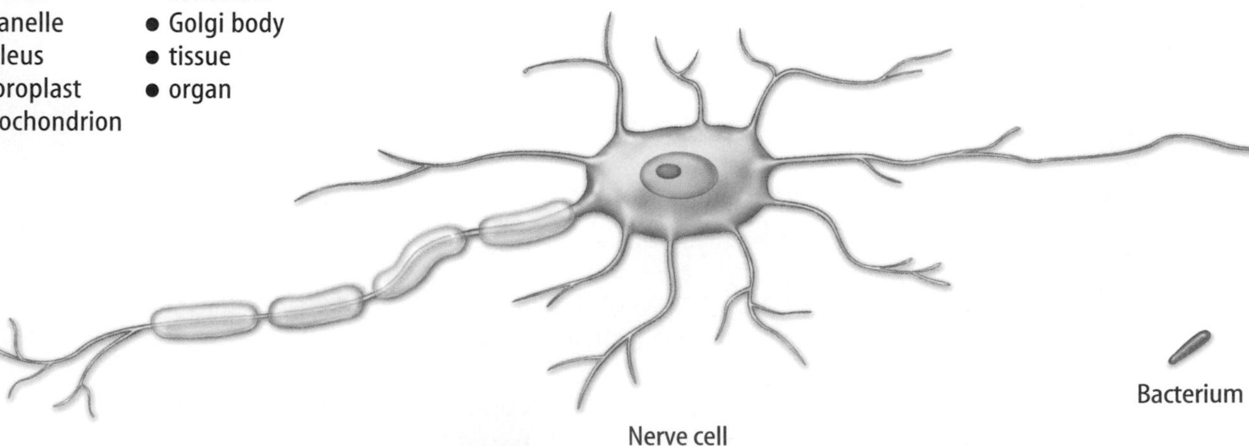

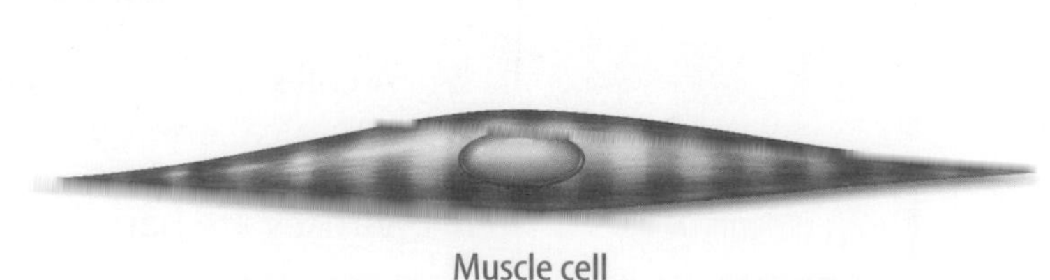

Figure 1 The shape of the cell can tell you something about its function. These cells are drawn 700 times their actual size.

Prokaryotic

Gel-like capsule
Cell wall
Cell membrane
Flagellum
Hereditary material
Ribosomes

Prokaryotic cell

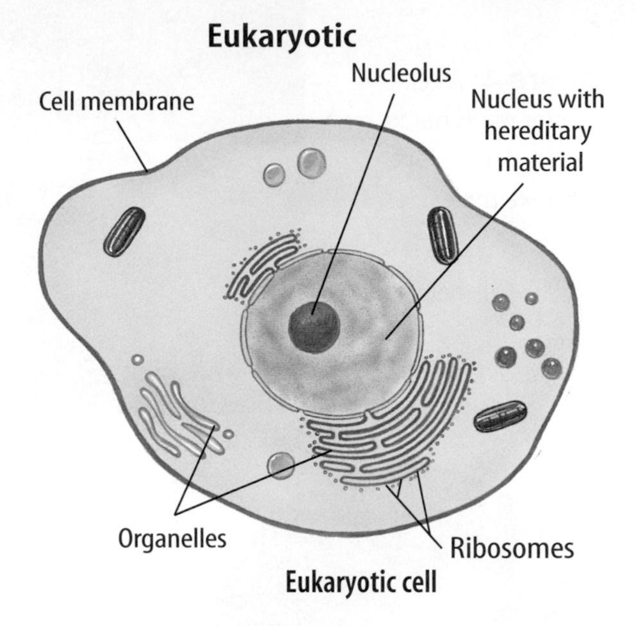

Eukaryotic cell

Cell Types Scientists have found that cells can be separated into two groups. One group has no membrane-bound structures inside the cell and the other group does, as shown in **Figure 2.** Cells without membrane-bound structures are called prokaryotic (proh KAYR ee yah tihk) cells. Cells with membrane-bound structures are called eukaryotic (yew KAYR ee yah tihk) cells.

Figure 2 Examine these drawings of cells. Prokaryotic cells are only found in one-celled organisms, such as bacteria. Protists, fungi, plants, and animals are made of eukaryotic cells. **Describe** *differences you see between them.*

Reading Check *Into what two groups can cells be separated?*

Cell Organization

Each cell in your body has a specific function. You might compare a cell to a busy delicatessen that is open 24 hours every day. Raw materials for the sandwiches are brought in often. Some food is eaten in the store, and some customers take their food with them. Sometimes food is prepared ahead of time for quick sale. Wastes are put into trash bags for removal or recycling. Similarly, your cells are taking in nutrients, secreting and storing chemicals, and breaking down substances 24 hours every day.

Cell Wall Just like a deli that is located inside the walls of a building, some cells are enclosed in a cell wall. The cells of plants, algae, fungi, and most bacteria are enclosed in a cell wall. **Cell walls** are tough, rigid outer coverings that protect the cell and give it shape.

A plant cell wall, as shown in **Figure 3,** mostly is made up of a substance called cellulose. The long, threadlike fibers of cellulose form a thick mesh that allows water and dissolved materials to pass through it. Cell walls also can contain pectin, which is used in jam and jelly, and lignin, which is a compound that makes cell walls rigid. Plant cells responsible for support have a lot of lignin in their walls.

Figure 3 The protective cell wall of a plant cell is outside the cell membrane.

Color-enhanced TEM Magnification: 9000×

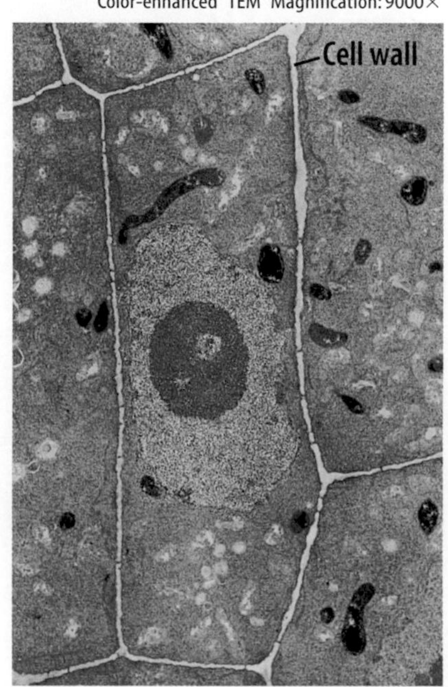

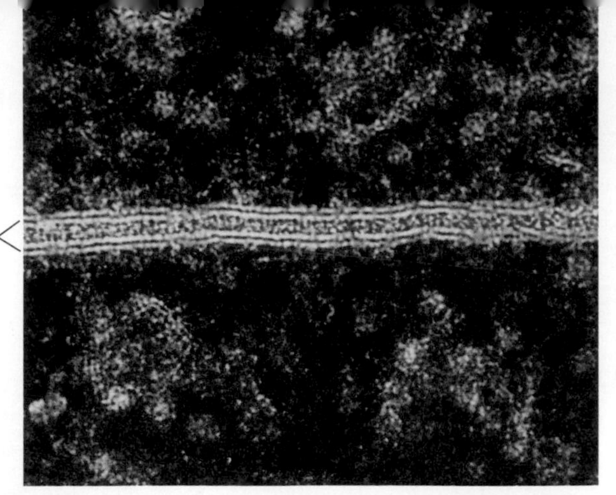

Cell membranes

Color-enhanced TEM Magnification: 125000×

Figure 4 A cell membrane is made up of a double layer of fatlike molecules.

Cell Membrane The protective layer around all cells is the cell membrane, as shown in **Figure 4.** If cells have cell walls, the cell membrane is inside of it. The cell membrane regulates interactions between the cell and the environment. Water is able to move freely into and out of the cell through the cell membrane. Food particles and some molecules enter and waste products leave through the cell membrane.

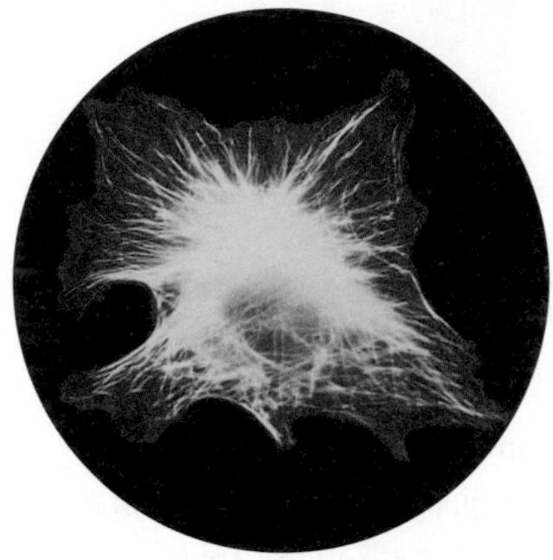

Stained LM Magnification: 700×

Figure 5 Cytoskeleton, a network of fibers in the cytoplasm, gives cells structure and helps them maintain shape.

Cytoplasm Cells are filled with a gelatinlike substance called cytoplasm that constantly flows inside the cell membrane. Many important chemical reactions occur within the cytoplasm.

Throughout the cytoplasm is a framework called the cytoskeleton, which helps the cell maintain or change its shape. Cytoskeletons enable some cells to move. An amoeba, a one-celled eukaryotic organism, moves by stretching and contracting its cytoskeleton. A cytoskeleton is made up of thin, hollow tubes of protein and thin, solid protein fibers, as shown in **Figure 5.** Proteins are organic molecules made up of amino acids.

Reading Check ***What is the function of the cytoskeleton?***

Most of a cell's life processes occur in the cytoplasm. Within the cytoplasm of eukaryotic cells are structures called **organelles.** Some organelles process energy and others manufacture substances needed by the cell or other cells. Certain organelles move materials, while others act as storage sites. Most organelles are surrounded by membranes. The nucleus is usually the largest organelle in a cell.

Modeling Cytoplasm

Procedure

1. Add 100 mL of **water** to a **clear container.**
2. Add **unflavored gelatin** and stir.
3. Shine a **flashlight** through the solution.

Analysis

1. Describe what you see.
2. How does a model help you understand what cytoplasm might be like?

Nucleus The nucleus is like the deli manager who directs the store's daily operations and passes on information to employees. The **nucleus,** shown in **Figure 6,** directs all cell activities and is separated from the cytoplasm by a membrane. Materials enter and leave the nucleus through openings in this membrane.

The nucleus contains the instructions for everything the cell does. These instructions are found on long, threadlike, hereditary material made of DNA. DNA is the chemical that contains the code for the cell's structure and activities. During cell division, the hereditary material coils tightly around proteins to form structures called chromosomes. A structure called a nucleolus also is found in the nucleus.

Figure 6 Refer to these diagrams of a typical animal cell (top) and plant cell (bottom) as you read about cell structures and their functions. **Determine** *which structures a plant cell has that are not found in animal cells.*

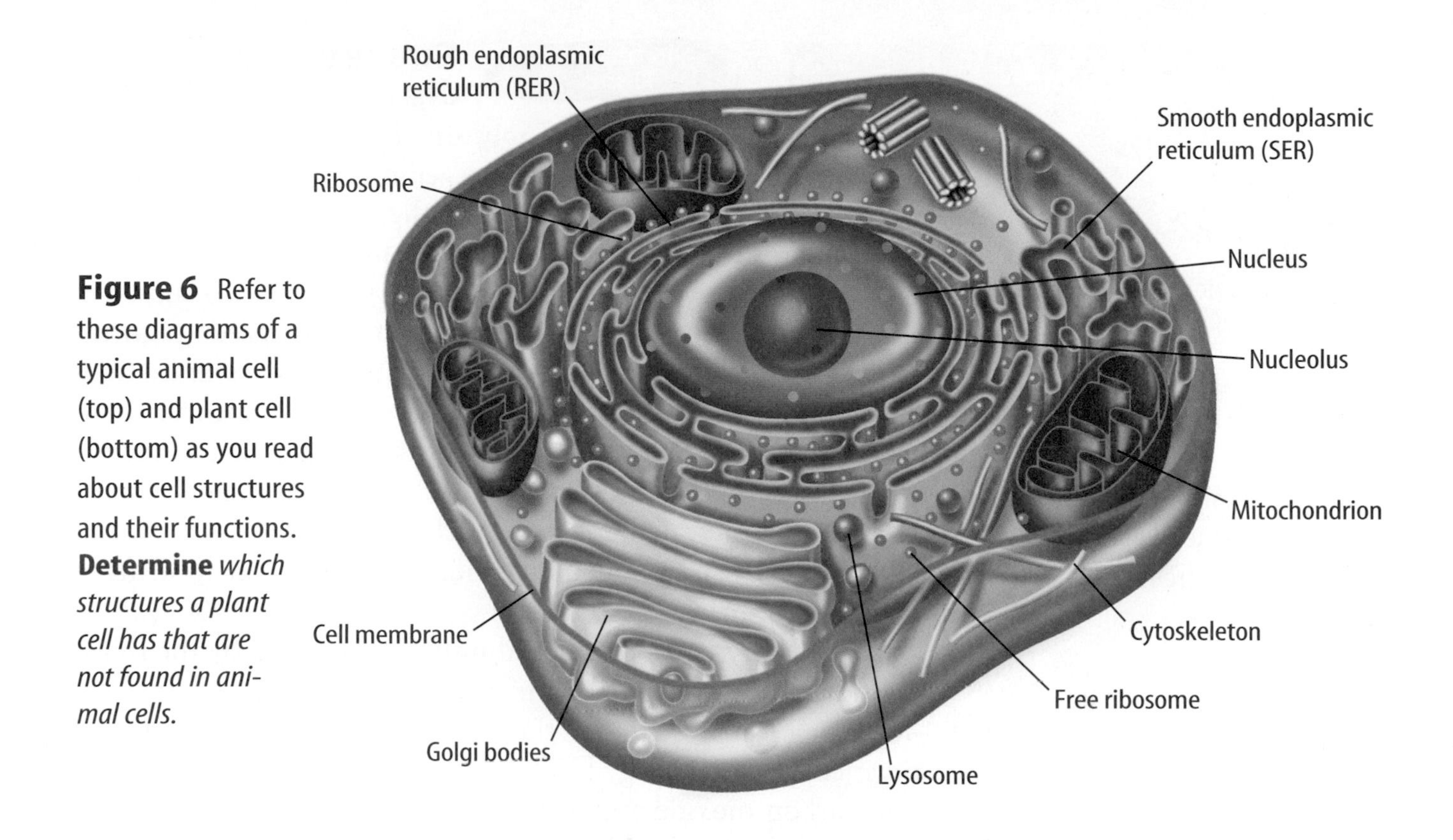

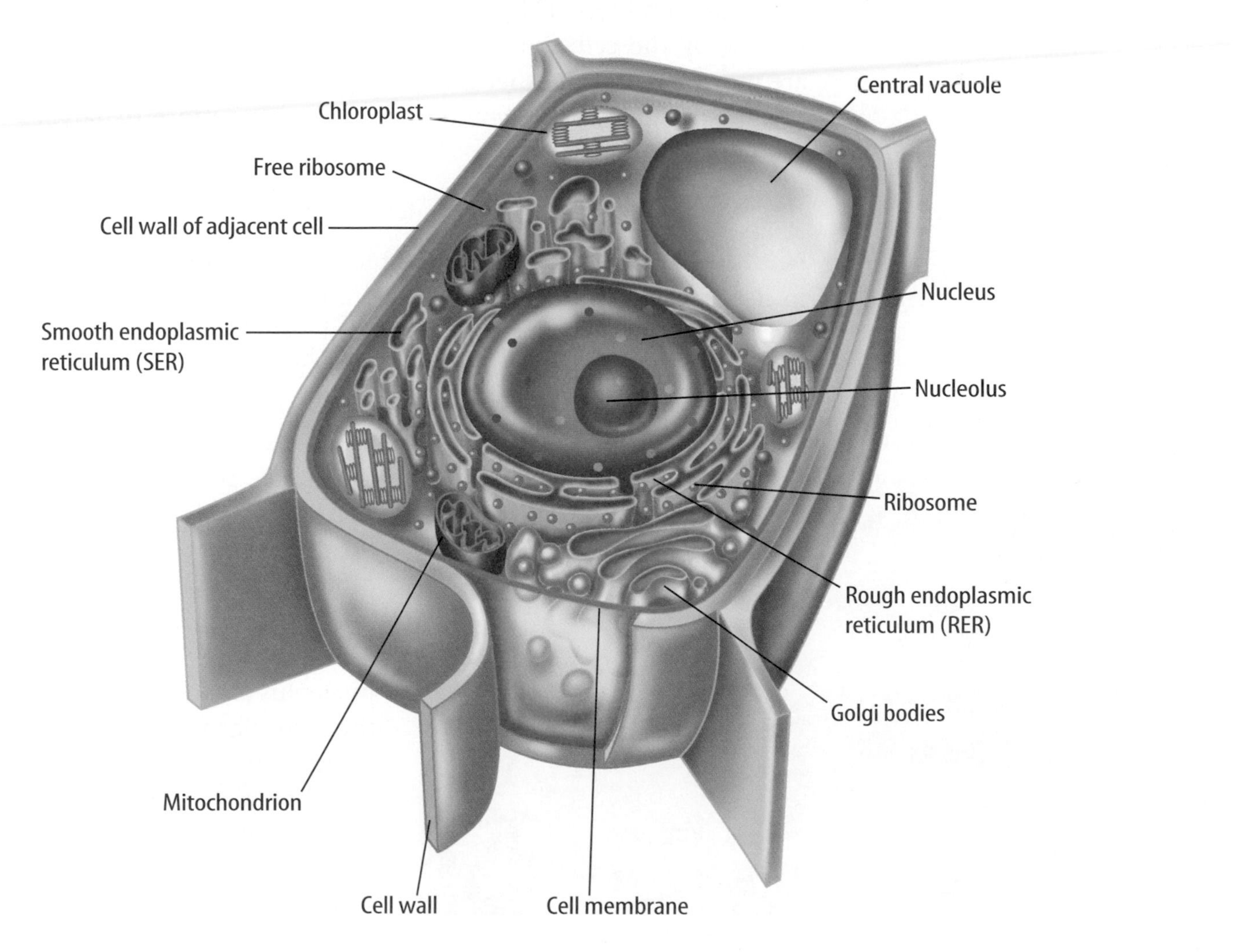

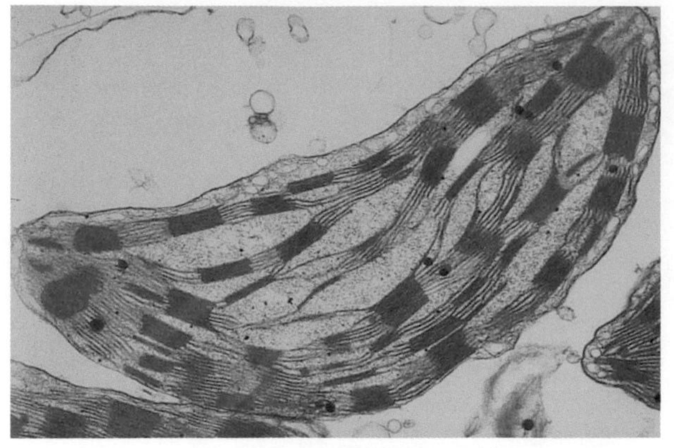

Color-enhanced TEM Magnification: 37000×

Figure 7 Chloroplasts are organelles that use light energy and make sugar from carbon dioxide and water.

Energy-Processing Organelles Cells require a continuous supply of energy to process food, make new substances, eliminate wastes, and communicate with each other. In some plant cells, food is made in green organelles in the cytoplasm called **chloroplasts** (KLOR uh plasts), as shown in **Figure 7.** Chloroplasts contain the green pigment chlorophyll, which gives many leaves and stems their green color. Chlorophyll captures light energy that is used to make a sugar called glucose. Glucose molecules store the captured light energy as chemical energy. Many cells, including animal cells, do not have chloroplasts for making food. They must get food from their environment.

The energy in food is stored until it is released by the mitochondria. **Mitochondria** (mi tuh KAHN dree uh) (singular, *mitochondrion*), such as the one shown in **Figure 8,** are organelles where energy is released from the breakdown of food into carbon dioxide and water. Just as the gas or electric company supplies fuel for the deli, a mitochondrion releases energy for use by the cell. Some types of cells, such as muscle cells, are more active than other cells. These cells have large numbers of mitochondria. Why would active cells have more or larger mitochondria?

Figure 8 Mitochondria are known as the powerhouses of the cell because they release energy that is needed by the cell from food.
Name *the cell types that might contain many mitochondria.*

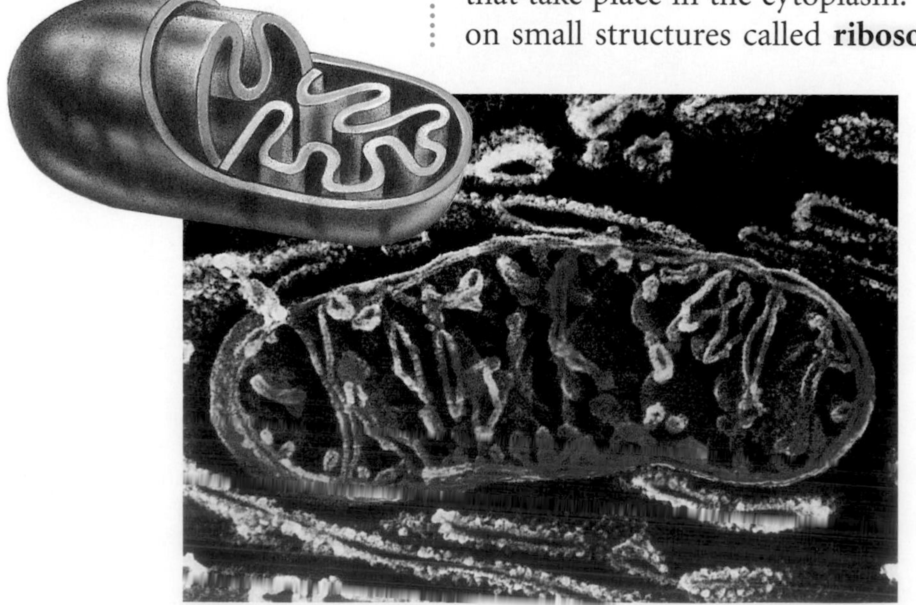

Color-enhanced SEM Magnification: [illegible]

Manufacturing Organelles One substance that takes part in nearly every cell activity is protein. Proteins are part of cell membranes. Other proteins are needed for chemical reactions that take place in the cytoplasm. Cells make their own proteins on small structures called **ribosomes.** Even though ribosomes are considered organelles, they are not membrane bound. Some ribosomes float freely in the cytoplasm; others are attached to the endoplasmic reticulum. Ribosomes are made in the nucleolus and move out into the cytoplasm. Ribosomes receive directions from the hereditary material in the nucleus on how, when, and in what order to make specific proteins.

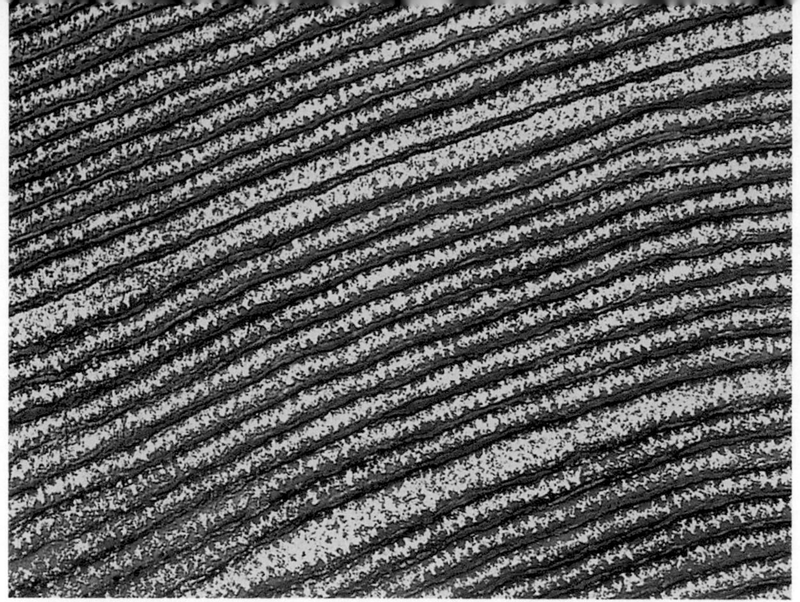
Color-enhanced TEM Magnification: 65000×

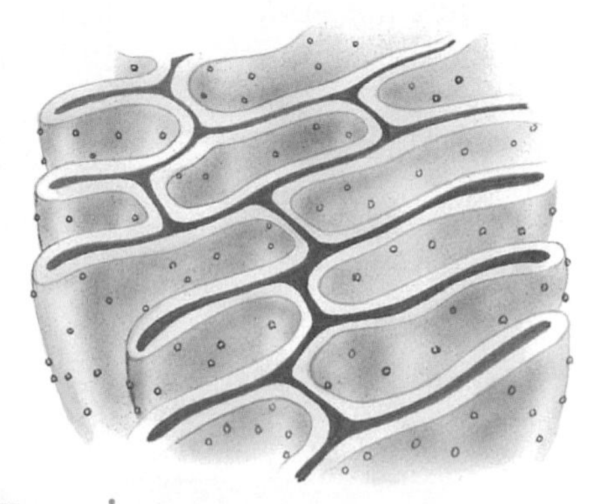
Figure 9 Endoplasmic reticulum (ER) is a complex series of membranes in the cytoplasm of the cell.
Infer *what smooth ER would look like.*

Processing, Transporting, and Storing Organelles

The **endoplasmic reticulum** (en duh PLAZ mihk • rih TIHK yuh lum) or ER, as shown in **Figure 9,** extends from the nucleus to the cell membrane. It is a series of folded membranes in which materials can be processed and moved around inside of the cell. The ER takes up a lot of space in some cells.

The endoplasmic reticulum may be "rough" or "smooth." ER that has no attached ribosomes is called smooth endoplasmic reticulum. This type of ER processes other cellular substances such as lipids that store energy. Ribsomes are attached to areas on the rough ER. There they carry out their job of making proteins that are moved out of the cell or used within the cell.

Reading Check ***What is the difference between rough ER and smooth ER?***

After proteins are made in a cell, they are transferred to another type of cell organelle called the Golgi (GAWL jee) bodies. The **Golgi bodies,** as shown ion **Figure 10,** are stacked, flattened membranes. The Golgi bodies sort proteins and other cellular substances and package them into membrane-bound structures called vesicles. The vesicles deliver cellular substances to areas inside the cell. They also carry cellular substances to the cell membrane where they are released to the outside of the cell.

Just as a deli has refrigerators for temporary storage of some of its foods and ingredients, cells have membrane-bound spaces called vacuoles for the temporary storage of materials. A vacuole can store water, waste products, food, and other cellular materials. In plant cells, the vacuole may make up most of the cell's volume.

Figure 10 The Golgi body packages materials and moves them to the outside of the cell.
Explain *why materials are removed from the cell.*

Color-enhanced TEM Magnification: 28000×
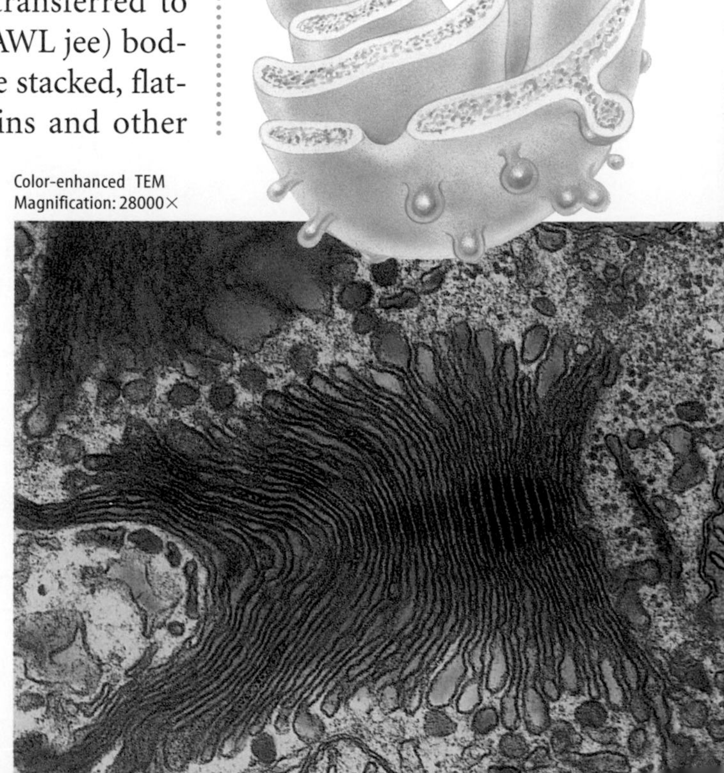

Recycling Just like a cell, you can recycle materials. Paper, plastics, aluminum, and glass are materials that can be recycled into usable items. Make a promotional poster to encourage others to recycle.

Recycling Organelles Active cells break down and recycle substances. Organelles called lysosomes (LI suh sohmz) contain digestive chemicals that help break down food molecules, cell wastes, and worn-out cell parts. In a healthy cell, chemicals are released into vacuoles only when needed. The lysosome's membrane prevents the digestive chemicals inside from leaking into the cytoplasm and destroying the cell. When a cell dies, a lysosome's membrane disintegrates. This releases digestive chemicals that allow the quick breakdown of the cell's contents.

What is the function of the lysosome's membrane?

Applying Math — Calculate a Ratio

CELL RATIO Assume that a cell is like a cube with six equal sides. Find the ratio of surface area to volume for a cube that is 4 cm high.

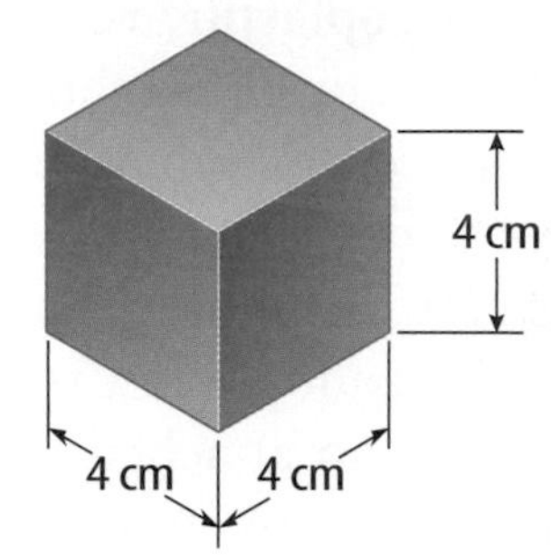

Solution

1 *This is what you know:* A cube has 6 equal sides of 4 cm × 4 cm.

2 *This is what you need to find out:* What is the ratio (R) of surface area to volume for the cube?

3 *These are the equations you use:*
- surface area (A) = width × length × 6
- volume (V) = length × width × height
- $R = A/V$

4 *This is the procedure you need to use:*
- Substitute in known values and solve the equations

$A = 4\text{ cm} \times 4\text{ cm} \times 6 = 96\text{ cm}^2$

$V = 4\text{ cm} \times 4\text{ cm} \times 4\text{ cm} = 64\text{ cm}^3$

$R = 96\text{ cm}^2/64\text{ cm}^3 = 1.5\text{ cm}^2/\text{cm}^3$

5 *Check your answer:* Multiply the ratio by the volume. Did you calculate the surface area?

Practice Problems

1. Calculate the ratio of surface area to volume for a cube that is 2 cm high. What happens to this ratio as the size of the cube decreases?
2. If a 4-cm cube doubled just one of its dimensions, what would happen to the ratio of surface area to volume?

For more practice, visit booka.msscience.com/math_practice

From Cell to Organism

Many one-celled organisms perform all their life functions by themselves. Cells in a many-celled organism, however, do not work alone. Each cell carries on its own life functions while depending in some way on other cells in the organism.

In **Figure 11,** you can see cardiac muscle cells grouped together to form a tissue. A **tissue** is a group of similar cells that work together to do one job. Each cell in a tissue does its part to keep the tissue alive.

Tissues are organized into organs. An **organ** is a structure made up of two or more different types of tissues that work together. Your heart is an organ made up of cardiac muscle tissue, nerve tissue, and blood tissues. The cardiac muscle tissue contracts, making the heart pump. The nerve tissue brings messages from the brain that tell the heart how fast to beat. The blood tissue is carried from the heart to other organs of the body.

What types of tissues make up your heart?

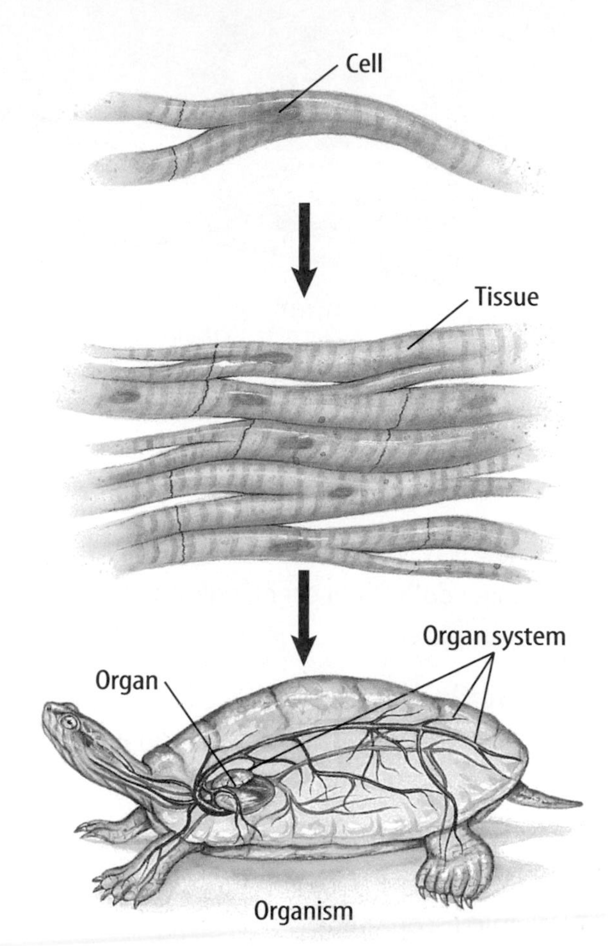

Figure 11 In a many-celled organism, cells are organized into tissues, tissues into organs, organs into systems, and systems into an organism.

A group of organs working together to perform a certain function is an organ system. Your heart, arteries, veins, and capillaries make up your cardiovascular system. In a many-celled organism, several systems work together in order to perform life functions efficiently. Your nervous, circulatory, respiratory, muscular, and other systems work together to keep you alive.

section 1 review

Summary

Common Cell Traits

- All cells have an outer covering called a cell membrane.
- Cells can be classified as prokaryotic or eukaryotic.

Cell Organization

- Each cell in your body has a specific function.
- Most of a cell's life processes occur in the cytoplasm.

From Cell to Organism

- In a many-celled organism, several systems work together to perform life functions.

Self Check

1. **Explain** why the nucleus is important in the life of a cell.
2. **Determine** why digestive enzymes in a cell are enclosed in a membrane-bound organelle.
3. **Discuss** how cells, tissues, organs, and organ systems are related.
4. **Think Critically** How is the cell of a one-celled organism different from the cells in many-celled organisms?

Applying Skills

5. **Interpret Scientific Illustrations** Examine **Figure 6.** Make a list of differences and similarities between the animal cell and the plant cell.

Comparing Cells

If you compared a goldfish to a rose, you would find them unlike each other. Are their individual cells different also?

Real-World Question

How do human cheek cells and plant cells compare?

Goals

- **Compare and contrast** an animal cell and a plant cell.

Materials

microscope
microscope slide
coverslip
forceps
tap water
dropper
Elodea plant
prepared slide of human cheek cells

Safety Precautions

Procedure

1. Copy the data table in your Science Journal. Check off the cell parts as you observe them.

Cell Observations		
Cell Part	Cheek	*Elodea*
Cytoplasm		
Nucleus		
Chloroplasts	Do not write in this book.	
Cell wall		
Cell membrane		

2. Using forceps, make a wet-mount slide of a young leaf from the tip of an *Elodea* plant.
3. **Observe** the leaf on low power. Focus on the top layer of cells.
4. Switch to high power and focus on one cell. In the center of the cell is a membrane-bound organelle called the central vacuole. Observe the chloroplasts—the green, disk-shaped objects moving around the central vacuole. Try to find the cell nucleus. It looks like a clear ball.
5. **Draw** the *Elodea* cell. Label the cell wall, cytoplasm, chloroplasts, central vacuole, and nucleus. Return to low power and remove the slide. Properly dispose of the slide.
6. **Observe** the prepared slide of cheek cells under low power.
7. Switch to high power and observe the cell nucleus. Draw and label the cell membrane, cytoplasm, and nucleus. Return to low power and remove the slide.

Conclude and Apply

1. **Compare and contrast** the shapes of the cheek cell and the *Elodea* cell.
2. **Draw conclusions** about the differences between plant and animal cells.

Communicating Your Data

Draw the two kinds of cells on one sheet of paper. Use a green pencil to label the organelles found only in plants, a red pencil to label the organelles found only in animals, and a blue pencil to label the organelles found in both. **For more help, refer to the Science Skill Handbook.**

section 2

Viewing Cells

Magnifying Cells

The number of living things in your environment that you can't see is much greater than the number that you can see. Many of the things that you cannot see are only one cell in size. To see most cells, you need to use a microscope.

Trying to see separate cells in a leaf, like the ones in **Figure 12,** is like trying to see individual photos in a photo mosaic picture that is on the wall across the room. As you walk toward the wall, it becomes easier to see the individual photos. When you get right up to the wall, you can see details of each small photo. A microscope has one or more lenses that enlarge the image of an object as though you are walking closer to it. Seen through these lenses, the leaf appears much closer to you, and you can see the individual cells that carry on life processes.

as you read

What **You'll Learn**

- **Compare** the differences between the compound light microscope and the electron microscope.
- **Summarize** the discoveries that led to the development of the cell theory.

Why **It's Important**

Humans are like other living things because they are made of cells.

Review Vocabulary

magnify: to increase the size of something

New Vocabulary

- cell theory

Early Microscopes In the late 1500s, the first microscope was made by a Dutch maker of reading glasses. He put two magnifying glasses together in a tube and could see an image that was larger than that made by either lens alone.

In the mid 1600s, Antonie van Leeuwenhoek, a Dutch fabric merchant, made a simple microscope with a tiny glass bead for a lens, as shown in **Figure 13.** With it, he reported seeing things in pond water that no one had ever imagined. His microscope could magnify up to 270 times. Another way to say this is that his microscope could make the object appear 270 times larger than its actual size. Today you would say his lens had a power of 270×. Early compound microscopes were crude by today's standards. The lenses would make a larger image, but it wasn't always sharp or clear.

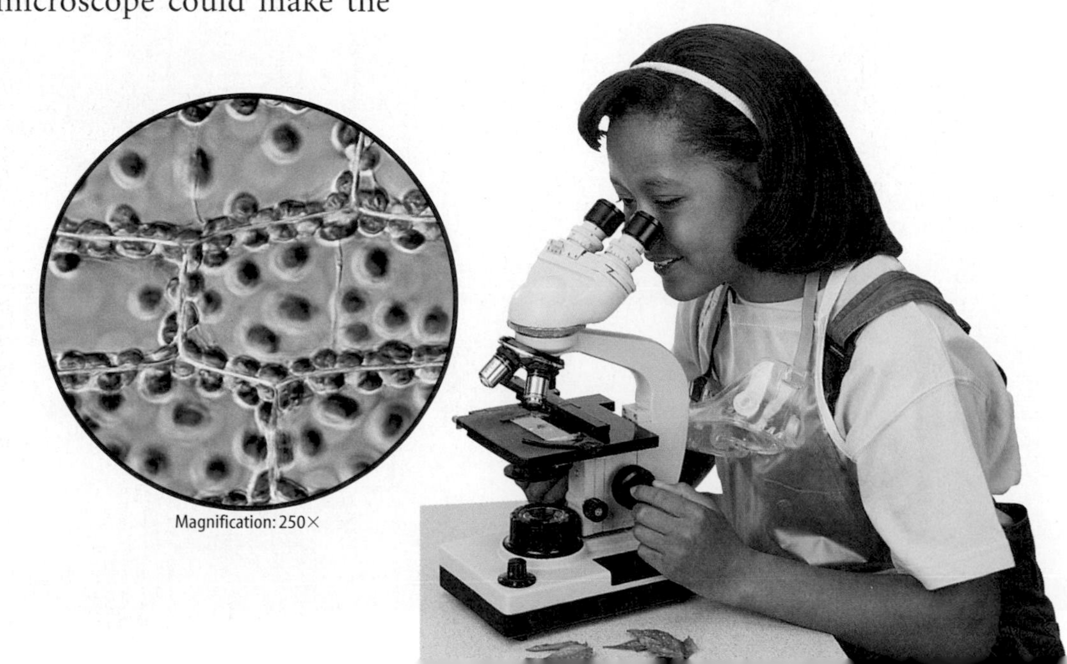

Figure 12 Individual cells become visible when a plant leaf is viewed using a microscope with enough magnifying power.

NATIONAL GEOGRAPHIC **VISUALIZING MICROSCOPES**

Figure 13

Microscopes give us a glimpse into a previously invisible world. Improvements have vastly increased their range of visibility, allowing researchers to study life at the molecular level. A selection of these powerful tools—and their magnification power—is shown here.

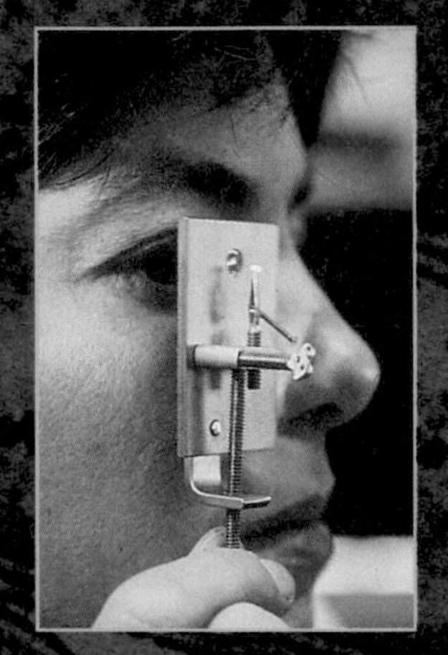

Up to 250× LEEUWENHOEK MICROSCOPE Held by a modern researcher, this historic microscope allowed Leeuwenhoek to see clear images of tiny freshwater organisms that he called "beasties."

Up to 2,000× BRIGHTFIELD / DARKFIELD MICROSCOPE The light microscope is often called the brightfield microscope because the image is viewed against a bright background. A brightfield microscope is the tool most often used in laboratories to study cells. Placing a thin metal disc beneath the stage, between the light source and the objective lenses, converts a brightfield microscope to a darkfield microscope. The image seen using a darkfield microscope is bright against a dark background. This makes details more visible than with a brightfield microscope. Below are images of a *Paramecium* as seen using both processes.

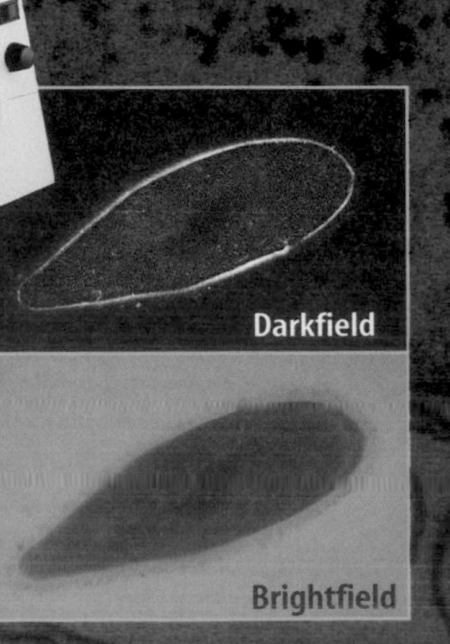

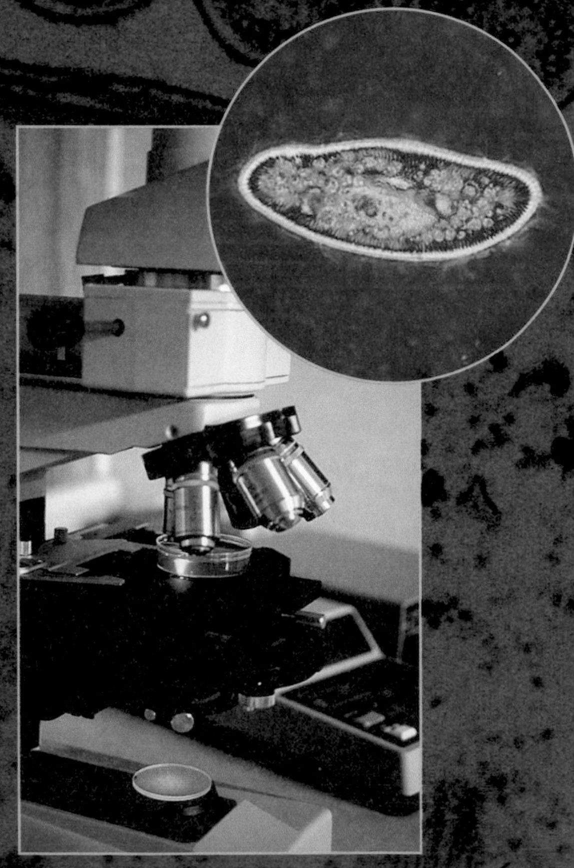

Up to 1,500× FLUORESCENCE MICROSCOPE This type of microscope requires that the specimen be treated with special fluorescent stains. When viewed through this microscope, certain cell structures or types of substances glow, as seen in the image of a *Paramecium* above.

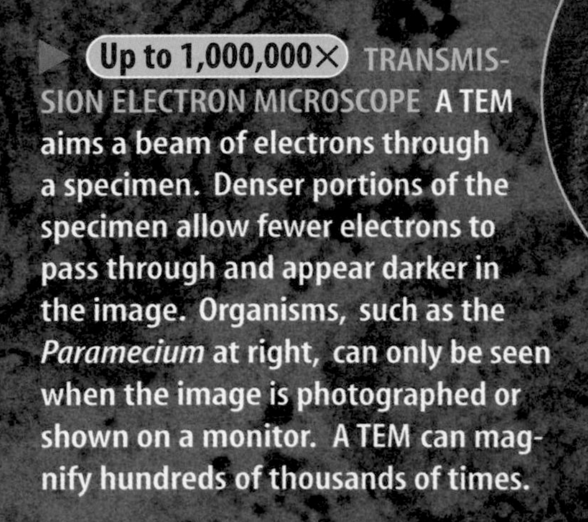

Up to 1,000,000× TRANSMISSION ELECTRON MICROSCOPE A TEM aims a beam of electrons through a specimen. Denser portions of the specimen allow fewer electrons to pass through and appear darker in the image. Organisms, such as the *Paramecium* at right, can only be seen when the image is photographed or shown on a monitor. A TEM can magnify hundreds of thousands of times.

Up to 1,500× PHASE-CONTRAST MICROSCOPE A phase-contrast microscope emphasizes slight differences in a specimen's capacity to bend light waves, thereby enhancing light and dark regions without the use of stains. This type of microscope is especially good for viewing living cells, like the *Paramecium* above left. The images from a phase-contrast microscope can only be seen when the specimen is photographed or shown on a monitor.

Up to 200,000× SCANNING ELECTRON MICROSCOPE An SEM sweeps a beam of electrons over a specimen's surface, causing other electrons to be emitted from the specimen. SEMs produce realistic, three-dimensional images, which can only be viewed as photographs or on a monitor, as in the image of the *Paramecium* at right. Here a researcher compares an SEM picture to a computer monitor showing an enhanced image.

Observing Magnified Objects

Procedure

1. Look at a **newspaper** through the curved side and through the flat bottom of an **empty, clear glass.**
2. Look at the newspaper through a **clear glass bowl filled with water** and then with a **magnifying lens.**

Analysis

In your Science Journal, compare how well you can see the newspaper through each of the objects.

Try at Home

Cell Biologist Microscopes are important tools for cell biologists as they research diseases. In your Science Journal, make a list of diseases for which you think cell biologists are trying to find effective drugs.

Modern Microscopes Scientists use different microscopes to study organisms, cells, and cell parts that are too small to be seen with just the human eye. Depending on how many lenses a microscope contains, it is called simple or compound. A simple microscope is similar to a magnifying lens. It has only one lens. A microscope's lens makes an enlarged image of an object and directs light toward your eye. The change in apparent size produced by a microscope is called magnification. Microscopes vary in powers of magnification. Some microscopes can make images of individual atoms.

The microscope you probably will use to study life science is a compound light microscope, similar to the one in the Reference Handbook at the back of this book. The compound light microscope has two sets of lenses—eyepiece lenses and objective lenses. The eyepiece lenses are mounted in one or two tubelike structures. Images of objects viewed through two eyepieces, or stereomicroscopes, are three-dimensional. Images of objects viewed through one eyepiece are not. Compound light microscopes usually have two to four movable objective lenses.

Magnification The powers of the eyepiece and objective lenses determine the total magnifications of a microscope. If the eyepiece lens has a power of 10× and the objective lens has a power of 43×, then the total magnification is 430× (10× times 43×). Some compound microscopes, like those in **Figure 13,** have more powerful lenses that can magnify an object up to 2,000 times its original size.

Electron Microscopes Things that are too small to be seen with other microscopes can be viewed with an electron microscope. Instead of using lenses to direct beams of light, an electron microscope uses a magnetic field in a vacuum to direct beams of electrons. Some electron microscopes can magnify images up to one million times. To see electron microscope images, they must be photographed or electronically produced.

Several kinds of electron microscopes have been invented, as shown in **Figure 13.** Scanning electron microscopes (SEM) produce a realistic, three-dimensional image. Only the surface of the specimen can be observed using an SEM. Transmission electron microscopes (TEM) produce a two-dimensional image of a thinly-sliced specimen. Details of cell parts can be examined using a TEM. Scanning tunneling microscopes (STM) are able to show the arrangement of atoms on the surface of a molecule. A metal probe is placed near the surface of the specimen and electrons flow from the tip. The hills and valleys of the specimen's surface are mapped.

Cell Theory

Table 1 The Cell Theory	
All organisms are made up of one or more cells.	An organism can be one cell or many cells like most plants and animals.
The cell is the basic unit of organization in organisms.	Even in complex organisms, the cell is the basic unit of structure and function.
All cells come from cells.	Most cells can divide to form two new, identical cells.

During the seventeenth century, scientists used their new invention, the microscope, to explore the newly discovered microscopic world. They examined drops of blood, scrapings from their own teeth, and other small things. Cells weren't discovered until the microscope was improved. In 1665, Robert Hooke cut a thin slice of cork and looked at it under his microscope. To Hooke, the cork seemed to be made up of empty little boxes, which he named cells.

In the 1830s, Matthias Schleiden used a microscope to study plants and concluded that all plants are made of cells. Theodor Schwann, after observing different animal cells, concluded that all animals are made up of cells. Eventually, they combined their ideas and became convinced that all living things are made of cells.

Several years later, Rudolf Virchow hypothesized that cells divide to form new cells. Virchow proposed that every cell came from a cell that already existed. His observations and conclusions and those of others are summarized in the **cell theory,** as described in **Table 1.**

Who first concluded that all animals are made of cells?

section 2 review

Summary

Magnifying Cells

- The powers of the eyepiece and objective lenses determine the total magnification of a microscope.
- An electron microscope uses a magnetic field in a vacuum to direct beams of electrons.

Development of the Cell Theory

- In 1665, Robert Hooke looked at a piece of cork under his microscope and called what he saw cells.
- The conclusions of Rudolf Virchow and those of others are summarized in the cell theory.

Self Check

1. **Determine** why the invention of the microscope was important in the study of cells.
2. **State** the cell theory.
3. **Compare** a simple and a compound light microscope.
4. **Explain** Virchow's contribution to the cell theory.
5. **Think Critically** Why would it be better to look at living cells than at dead cells?

Applying Math

6. **Solve One-Step Equations** Calculate the magnifications of a microscope that has an 8× eyepiece and 10× and 40× objectives.

section 3

Viruses

as you read

What You'll Learn

- **Explain** how a virus makes copies of itself.
- **Identify** the benefits of vaccines.
- **Investigate** some uses of viruses.

Why It's Important

Viruses infect nearly all organisms, usually affecting them negatively yet sometimes affecting them positively.

Review Vocabulary
disease: a condition that results from the disruption in function of one or more of an organism's normal processes

New Vocabulary
- virus
- host cell

What are viruses?

Cold sores, measles, chicken pox, colds, the flu, and AIDS are diseases caused by nonliving particles called viruses. A **virus** is a strand of hereditary material surrounded by a protein coating. Viruses don't have a nucleus or other organelles. They also lack a cell membrane. Viruses, as shown in **Figure 14,** have a variety of shapes. Because they are too small to be seen with a light microscope, they were discovered only after the electron microscope was invented. Before that time, scientists only hypothesized about viruses.

How do viruses multiply?

All viruses can do is make copies of themselves. However, they can't do that without the help of a living cell called a **host cell.** Crystalized forms of some viruses can be stored for years. Then, if they enter an organism, they can multiply quickly.

Once a virus is inside of a host cell, the virus can act in two ways. It can either be active or it can become latent, which is an inactive stage.

Figure 14 Viruses come in a variety of shapes.

Color-enhanced TEM Magnification: 160000×

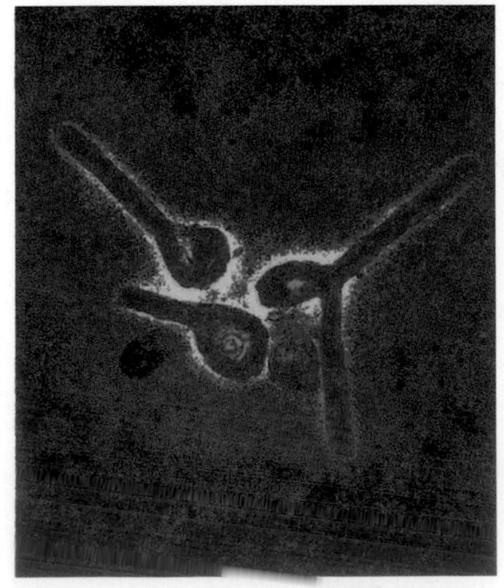

Filoviruses do not have uniform shapes. Some of these *Ebola* viruses have a loop at one end.

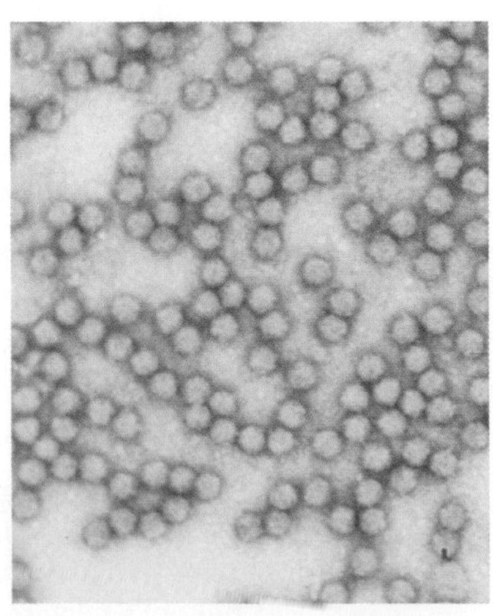

The potato leafroll virus, *Polervirus,* damages potato crops worldwide.

Color-enhanced SEM Magnification: 140000×

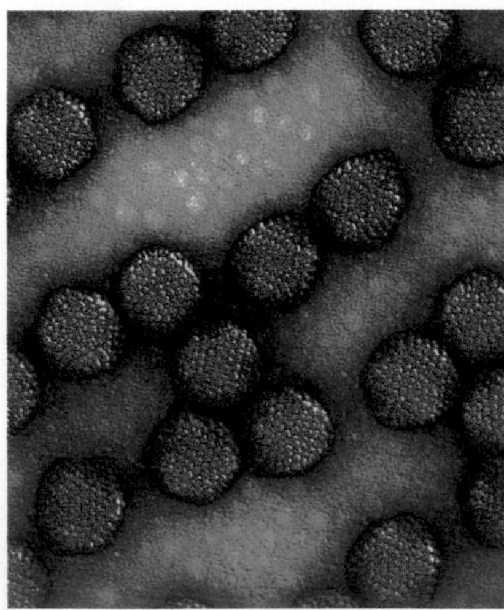

This is just one of the many adenoviruses that can cause the common cold.

Figure 15 An active virus multiplies and destroys the host cell.

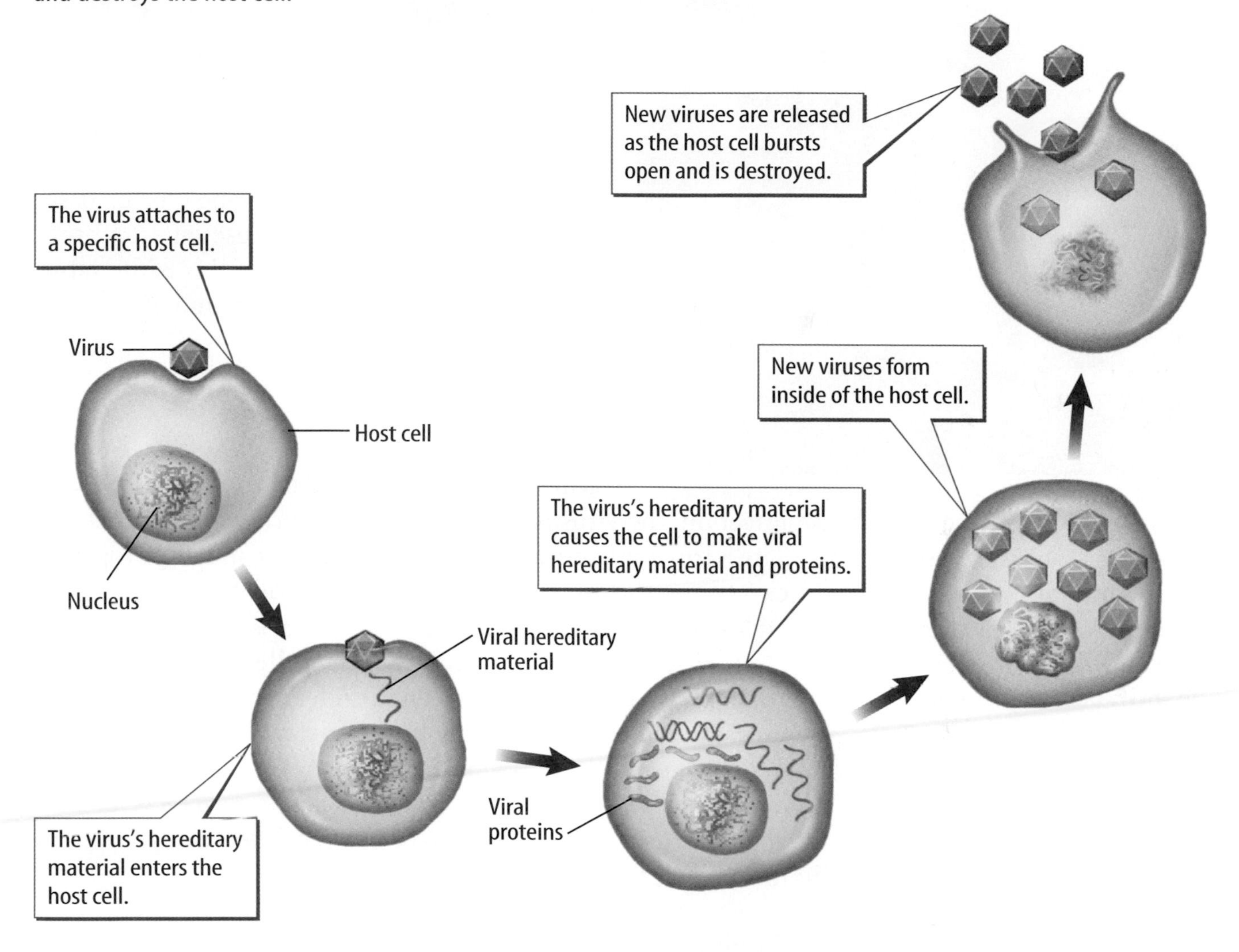

Active Viruses When a virus enters a cell and is active, it causes the host cell to make new viruses. This process destroys the host cell. Follow the steps in **Figure 15** to see one way that an active virus functions inside a cell.

Latent Viruses Some viruses can be latent. That means that after the virus enters a cell, its hereditary material can become part of the cell's hereditary material. It does not immediately make new viruses or destroy the cell. As the host cell reproduces, the viral DNA is copied. A virus can be latent for many years. Then, certain conditions, either inside or outside your body, cause the latent virus to become an active virus.

If you have had a cold sore on your lip, a latent virus in your body has become active. The cold sore is a sign that the virus is active and destroying cells in your lip. When the cold sore disappears, the virus has become latent again. The virus is still in your body's cells, but it is hiding and doing no apparent harm.

Topic: Virus Reactivation
Visit booka.msscience.com for Web links to information about viruses.

Activity In your Science Journal, list five stimuli that might activate a latent virus.

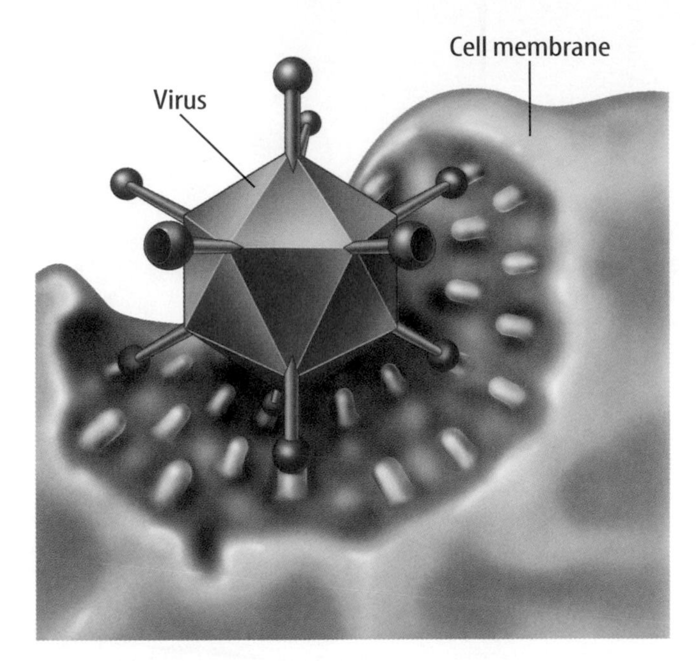

Figure 16 Viruses and the attachment sites of the host cell must match exactly. That's why most viruses infect only one kind of host cell.
Identify *diseases caused by viruses.*

How do viruses affect organisms?

Viruses attack animals, plants, fungi, protists, and bacteria. Some viruses can infect only specific kinds of cells. For instance, many viruses, such as the potato leafroll virus, are limited to one host species or to one type of tissue within that species. A few viruses affect a broad range of hosts. An example of this is the rabies virus. Rabies can infect humans and many other animal hosts.

A virus cannot move by itself, but it can reach a host's body in several ways. For example, it can be carried onto a plant's surface by the wind or it can be inhaled by an animal. In a viral infection, the virus first attaches to the surface of the host cell. The virus and the place where it attaches must fit together exactly, as shown in **Figure 16.** Because of this, most viruses attack only one kind of host cell.

Viruses that infect bacteria are called bacteriophages (bak TIHR ee uh fay jihz). They differ from other kinds of viruses in the way that they enter bacteria and release their hereditary material. Bacteriophages attach to a bacterium and inject their hereditary material. The entire cycle takes about 20 min, and each virus-infected cell releases an average of 100 viruses.

Fighting Viruses

Vaccines are used to prevent disease. A vaccine is made from weakened virus particles that can't cause disease anymore. Vaccines have been made to prevent many diseases, including measles, mumps, smallpox, chicken pox, polio, and rabies.

Reading Check *What is a vaccine?*

The First Vaccine Edward Jenner is credited with developing the first vaccine in 1796. He developed a vaccine for smallpox, a disease that was still feared in the early twentieth century. Jenner noticed that people who got a disease called cowpox didn't get smallpox. He prepared a vaccine from the sores of people who had cowpox. When injected into healthy people, the cowpox vaccine protected them from smallpox. Jenner didn't know he was fighting a virus. At that time, no one understood what caused disease or how the body fought disease.

Topic: Filoviruses
Visit booka.msscience.com for Web links to information about the virus family *Filoviridae.*

Activity Make a table that displays the virus name, location, and year of the initial outbreaks associated with the *Filoviridae* family.

Treating Viral Diseases Antibiotics treat bacterial infections but are not effective against viral diseases. One way your body can stop viral infections is by making interferons. Interferons are proteins that are produced rapidly by virus-infected cells and move to noninfected cells in the host. They cause the noninfected cells to produce protective substances.

Antiviral drugs can be given to infected patients to help fight a virus. A few drugs show some effectiveness against viruses but some have limited use because of their adverse side effects.

Preventing Viral Diseases Public health measures for preventing viral diseases include vaccinating people, improving sanitary conditions, quarantining patients, and controlling animals that spread disease. For example, annual rabies vaccinations of pets and farm animals protect them and humans from infection. To control the spread of rabies in wild animals such as coyotes and wolves, wildlife workers place bait containing an oral rabies vaccine, as shown in **Figure 17,** where wild animals will find it.

Figure 17 This oral rabies bait is being prepared for an aerial drop by the Texas Department of Health as part of their Oral Rabies Vaccination Program. This five-year program has prevented the expansion of rabies into Texas.

Research with Viruses

You might think viruses are always harmful. However, through research, scientists are discovering helpful uses for some viruses. One use, called gene transfer, substitutes normal hereditary material for a cell's defective hereditary material. The normal material is enclosed in viruses that "infect" targeted cells. The new hereditary material enters the cells and replaces the defective hereditary material. Using gene therapy, scientists hope to help people with genetic disorders and find a cure for cancer.

section 3 review

Summary

What are viruses?

- A virus is a strand of hereditary material surrounded by a protein coating.

How do viruses multiply?

- An active virus immediately destroys the host cell but a latent virus does not.

Fighting Viruses and Research with Viruses

- Antiviral drugs can be given to infected patients to help fight a virus.
- Scientists are discovering helpful uses for some viruses.

Self Check

1. **Describe** how viruses multiply.
2. **Explain** how vaccines are beneficial.
3. **Determine** how some viruses might be helpful.
4. **Discuss** how viral diseases might be prevented.
5. **Think Critically** Explain why a doctor might not give you any medication if you have a viral disease.

Applying Skills

6. **Concept Map** Make an events-chain concept map to show what happens when a latent virus becomes active.

Design Your Own

Comparing Light Microscopes

Goals

- **Learn** how to correctly use a stereomicroscope and a compound light microscope.
- **Compare** the uses of the stereomicroscope and compound light microscope.

Possible Materials

compound light microscope
stereomicroscope
items from the classroom—include some living or once-living items (8)
microscope slides and coverslips
plastic petri dishes
distilled water
dropper

Safety Precautions

Real-World Question

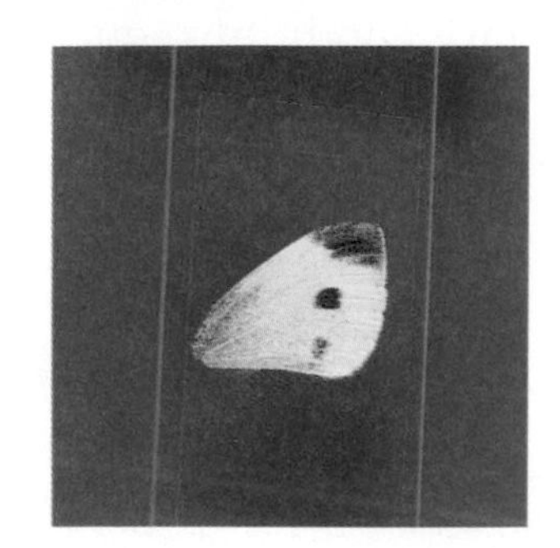

You're a technician in a police forensic laboratory. You use a stereomicroscope and a compound light microscope in the laboratory. A detective just returned from a crime scene with bags of evidence. You must examine each piece of evidence under a microscope. How do you decide which microscope is the best tool to use? Will all of the evidence that you've collected be viewable through both microscopes?

Form a Hypothesis

Compare the items to be examined under the microscopes. Form a hypothesis to predict which microscope will be used for each item and explain why.

Test Your Hypothesis

Make a Plan

1. As a group, decide how you will test your hypothesis.
2. **Describe** how you will carry out this experiment using a series of specific steps. Make sure the steps are in a logical order. Remember that you must place an item in the bottom of a plastic petri dish to examine it under the stereomicroscope and you must make a wet mount of any item to be examined under the compound light microscope. For more help, see the Reference Handbook.
3. If you need a data table or an observation table, design one in your Science Journal.

Follow Your Plan

1. Make sure your teacher approves the objects you'll examine, your plan, and your data table before you start.
2. Carry out the experiment.
3. While doing the experiment, record your observations and complete the data table.

Analyze Your Data

1. **Compare** the items you examined with those of your classmates.
2. **Classify** the eight items you observed based on this experiment.

Conclude and Apply

1. **Infer** which microscope a scientist might use to examine a blood sample, fibers, and live snails.
2. **List** five careers that require people to use a stereomicroscope. List five careers that require people to use a compound light microscope. Enter the lists in your Science Journal.
3. **Infer** how the images would differ if you examined an item under a compound light microscope and a stereomicroscope.
4. **Determine** which microscope is better for looking at large, or possibly live, items.

Communicating Your Data

In your Science Journal, **write** a short description of an imaginary crime scene and the evidence found there. Sort the evidence into two lists—items to be examined under a stereomicroscope and items to be examined under a compound light microscope. **For more help, refer to the Science Skill Handbook.**

TIME SCIENCE AND HISTORY

SCIENCE CAN CHANGE THE COURSE OF HISTORY!

Cobb Against Cancer

This colored scanning electron micrograph (SEM) shows two breast cancer cells in the final stage of cell division.

Jewel Plummer Cobb is a cell biologist who did important background research on the use of drugs against cancer in the 1950s. She removed cells from cancerous tumors and cultured them in the lab. Then, in a controlled study, she tried a series of different drugs against batches of the same cells. Her goal was to find the right drug to cure each patient's particular cancer. Cobb never met that goal, but her research laid the groundwork for modern chemotherapy—the use of chemicals to treat cancer.

Jewel Cobb also influenced science in another way. She was a role model, especially in her role as dean or president of several universities. Cobb promoted equal opportunity for students of all backgrounds, especially in the sciences.

Light Up a Cure

Vancouver, British Columbia 2000. While Cobb herself was only able to infer what was going on inside a cell from its reactions to various drugs, her work has helped others go further. Building on Cobb's work, Professor Julia Levy and her research team at the University of British Columbia actually go inside cells, and even organelles, to work against cancer. One technique they are pioneering is the use of light to guide cancer drugs to the right cells. First, the patient is given a chemotherapy drug that reacts to light. Then, a fiber optic tube is inserted into the tumor. Finally, laser light is passed through the tube, which activates the light-sensitive drug—but only in the tumor itself. This will hopefully provide a technique to keep healthy cells healthy while killing sick cells.

Write Report on Cobb's experiments on cancer cells. What were her dependent and independent variables? What would she have used as a control? What sources of error did she have to guard against? Answer the same questions about Levy's work.

Reviewing Main Ideas

Section 1 Cell Structure

1. Prokaryotic and eukaryotic are the two cell types.
2. The DNA in the nucleus controls cell functions.
3. Organelles such as mitochondria and chloroplasts process energy.
4. Most many-celled organisms are organized into tissues, organs, and organ systems.

Section 2 Viewing Cells

1. A simple microscope has just one lens. A compound light microscope has an eyepiece and objective lenses.
2. To calculate the magnification of a microscope, multiply the power of the eyepiece by the power of the objective lens.
3. According to the cell theory, the cell is the basic unit of life. Organisms are made of one or more cells, and all cells come from other cells.

Section 3 Viruses

1. A virus is a structure containing hereditary material surrounded by a protein coating.
2. A virus can make copies of itself only when it is inside a living host cell.

Visualizing Main Ideas

Copy and complete the following concept map of the basic units of life.

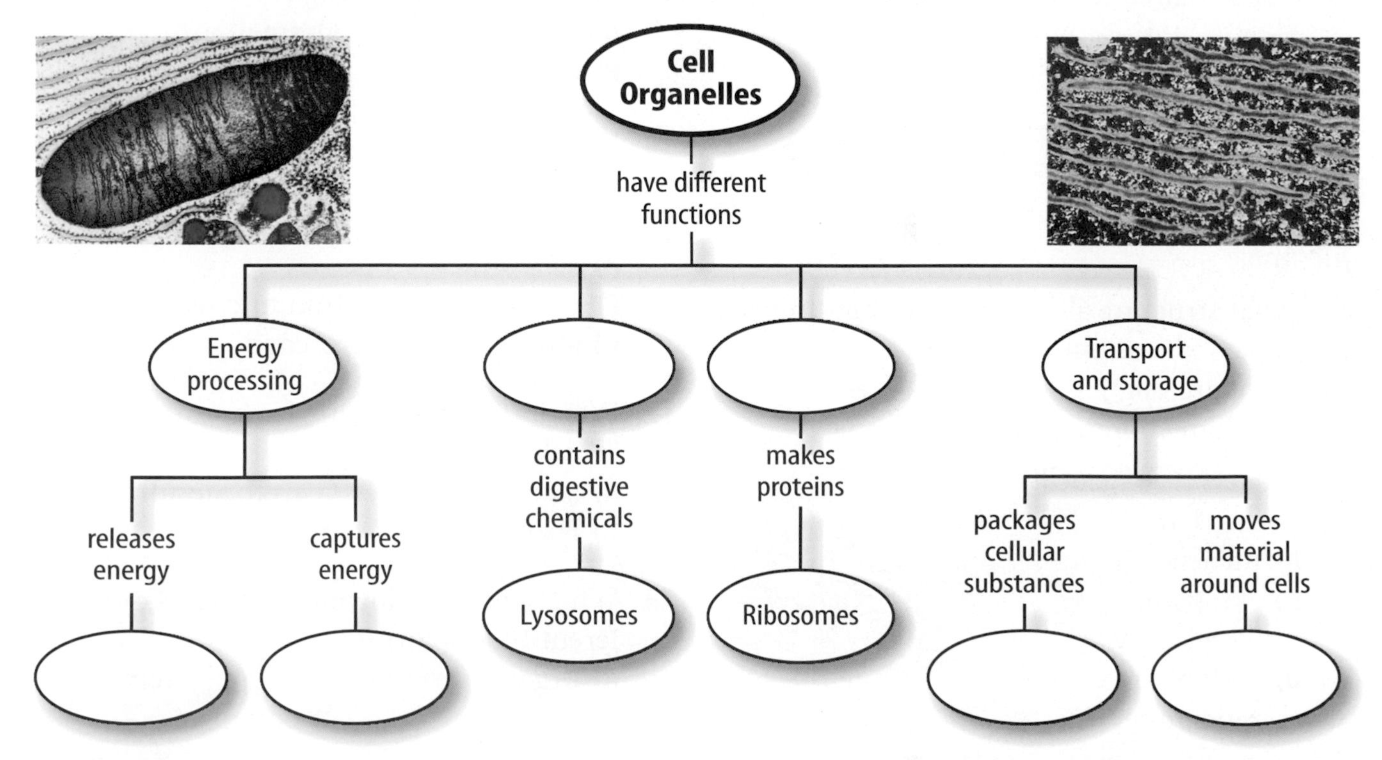

Using Vocabulary

cell membrane p. 40	host cell p. 54
cell theory p. 53	mitochondrion p. 44
cell wall p. 41	nucleus p. 42
chloroplast p. 44	organ p. 47
cytoplasm p. 40	organelle p. 42
endoplasmic reticulum p. 45	ribosome p. 44
Golgi body p. 45	tissue p. 47
	virus p. 54

Using the vocabulary words, give an example of each of the following.

1. found in every organ
2. smaller than one cell
3. a plant-cell organelle
4. part of every cell
5. powerhouse of a cell
6. used by biologists
7. contains hereditary material
8. a structure that surrounds the cell
9. can be damaged by a virus
10. made up of cells

Checking Concepts

Choose the word or phrase that best answers the question.

11. What structure allows only certain things to pass in and out of the cell?
    A) cytoplasm **C)** ribosomes
    B) cell membrane **D)** Golgi body

12. What is the organelle to the right?
    A) nucleus
    B) cytoplasm
    C) Golgi body
    D) endoplasmic reticulum

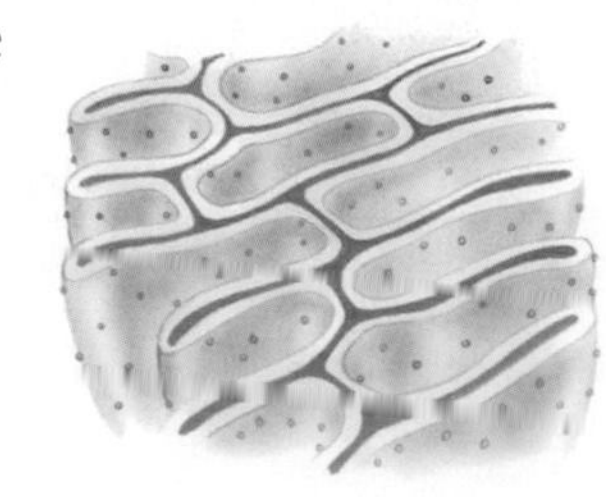

Use the illustration below to answer question 13.

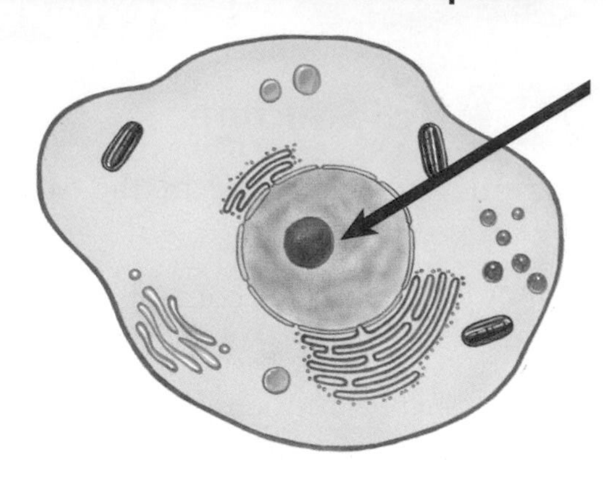

13. In the figure above, what is the function of the structure that the arrow is pointing to?
    A) recycles old cell parts
    B) controls cell activities
    C) protection
    D) releases energy

14. Which scientist gave the name *cells* to structures he viewed?
    A) Hooke **C)** Schleiden
    B) Schwann **D)** Virchow

15. Which of the following is a viral disease?
    A) tuberculosis **C)** smallpox
    B) anthrax **D)** tetanus

16. Which microscope can magnify up to a million times?
    A) compound light microscope
    B) stereomicroscope
    C) transmission electron microscope
    D) atomic force microscope

17. Which of the following is part of a bacterial cell?
    A) a cell wall **C)** mitochondria
    B) lysosomes **D)** a nucleus

18. Which of the following do groups of different tissues form?
    A) organ **C)** organ system
    B) organelle **D)** organism

Science Online booka.msscience.com/vocabulary_puzzlemaker

Thinking Critically

19. **Infer** why it is difficult to treat a viral disease.
20. **Explain** which type of microscope would be best to view a piece of moldy bread.
21. **Predict** what would happen to a plant cell that suddenly lost its chloroplasts.
22. **Predict** what would happen if the animal cell shown to the right didn't have ribosomes.

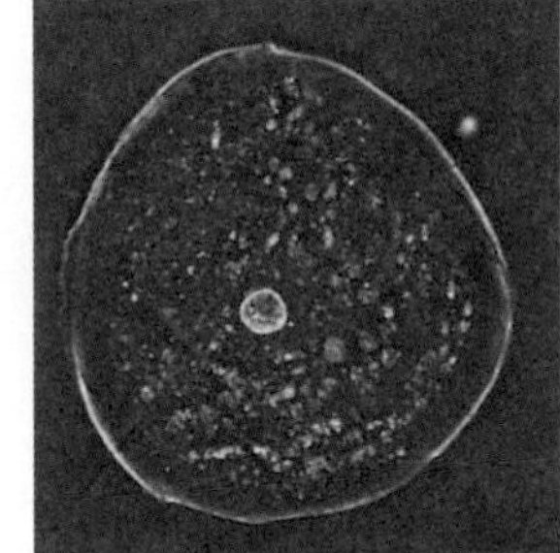

23. **Determine** how you would decide whether an unknown cell was an animal cell, a plant cell, or a bacterial cell.
24. **Concept Map** Make an events-chain concept map of the following from simple to complex: *small intestine, circular muscle cell, human,* and *digestive system.*
25. **Interpret Scientific Illustrations** Use the illustrations in **Figure 1** to describe how the shape of a cell is related to its function.

Use the table below to answer question 26.

Cell Structures

Structure	Prokaryotic Cell	Eukaryotic Cell
Cell membrane		Yes
Cytoplasm	Yes	
Nucleus		Yes
Endoplasmic reticulum		
Golgi bodies		

26. **Compare and Contrast** Copy and complete the table above.
27. **Make a Model** Make and illustrate a time line about the development of the cell theory. Begin with the development of the microscope and end with Virchow. Include the contributions of Leeuwenhoek, Hooke, Schleiden, and Schwann.

Performance Activities

28. **Model** Use materials that resemble cell parts or represent their functions to make a model of a plant cell or an animal cell. Include a cell-parts key.
29. **Poster** Make a poster about the history of vaccinations. Contact your local Health Department for current information.

Applying Math

Use the illustration below to answer question 30.

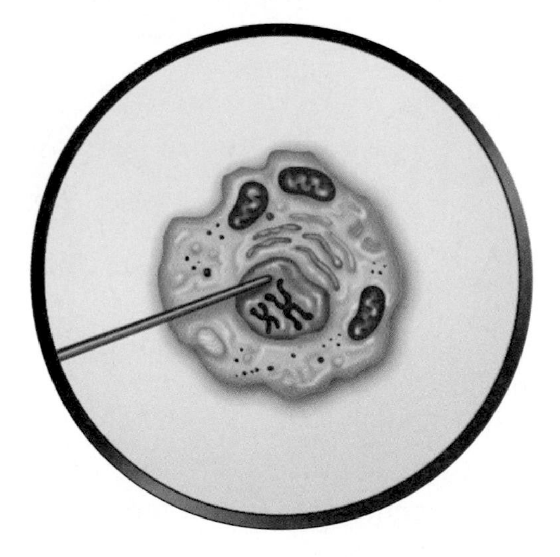

30. **Cell Width** If the pointer shown above with the cell is 10 micrometers (μm) in length, then about how wide is this cell?
    A) 20 μm **C)** 5 μm
    B) 10 μm **D)** 0.1 μm
31. **Magnification** Calculate the magnification of a microscope with a 20× eyepiece and a 40× objective.

chapter 2 Standardized Test Practice

Part 1 Multiple Choice

Record your answers on the answer sheet provided by your teacher or on a sheet of paper.

1. What do a bacterial cell, a plant cell, and a nerve cell have in common?

A. cell wall and nucleus
B. cytoplasm and cell membrane
C. endoplasmic reticulum
D. flagella

2. Which is not a function of an organelle?

A. cell shape and movement
B. energy release
C. chemical transfer
D. chemical storage

Use the images below to answer question 3.

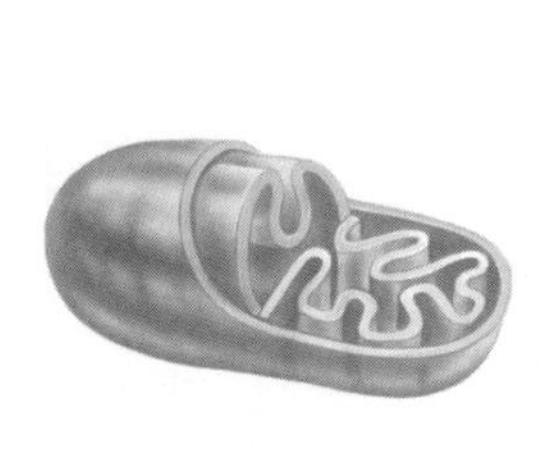

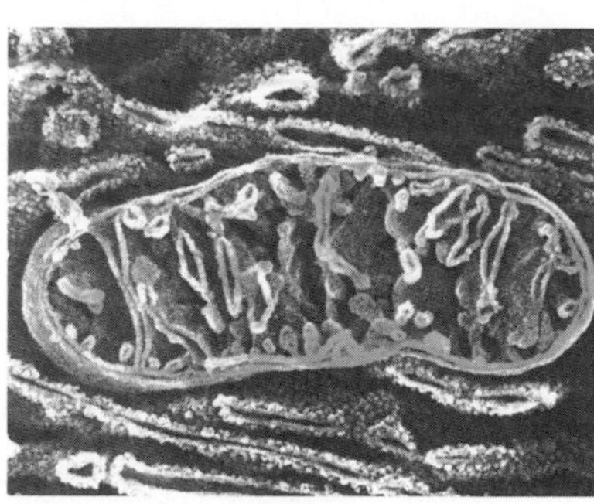

3. What is the primary function of this organelle?

A. capturing light energy
B. directing cell processes
C. releasing energy stored in food
D. making proteins

4. Which organelles receive the directions from the DNA in the nucleus about which proteins to make?

A. ribosomes
B. endoplasmic reticulum
C. Golgi bodies
D. cell wall

5. Why is a virus not considered a living cell?

A. It has a cell wall.
B. It has hereditary material.
C. It has no organelles.
D. It cannot multiply.

Use the illustration below to answer questions 6 and 7.

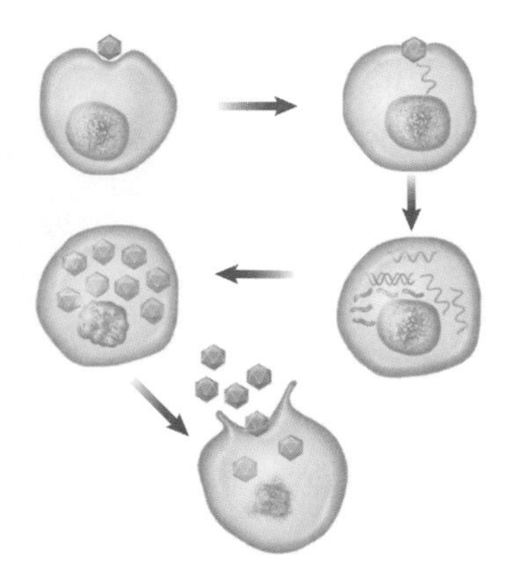

6. What does the diagram above represent?

A. cell reproduction
B. bacterial reproduction
C. active virus multiplication
D. vaccination

7. What does the largest circular structure represent?

A. a host cell
B. a ribosome
C. a vacuole
D. the nucleus

8. Where do most of a cell's life processes occur?

A. nucleus
B. cell wall
C. organ
D. cytoplasm

9. What is a group of similar cells that work together?

A. tissue
B. organ
C. organ system
D. organism

Test-Taking Tip

Read Carefully Read each question carefully for full understanding.

Part 2 Short Response/Grid In

Record your answers on the answer sheet provided by your teacher or on a sheet of paper.

10. Compare and contrast the cell wall and the cell membrane.

11. How would a cell destroy or breakdown a harmful chemical which entered the cytoplasm?

12. How does your body stop viral infections? What are other ways of protection against viral infections?

13. Where is cellulose found in a cell and what is its function?

Use the following table to answer question 14.

Organelle	Function
	Directs all cellular activities
Mitochondria	
	Captures light energy to make glucose
Ribosomes	

14. Copy and complete the table above with the appropriate information.

15. How are Golgi bodies similar to a packaging plant?

16. Why does a virus need a host cell?

17. Give an example of an organ system and list the organs in it.

18. Compare and contrast the energy processing organelles.

19. Describe the structure of viruses.

20. How do ribosomes differ from other cell structures found in the cytoplasm?

21. What kind of microscope uses a series of lenses to magnify?

Part 3 Open Ended

Record your answers on a sheet of paper.

22. Name three different types of microscopes and give uses for each.

23. Some viruses, like the common cold, only make the host organism sick, but other viruses, like *Ebola,* are deadly to the host organism. Which of these strategies is more effective for replication and transmission of the virus to new host organisms? Which type of virus would be easier to study and develop a vaccine against?

24. Discuss the importance of the cytoplasm.

25. Explain how Hooke, Schleiden, and Schwann contributed to the cell theory.

Use the illustration below to answer question 26.

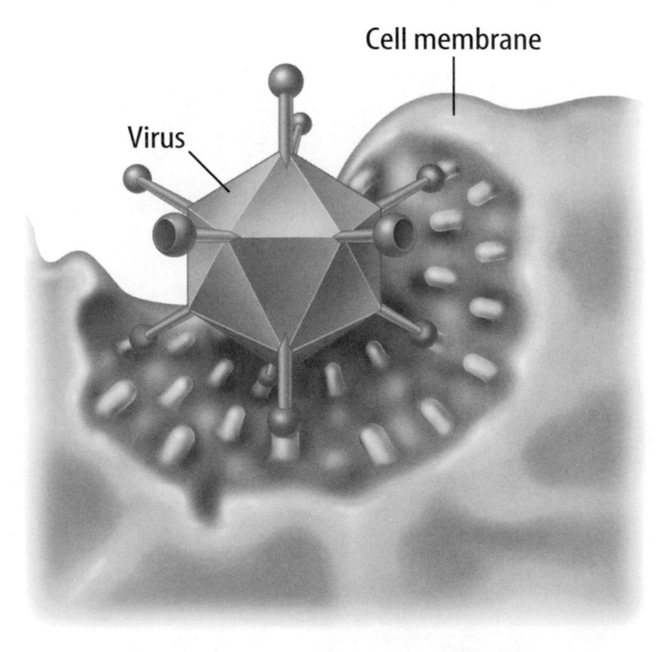

26. What interaction is taking place in the illustration above? What are two possible outcomes of this interaction?

27. Describe how the first vaccine was developed.

chapter 3

Cell Processes

The BIG Idea

Each cell undergoes processes that ensure its survival and often, the survival of other organisms.

SECTION 1
Chemistry of Life
Main Idea All organisms require certain elements that combine and form countless substances needed for life.

SECTION 2
Moving Cellular Material
Main Idea A cell can survive only if substances can move within the cell and pass through its cell membrane.

SECTION 3
Energy for Life
Main Idea All cells require and use energy.

The Science of Gardening

Growing a garden is hard work for both you and the plants. Like you, plants need water and food for energy. How plants get food and water is different from you. Understanding how living things get the energy they need to survive will make a garden seem like much more than just plants and dirt.

Science Journal Describe two ways in which you think plants get food for energy.

Start-Up Activities

Why does water enter and leave plant cells?

If you forget to water a plant, it will wilt. After you water the plant, it probably will straighten up and look healthier. In the following lab, find out how water causes a plant to wilt and straighten.

1. Label a small bowl *Salt Water.* Pour 250 mL of water into the bowl. Then add 15 g of salt to the water and stir.
2. Pour 250 mL of water into another small bowl.
3. Place two carrot sticks into each bowl. Also, place two carrot sticks on the lab table.
4. After 30 min, remove the carrot sticks from the bowls and keep them next to the bowl they came from. Examine all six carrot sticks, then describe them in your Science Journal.
5. **Think Critically** Write a paragraph in your Science Journal that describes what would happen if you moved the carrot sticks from the plain water to the lab table, the ones from the salt water into the plain water, and the ones from the lab table into the salt water for 30 min. Now move the carrot sticks as described and write the results in your Science Journal.

FOLDABLES™ Study Organizer

How Living Things Survive Make the following vocabulary Foldable to help you understand the chemistry of living things and how energy is obtained for life.

STEP 1 **Fold** a vertical sheet of notebook paper from side to side.

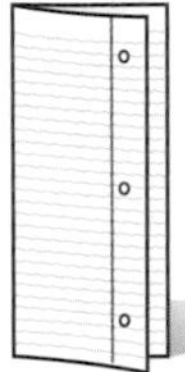

STEP 2 **Cut** along every third line of only the top layer to form tabs.

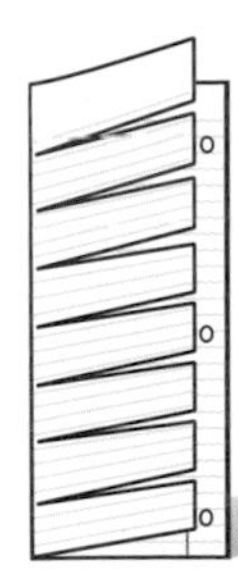

Build Vocabulary As you read this chapter, list the vocabulary words about cell processes on the tabs. As you learn the definitions, write them under the tab for each vocabulary word. Write a sentence about one of the cell processes using the vocabulary word on the tab.

Preview this chapter's content and activities at booka.msscience.com

Get Ready to Read

New Vocabulary

1 Learn It! What should you do if you find a word you don't know or understand? Here are some suggested strategies:

1. Use context clues (from the sentence or the paragraph) to help you define it.
2. Look for prefixes, suffixes, or root words that you already know.
3. Write it down and ask for help with the meaning.
4. Guess at its meaning.
5. Look it up in the glossary or a dictionary.

2 Practice It! Look at the term *inorganic compounds* in the following passage. See how context clues can help you understand its meaning.

Context Clue
Inorganic compounds contain elements other than carbon and have fewer atoms than organic compounds.

Context Clue
Many elements needed for life come from inorganic compounds.

Context Clue
Water is one of the most important inorganic compounds for living things.

Most **inorganic compounds** are made from elements other than carbon. Generally, inorganic molecules contain fewer atoms than organic molecules. Inorganic compounds are the sources for many elements needed by living things. For example, plants take up inorganic compounds from the soil. These inorganic compounds can contain the elements nitrogen, phosphorus, and sulfur. Many foods that you eat contain inorganic compounds. **Table 3** shows some of the inorganic compounds that are important to you. One of the most important inorganic compounds for living things is water.

—from page 73

3 Apply It! Make a vocabulary bookmark with a strip of paper. As you read, keep track of words you do not know or want to learn more about.

Read a paragraph containing a vocabulary word from beginning to end. Then, go back to determine the meaning of the word.

Target Your Reading

Use this to focus on the main ideas as you read the chapter.

1. **Before you read** the chapter, respond to the statements below on your worksheet or on a numbered sheet of paper.
   - Write an **A** if you **agree** with the statement.
   - Write a **D** if you **disagree** with the statement.

2. **After you read** the chapter, look back to this page to see if you've changed your mind about any of the statements.
   - If any of your answers changed, explain why.
   - Change any false statements into true statements.
   - Use your revised statements as a study guide.

Before You Read A or D		Statement	After You Read A or D
	1	Osmosis is the movement of water into and out of a cell.	
	2	All substances can easily pass through a cell's cell membrane.	
	3	Photosynthesis produces oxygen and a sugar.	
	4	Proteins are organic compounds that are important for storing energy.	
	5	Ions play an important role in many life processes.	
	6	Diffusion continues until equilibrium occurs.	
	7	Matter is anything that has mass and takes up space.	
	8	Only plant cells can transform energy.	
	9	Water is the most abundant compound in a cell.	
	10	Cellular respiration requires oxygen and releases energy for a cell.	

Science Online
Print out a worksheet of this page at booka.msscience.com

section 1

Chemistry of Life

as you read

What You'll Learn

- **List** the differences among atoms, elements, molecules, and compounds.
- **Explain** the relationship between chemistry and life science.
- **Discuss** how organic compounds are different from inorganic compounds.

Why It's Important

You grow because of chemical reactions in your body.

Review Vocabulary
cell: the smallest unit of a living thing that can perform the functions of life

New Vocabulary
- mixture
- organic compound
- enzyme
- inorganic compound

The Nature of Matter

Think about everything that surrounds you—chairs, books, clothing, other students, and air. What are all these things made up of? You're right if you answer "matter and energy." Matter is anything that has mass and takes up space. Energy is anything that brings about change. Everything in your environment, including you, is made of matter. Energy can hold matter together or break it apart. For example, the food you eat is matter that is held together by chemical energy. When food is cooked, energy in the form of thermal energy can break some of the bonds holding the matter in food together.

Atoms Whether it is solid, liquid, or gas, matter is made of atoms. **Figure 1** shows a model of an oxygen atom. At the center of an atom is a nucleus that contains protons and neutrons. Although they have nearly equal masses, a proton has a positive charge and a neutron has no charge. Outside the nucleus are electrons, each of which has a negative charge. It takes about 1,837 electrons to equal the mass of one proton. Electrons are important because they are the part of the atom that is involved in chemical reactions. Look at **Figure 1** again and you will see that an atom is mostly empty space. Energy holds the parts of an atom together.

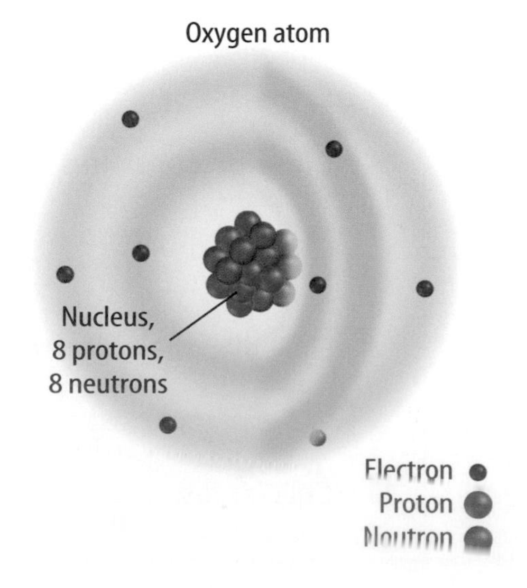

Figure 1 An oxygen atom model shows the placement of electrons, protons, and neutrons.

Table 1 Elements in the Human Body

Symbol	Element	Percent
O	Oxygen	65.0
C	Carbon	18.5
H	Hydrogen	9.5
N	Nitrogen	3.2
Ca	Calcium	1.5
P	Phosphorus	1.0
K	Potassium	0.4
S	Sulfur	0.3
Na	Sodium	0.2
Cl	Chlorine	0.2
Mg	Magnesium	0.1
	Other elements	0.1

Elements When something is made up of only one kind of atom, it is called an element. An element can't be broken down into a simpler form by chemical reactions. The element oxygen is made up of only oxygen atoms, and hydrogen is made up of only hydrogen atoms. Scientists have given each element its own one- or two-letter symbol.

All elements are arranged in a chart known as the periodic table of elements. You can find this table at the back of this book. The table provides information about each element including its mass, how many protons it has, and its symbol.

Everything is made up of elements. Most things, including all living things, are made up of a combination of elements. Few things exist as pure elements. **Table 1** lists elements that are in the human body. What two elements make up most of your body?

Six of the elements listed in the table are important because they make up about 99 percent of living matter. The symbols for these elements are S, P, O, N, C, and H. Use **Table 1** to find the names of these elements.

Reading Check *What types of things are made up of elements?*

Figure 2 The words *atoms, molecules,* and *compounds* are used to describe substances.
Explain *how these terms are related to each other.*

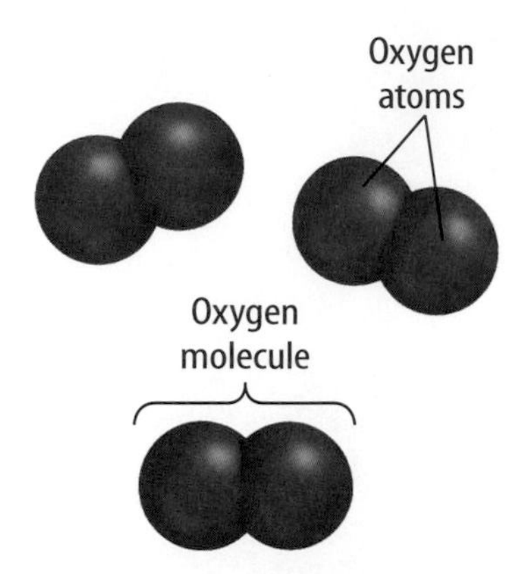

A Some elements, like oxygen, occur as molecules. These molecules contain atoms of the same element bonded together.

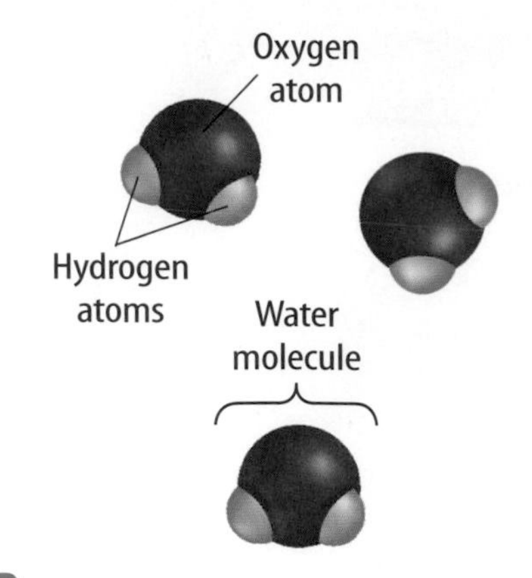

B Compounds also are composed of molecules. Molecules of compounds contain atoms of two or more different elements bonded together, as shown by these water molecules.

Compounds and Molecules

Suppose you make a pitcher of lemonade using a powdered mix and water. The water and the lemonade mix, which is mostly sugar, contain the elements oxygen and hydrogen. Yet, in one, they are part of a nearly tasteless liquid—water. In the other they are part of a sweet solid—sugar. How can the same elements be part of two materials that are so different? Water and sugar are compounds. Compounds are made up of two or more elements in exact proportions. For example, pure water, whether one milliliter of it or one million liters, is always made up of hydrogen atoms bonded to oxygen atoms in a ratio of two hydrogen atoms to one oxygen atom. Compounds have properties different from the elements they are made of. There are two types of compounds—molecular compounds and ionic compounds.

Molecular Compounds The smallest part of a molecular compound is a molecule. A molecule is a group of atoms held together by the energy of chemical bonds, as shown in **Figure 2.** When chemical reactions occur, chemical bonds break, atoms are rearranged, and new bonds form. The molecules produced are different from those that began the chemical reaction.

Molecular compounds form when different atoms share their outermost electrons. For example, two atoms of hydrogen each can share one electron on one atom of oxygen to form one molecule of water, as shown in **Figure 2B.** Water does not have the same properties as oxygen and hydrogen. Under normal conditions on Earth, oxygen and hydrogen are gases. Yet, water can be a liquid, a solid, or a gas. When hydrogen and oxygen combine, changes occur and a new substance forms.

Ions Atoms also combine because they've become positively or negatively charged. Atoms are usually neutral—they have no overall electric charge. When an atom loses an electron, it has more protons than electrons, then it is positively charged. When an atom gains an electron, it has more electrons than protons, then it is negatively charged. Electrically charged atoms—positive or negative—are called ions.

Ionic Compounds Ions of opposite charges attract one another to form electrically neutral compounds called ionic compounds. Table salt is made of sodium (Na) and chlorine (Cl) ions, as shown in **Figure 3B.** When they combine, a chlorine atom gains an electron from a sodium atom. The chlorine atom becomes a negatively charged ion, and the sodium atom becomes a positively charged ion. These oppositely charged ions attract each other and form the ionic compound sodium chloride, NaCl.

Ions are important in many life processes that take place in your body and in other organisms. For example, messages are sent along your nerves as potassium and sodium ions move in and out of nerve cells. Calcium ions are important in causing your muscles to contract. Ions also are involved in the transport of oxygen by your blood. The movement of some substances into and out of a cell would not be possible without ions.

LM Magnification: 8×

A Magnified crystals of salt look like this.

Na^+ Cl^-

B A salt crystal is held together by the attractions between sodium ions and chlorine ions.

Figure 3 Table salt crystals are held together by ionic bonds.

Mixtures

Some substances, such as a combination of sugar and salt, can't change each other or combine chemically. A **mixture** is a combination of substances in which individual substances retain their own properties. Mixtures can be solids, liquids, gases, or any combination of them.

✔ Reading Check *Why is a combination of sugar and salt said to be a mixture?*

Most chemical reactions in living organisms take place in mixtures called solutions. You've probably noticed the taste of salt when you perspire. Sweat is a solution of salt and water. In a solution, two or more substances are mixed evenly. A cell's cytoplasm is a solution of dissolved molecules and ions.

Living things also contain mixtures called suspensions. A suspension is formed when a liquid or a gas has another substance evenly spread throughout it. Unlike solutions, the substances in a suspension eventually sink to the bottom. If blood, shown in **Figure 4,** is left undisturbed, the red blood cells and white blood cells will sink gradually to the bottom. However, the pumping action of your heart constantly moves your blood and the blood cells remain suspended.

Figure 4 When a test tube of whole blood is left standing, the blood cells sink in the watery plasma.

Table 2 Organic Compounds Found in Living Things

	Carbohydrates	Lipids	Proteins	Nucleic Acids
Elements	carbon, hydrogen, and oxygen	carbon, oxygen, hydrogen, and phosphorus	carbon, oxygen, hydrogen, nitrogen, and sulfur	carbon, oxygen, hydrogen, nitrogen, and phosphorus
Examples	sugars, starch, and cellulose	fats, oils, waxes, phospholipids, and cholesterol	enzymes, skin, and hair	DNA and RNA
Function	supply energy for cell processes; form plant structures; short-term energy storage	store large amounts of energy long term; form boundaries around cells	regulate cell processes and build cell structures	carry hereditary information; used to make proteins

Organic Compounds

You and all living things are made up of compounds that are classified as organic or inorganic. Rocks and other nonliving things contain inorganic compounds, but most do not contain large amounts of organic compounds. **Organic compounds** always contain carbon and hydrogen and usually are associated with living things. One exception would be nonliving things that are products of living things. For example, coal contains organic compounds because it was formed from dead and decaying plants. Organic molecules can contain hundreds or even thousands of atoms that can be arranged in many ways. **Table 2** compares the four groups of organic compounds that make up all living things—carbohydrates, lipids, proteins, and nucleic acids.

Topic: Air Quality
Visit booka.msscience.com for Web links to information about air quality.

Activity Organic compounds such as soot, smoke, and ash can affect air quality. Look up the air quality forecast for today. List three locations where the air quality forecast is good, and three locations where it is unhealthy.

Carbohydrates Carbohydrates are organic molecules that supply energy for cell processes. Sugars and starches are carbohydrates that cells use for energy. Some carbohydrates also are important parts of cell structures. For example, a carbohydrate called cellulose is an important part of plant cells.

Lipids Another type of organic compound found in living things is a lipid. Lipids do not mix with water. Lipids such as fats and oils store and release even larger amounts of energy than carbohydrates do. One type of lipid, the phospholipid, is a major part of cell membranes.

What are three types of lipids?

Proteins Organic compounds called proteins have many important functions in living organisms. They are made up of smaller molecules called amino acids. Proteins are the building blocks of many structures in organisms. Your muscles contain large amounts of protein. Proteins are scattered throughout cell membranes. Certain proteins called **enzymes** regulate nearly all chemical reactions in cells.

Nucleic Acids Large organic molecules that store important coded information in cells are called nucleic acids. One nucleic acid, deoxyribonucleic (dee AHK sih ri boh noo klee ihk) acid, or DNA is the genetic material found in all cells at some point in their lives. It carries information that directs each cell's activities. Another nucleic acid, ribonucleic (ri boh noo klee ihk) acid, or RNA, is needed to make enzymes and other proteins.

Inorganic Compounds

Most **inorganic compounds** are made from elements other than carbon. Generally, inorganic molecules contain fewer atoms than organic molecules. Inorganic compounds are the source for many elements needed by living things. For example, plants take up inorganic compounds from the soil. These inorganic compounds can contain the elements nitrogen, phosphorus, and sulfur. Many foods that you eat contain inorganic compounds. **Table 3** shows some of the inorganic compounds that are important to you. One of the most important inorganic compounds for living things is water.

Table 3 Some Inorganic Compounds Important in Humans

Compound	Use in Body
Water	makes up most of the blood; most chemical reactions occur in water
Calcium phosphate	gives strength to bones
Hydrochloric acid	helps break down foods in the stomach
Sodium bicarbonate	helps the digestion of food to occur
Salts containing sodium, chlorine, and potassium	important in sending messages along nerves

Observing How Enzymes Work

Procedure

1. Get two small cups of **prepared gelatin** from your teacher. Do not eat or drink anything in lab.
2. On the gelatin in one of the cups, place a piece of **fresh pineapple.**
3. Let both cups stand undisturbed overnight.
4. Observe what happens to the gelatin.

Analysis

1. What effect did the piece of fresh pineapple have on the gelatin?
2. What does fresh pineapple contain that caused it to have the effect on the gelatin you observed?
3. Why do the preparation directions on a box of gelatin dessert tell you not to mix it with fresh pineapple?

Importance of Water Some scientists hypothesize that life began in the water of Earth's ancient oceans. Chemical reactions might have occurred that produced organic molecules. Similar chemical reactions can take place in cells in your body.

Living things are composed of more than 50 percent water and depend on water to survive. You can live for weeks without food but only for a few days without water. **Figure 5** shows where water is found in your body. Although seeds and spores of plants, fungi, and bacteria can exist without water, they must have water if they are to grow and reproduce. All the chemical reactions in living things take place in water solutions, and most organisms use water to transport materials through their bodies. For example, many animals have blood that is mostly water and moves materials. Plants use water to move minerals and sugars between the roots and leaves.

Applying Math — Solve an Equation

CALCULATE THE IMPORTANCE OF WATER All life on Earth depends on water for survival. Water is the most vital part of humans and other animals. It is required for all of the chemical processes that keep us alive. At least 60 percent of an adult human body consists of water. If an adult man weighs 90 kg, how many kilograms of water does his body contain?

Solution

1 *This is what you know:*
- adult human body = 60% water
- man = 90 kg

2 *This is what you need to find:* How many kilograms of water does the adult man have?

3 *This is the procedure you need to use:*
- Set up the ratio: $60/100 = x/90$.
- Solve the equation for x: $(60 \times 90)/100$.
- The adult man has 54 kg of water.

4 *Check your answer:* Divide your answer by 90, then multiply by 100. You should get 60%.

Practice Problems

1. A human body at birth consists of 78 percent water. This gradually decreases to 60 percent in an adult. Assume a baby weighed 3.2 kg at birth and grew into an adult weighing 95 kg. Calculate the approximate number of kilograms of water the human gained.
2. Assume an adult woman weighs 65 kg and an adult man weighs 90 kg. Calculate how much more water, in kilograms, the man has compared to the woman.

For more practice, visit bookA.msscience.com/math_practice

INTEGRATE Physics

Characteristics of Water The atoms of a water molecule are arranged in such a way that the molecule has areas with different charges. Water molecules are like magnets. The negative part of a water molecule is attracted to the positive part of another water molecule just like the north pole of a magnet is attracted to the south pole of another magnet. This attraction, or force, between water molecules is why a film forms on the surface of water. The film is strong enough to support small insects because the forces between water molecules are stronger than the force of gravity on the insect.

When thermal energy is added to any substance, its molecules begin to move faster. Because water molecules are so strongly attracted to each other, the temperature of water changes slowly. The large percentage of water in living things acts like an insulator. The water in a cell helps keep its temperature constant, which allows life-sustaining chemical reactions to take place.

You've seen ice floating on water. When water freezes, ice crystals form. In the crystals, each water molecule is spaced at a certain distance from all the others. Because this distance is greater in frozen water than in liquid water, ice floats on water. Bodies of water freeze from the top down. The floating ice provides insulation from extremely cold temperatures and allows living things to survive in the cold water under the ice.

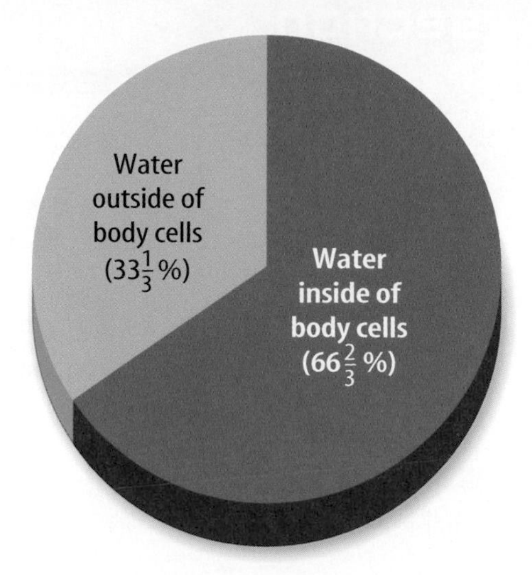

Figure 5 About two-thirds of your body's water is located within your body's cells. Water helps maintain the cells' shapes and sizes. One-third of your body's water is outside of your body's cells.

section 1 review

Summary

The Nature of Matter

- Atoms are made up of protons, neutrons, and electrons.
- Elements are made up of only one kind of atom.
- Compounds are made up of two or more elements.

Mixtures

- Solutions are made of two or more substances and are mixed evenly, whereas substances in suspension eventually will sink to the bottom.

Organic Compounds

- All living things contain organic compounds.

Inorganic Compounds

- Water is one of the most important inorganic compounds for living things.

Self Check

1. **Compare and contrast** atoms and molecules.
2. **Describe** the differences between an organic and an inorganic compound. Given an example of each type of compound.
3. **List** the four types of organic compounds found in all living things.
4. **Infer** why life as we know it depends on water.
5. **Think Critically** If you mix salt, sand, and sugar with water in a small jar, will the resulting mixture be a suspension, a solution, or both?

Applying Skills

6. **Interpret** Carefully observe **Figure 1** and determine how many protons, neutrons, and electrons an atom of oxygen has.

section 2

Moving Cellular Materials

as you read

What You'll Learn

- **Describe** the function of a selectively permeable membrane.
- **Explain** how the processes of diffusion and osmosis move molecules in living cells.
- **Explain** how passive transport and active transport differ.

Why It's Important

Cell membranes control the substances that enter and leave the cells in your body.

Review Vocabulary
cytoplasm: constantly moving gel-like mixture inside the cell membrane that contains hereditary material and is the location of most of a cell's life process

New Vocabulary

- passive transport
- diffusion
- equilibrium
- osmosis
- active transport
- endocytosis
- exocytosis

Passive Transport

"Close that window. Do you want to let in all the bugs and leaves?" How do you prevent unwanted things from coming through the window? As seen in **Figure 6,** a window screen provides the protection needed to keep unwanted things outside. It also allows some things to pass into or out of the room like air, unpleasant odors, or smoke.

Cells take in food, oxygen, and other substances from their environments. They also release waste materials into their environments. A cell has a membrane around it that works for a cell like a window screen does for a room. A cell's membrane is selectively permeable (PUR mee uh bul). It allows some things to enter or leave the cell while keeping other things outside or inside the cell. The window screen also is selectively permeable based on the size of its openings.

Things can move through a cell membrane in several ways. Which way things move depends on the size of the molecules or particles, the path taken through the membrane, and whether or not energy is used. The movement of substances through the cell membrane without the input of energy is called **passive transport.** Three types of passive transport can occur. The type depends on what is moving through the cell membrane.

Figure 6 A cell membrane, like a screen, will let some things through more easily than others. Air gets through a screen, but insects are kept out.

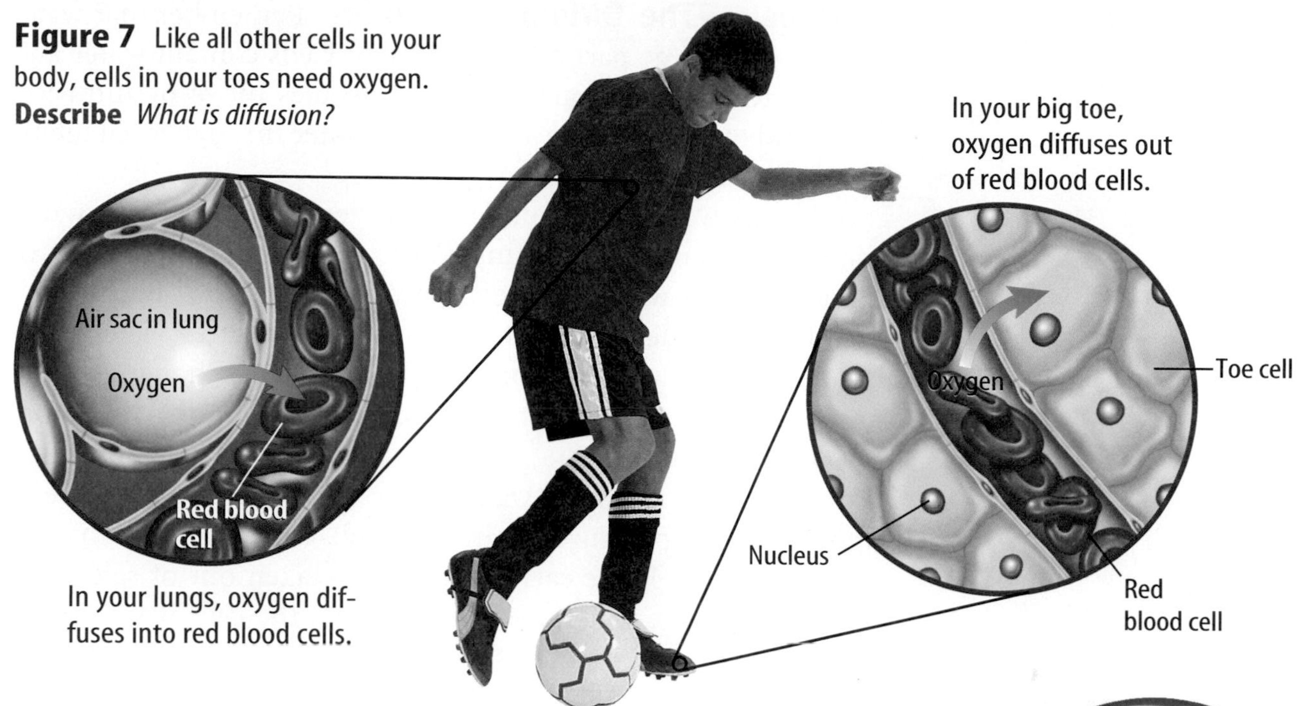

Figure 7 Like all other cells in your body, cells in your toes need oxygen.
Describe *What is diffusion?*

Diffusion Molecules in solids, liquids, and gases move constantly and randomly. You might smell perfume when you sit near or as you walk past someone who is wearing it. This is because perfume molecules randomly move throughout the air. This random movement of molecules from an area where there is relatively more of them into an area where there is relatively fewer of them is called **diffusion.** Diffusion is one type of cellular passive transport. Molecules of a substance will continue to move from one area into another until the relative number of these molecules is equal in the two areas. When this occurs, **equilibrium** is reached and diffusion stops. After equilibrium occurs, it is maintained because molecules continue to move.

Reading Check *What is equilibrium?*

Every cell in your body uses oxygen. When you breathe, how does oxygen get from your lungs to cells in your big toe? Oxygen is carried throughout your body in your blood by the red blood cells. When your blood is pumped from your heart to your lungs, your red blood cells do not contain much oxygen. However, your lungs have more oxygen molecules than your red blood cells do, so the oxygen molecules diffuse into your red blood cells from your lungs, as shown in **Figure 7.** When the blood reaches your big toe, there are more oxygen molecules in your red blood cells than in your big toe cells. The oxygen diffuses from your red blood cells and into your big toe cells, as shown also in **Figure 7.**

Mini LAB

Observing Molecule Movement

Procedure

1. Use two clean glasses of equal size. Label one *Hot,* then fill it until half full with **very warm water.** Label the other *Cold,* then fill it until half full with **cold water. WARNING:** *Do not use boiling hot water.*
2. Add one drop of **food coloring** to each glass. Carefully release the drop just at the water's surface to avoid splashing the water.
3. Immediately observe the water in the glasses. Record your observations and again after 15 min.

Analysis

What is the relationship between temperature and molecule movement?

Try at Home

Osmosis—The Diffusion of Water Remember that water makes up a large part of living matter. Cells contain water and are surrounded by water. Water molecules move by diffusion into and out of cells. The diffusion of water through a cell membrane is called **osmosis.**

If cells weren't surrounded by water that contains few dissolved substances, water inside of cells would diffuse out of them. This is why water left the carrot cells in this chapter's Launch Lab. Because there were relatively fewer water molecules in the salt solution around the carrot cells than in the carrot cells, water moved out of the cells and into the salt solution.

Losing water from a plant cell causes its cell membrane to come away from its cell wall, as shown on the left in **Figure 8.** This reduces pressure against its cell wall, and a plant cell becomes limp. If the carrot sticks were taken out of salt water and put in pure water, the water around the cells would move into them and they would fill with water. Their cell membranes would press against their cell walls, as shown on the right in **Figure 8,** pressure would increase, and the cells would become firm. That is why the carrot sticks would be crisp again.

Reading Check *Why do carrots in salt water become limp?*

Osmosis also takes place in animal cells. If animal cells were placed in pure water, they too would swell up. However, animal cells are different from plant cells. Just like an overfilled water balloon, animal cells will burst if too much water enters the cell.

Figure 8 Cells respond to differences between the amount of water inside and outside the cell.
Define *What is osmosis?*

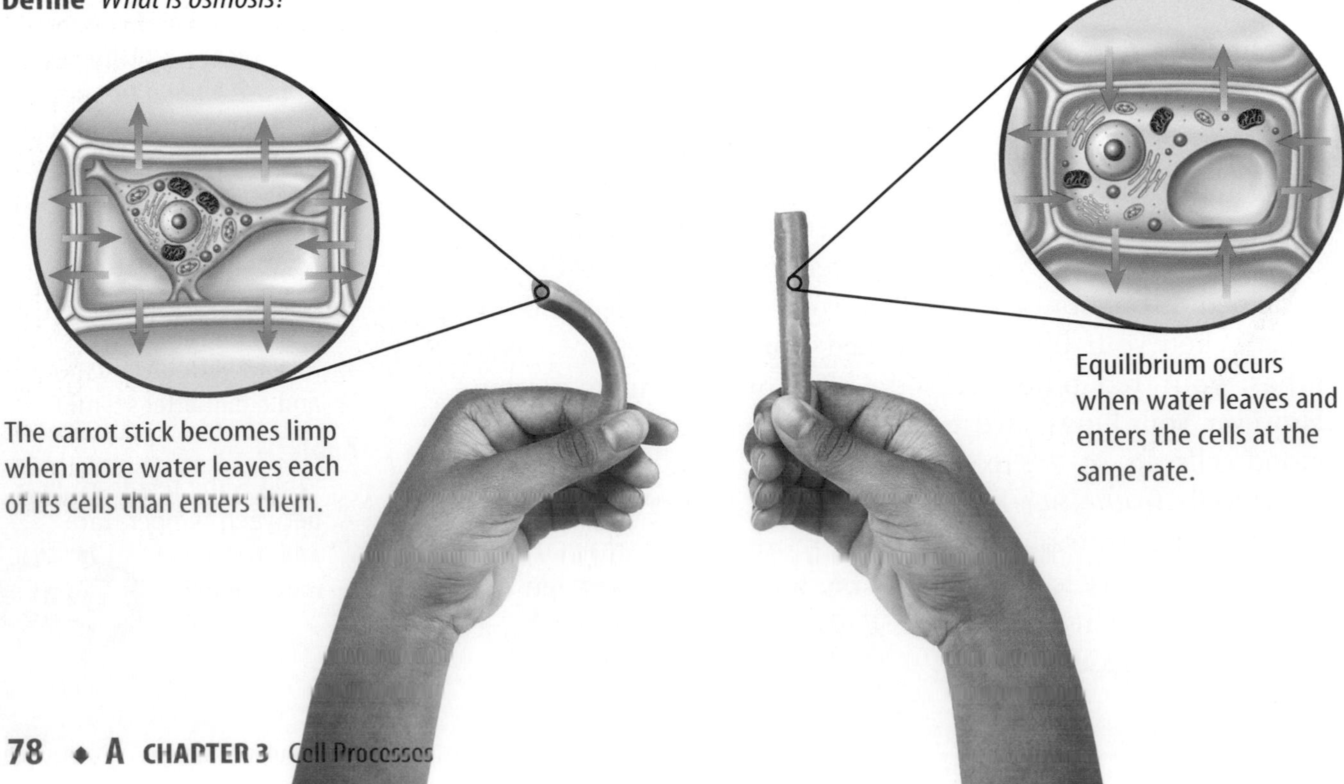

Facilitated Diffusion Cells take in many substances. Some substances pass easily through the cell membrane by diffusion. Other substances, such as sugar molecules, are so large that they can enter the cell only with the help of molecules in the cell membrane called transport proteins. This process, a type of passive transport, is known as facilitated diffusion. Have you ever used the drive through at a fast-food restaurant to get your meal? The transport proteins in the cell membrane are like the drive-through window at the restaurant. The window lets you get food out of the restaurant and put money into the restaurant. Similarly, transport proteins are used to move substances into and out of the cell.

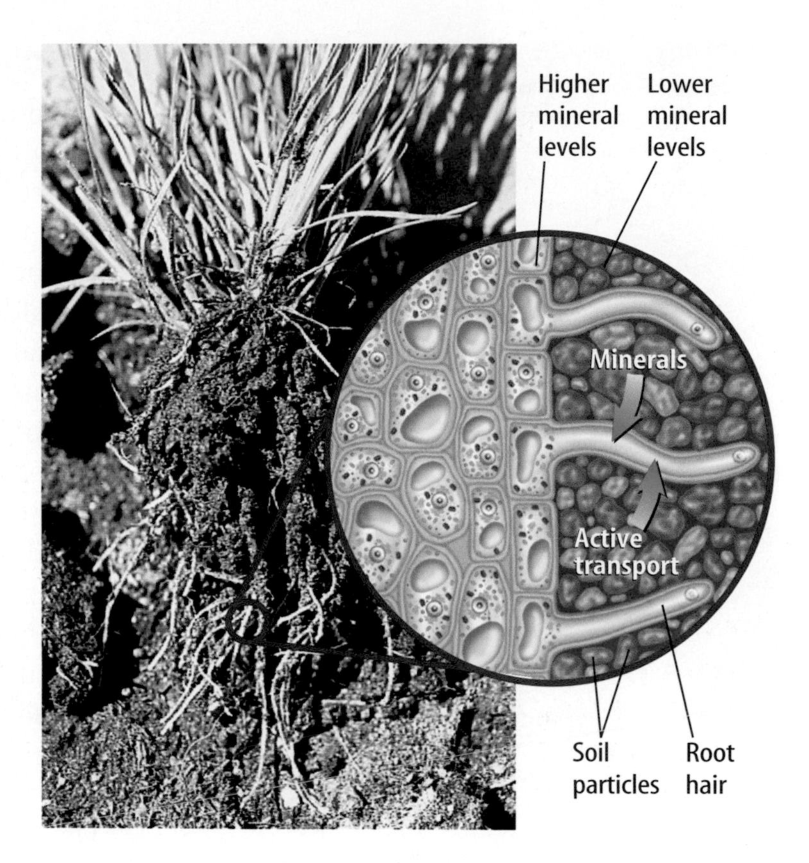

Figure 9 Some root cells have extensions called root hairs that may be 5 mm to 8 mm long. Minerals are taken in by active transport through the cell membranes of root hairs.

Active Transport

Imagine that a football game is over and you leave the stadium. As soon as you get outside of the stadium, you remember that you left your jacket on your seat. Now you have to move against the crowd coming out of the stadium to get back in to get your jacket. Which required more energy—leaving the stadium with the crowd or going back to get your jacket? Something similar to this happens in cells.

Sometimes, a substance is needed inside a cell even though the amount of that substance inside the cell is already greater than the amount outside the cell. For example, root cells require minerals from the soil. The roots of the plant in **Figure 9** already might contain more of those mineral molecules than the surrounding soil does. The tendency is for mineral molecules to move out of the root by diffusion or facilitated diffusion. But they need to move back across the cell membrane and into the cell just like you had to move back into the stadium. When an input of energy is required to move materials through a cell membrane, **active transport** takes place.

Active transport involves transport proteins, just as facilitated diffusion does. In active transport, a transport protein binds with the needed particle and cellular energy is used to move it through the cell membrane. When the particle is released, the transport protein can move another needed particle through the membrane.

Transport Proteins Your health depends on transport proteins. Sometimes transport proteins are missing or do not function correctly. What would happen if proteins that transport cholesterol across membranes were missing? Cholesterol is an important lipid used by your cells. Write your ideas in your Science Journal.

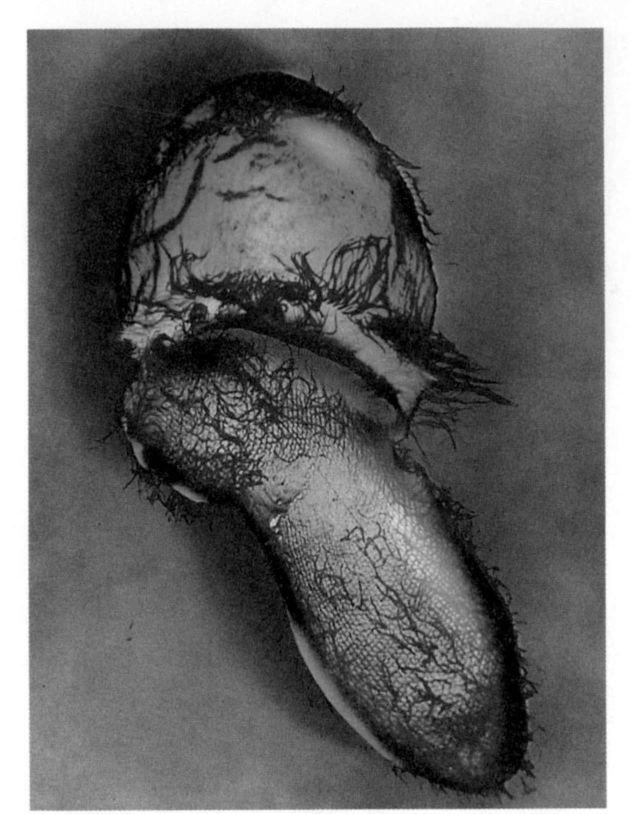

Color-enhanced TEM Magnification: 1,400×

Figure 10 One-celled organisms like this egg-shaped one can take in other one-celled organisms using endocytosis.

Endocytosis and Exocytosis

Some molecules and particles are too large to move by diffusion or to use the cell membrane's transport proteins. Large protein molecules and bacteria, for example, can enter a cell when they are surrounded by the cell membrane. The cell membrane folds in on itself, enclosing the item in a sphere called a vesicle. Vesicles are transport and storage structures in a cell's cytoplasm. The sphere pinches off, and the resulting vesicle enters the cytoplasm. A similar thing happens when you poke your finger into a partially inflated balloon. Your finger is surrounded by the balloon in much the same way that the protein molecule is surrounded by the cell membrane. This process of taking substances into a cell by surrounding it with the cell membrane is called **endocytosis** (en duh si TOH sus). Some one-celled organisms, as shown in **Figure 10,** take in food this way.

The contents of a vesicle can be released by a cell using the process called **exocytosis** (ek soh si TOH sus). Exocytosis occurs in the opposite way that endocytosis does. A vesicle's membrane fuses with a cell's membrane, and the vesicle's contents are released. Cells in your stomach use this process to release chemicals that help digest food. The ways that materials can enter or leave a cell are summarized in **Figure 11.**

section 2 review

Summary

Passive Transport

- Cells take in substances and release waste through their cell membranes.
- Facilitated diffusion and osmosis are types of passive transport.

Active Transport

- Transport proteins are involved in active transport.
- Transport proteins can be reused many times.

Endocytosis and Exocytosis

- Vesicles are formed when a cell takes in a substance by endocytosis.
- Contents of a vesicle are released to the outside of a cell by exocytosis.

Self Check

1. **Describe** how cell membranes are selectively permeable.
2. **Compare and contrast** the processes of osmosis and diffusion.
3. **Infer** why endocytosis and exocytosis are important processes to cells.
4. **Think Critically** Why are fresh fruits and vegetables sprinkled with water at produce markets?

Applying Skills

5. **Communicate** Seawater is saltier than tap water. Explain why drinking large amounts of seawater would be dangerous for humans.

booka.msscience.com/self_check_quiz

NATIONAL GEOGRAPHIC

VISUALIZING CELL MEMBRANE TRANSPORT

Figure 11

A flexible yet strong layer, the cell membrane is built of two layers of lipids (gold) pierced by protein "passageways" (purple). Molecules can enter or exit the cell by slipping between the lipids or through the protein passageways. Substances that cannot enter or exit the cell in these ways may be surrounded by the membrane and drawn into or expelled from the cell.

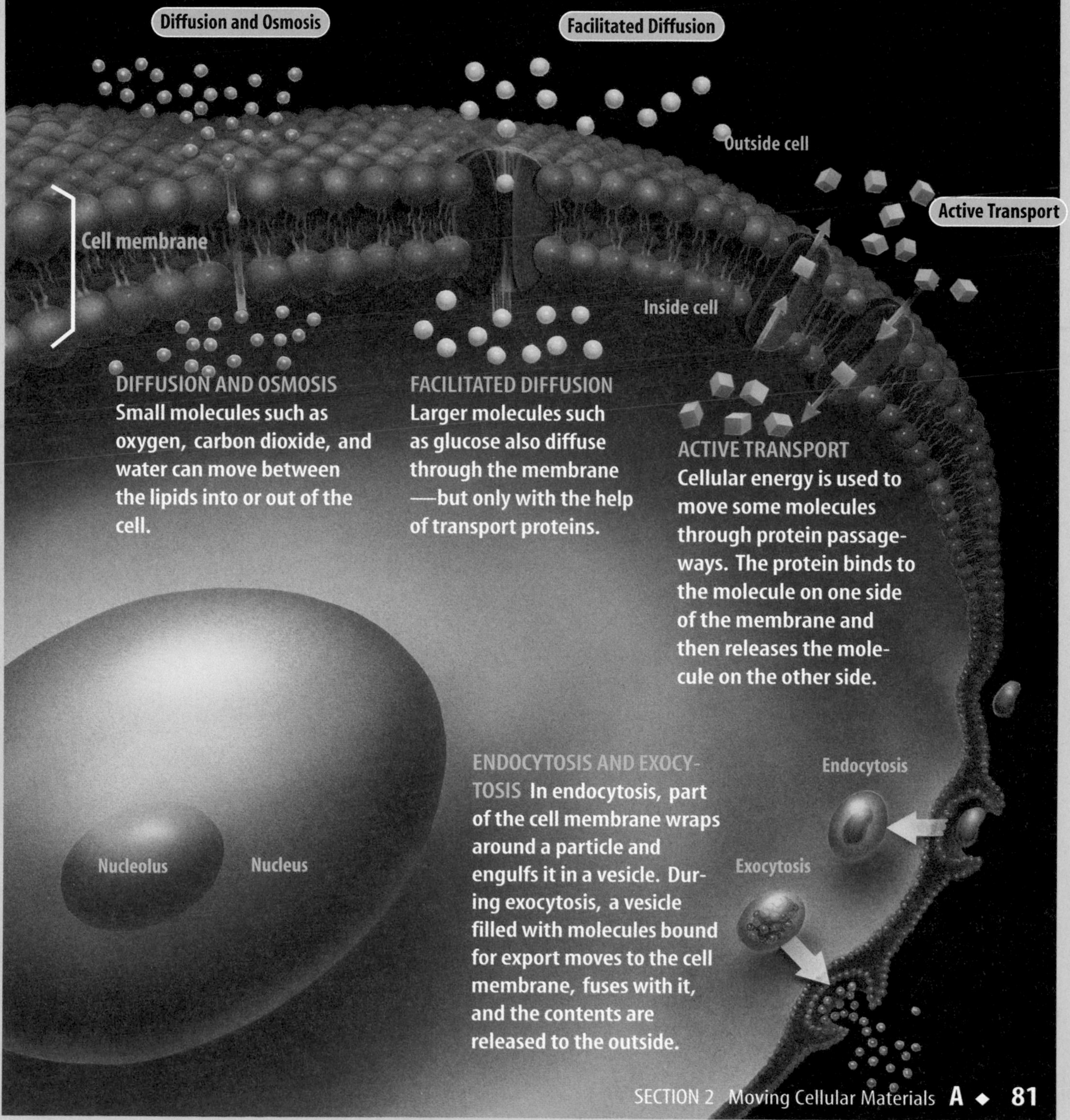

DIFFUSION AND OSMOSIS Small molecules such as oxygen, carbon dioxide, and water can move between the lipids into or out of the cell.

FACILITATED DIFFUSION Larger molecules such as glucose also diffuse through the membrane —but only with the help of transport proteins.

ACTIVE TRANSPORT Cellular energy is used to move some molecules through protein passageways. The protein binds to the molecule on one side of the membrane and then releases the molecule on the other side.

ENDOCYTOSIS AND EXOCYTOSIS In endocytosis, part of the cell membrane wraps around a particle and engulfs it in a vesicle. During exocytosis, a vesicle filled with molecules bound for export moves to the cell membrane, fuses with it, and the contents are released to the outside.

Observing Osmosis

It is difficult to observe osmosis in cells because most cells are so small. However, a few cells can be seen without the aid of a microscope. Try this lab to observe osmosis.

Real-World Question

How does osmosis occur in an egg cell?

Materials

unshelled egg*
balance
spoon
distilled water (250 mL)
light corn syrup (250 mL)
500-mL container

an egg whose shell has been dissolved by vinegar

Goals

- **Observe** osmosis in an egg cell.
- **Determine** what affects osmosis.

Safety Precautions

WARNING: *Eggs may contain bacteria. Avoid touching your face.*

Procedure

1. Copy the table below into your Science Journal and use it to record your data.

Egg Mass Data

	Beginning Egg Mass	Egg Mass After Two Days
Distilled water	Do not write in this book.	
Corn syrup		

2. Obtain an unshelled egg from your teacher. Handle the egg gently. Use a balance to find the egg's mass and record it in the table.

3. Place the egg in the container and add enough distilled water to cover it.
4. **Observe** the egg after 30 min, one day, and two days. After each observation, record the egg's appearance in your Science Journal.
5. After day two, remove the egg with a spoon and allow it to drain. Find the egg's mass and record it in the table.
6. Empty the container, then put the egg back in. Now add enough corn syrup to cover it. Repeat steps 4 and 5.

Conclude and Apply

1. **Explain** the difference between what happened to the egg in water and in corn syrup.
2. **Calculate** the mass of water that moved into and out of the egg.
3. **Hypothesize** why you used an unshelled egg for this investigation.
4. **Infer** what part of the egg controlled water's movement into and out of the egg.

Communicating Your Data

Compare your conclusions with those of other students in your class. **For more help, refer to the Science Skill Handbook.**

section 3

Energy for Life

Trapping and Using Energy

Think of all the energy that players use in a soccer game. Where does the energy come from? The simplest answer is "from the food they eat." The chemical energy stored in food molecules is changed inside of cells into forms needed to perform all the activities necessary for life. In every cell, these changes involve chemical reactions. All of the activities of an organism involve chemical reactions in some way. The total of all chemical reactions in an organism is called **metabolism.**

The chemical reactions of metabolism need enzymes. What do enzymes do? Suppose you are hungry and decide to open a can of spaghetti. You use a can opener to open the can. Without a can opener, the spaghetti is unusable. The can of spaghetti changes because of the can opener, but the can opener does not change. The can opener can be used again later to open more cans of spaghetti. Enzymes in cells work something like can openers. The enzyme, like the can opener, causes a change, but the enzyme is not changed and can be used again, as shown in **Figure 12.** Unlike the can opener, which can only cause things to come apart, enzymes also can cause molecules to join. Without the right enzyme, a chemical reaction in a cell cannot take place. Each chemical reaction in a cell requires a specific enzyme.

as you read

What You'll Learn

- **List** the differences between producers and consumers.
- **Explain** how the processes of photosynthesis and cellular respiration store and release energy.
- **Describe** how cells get energy from glucose through fermentation.

Why It's Important

Because of photosynthesis and cellular respiration, you use the Sun's energy.

Review Vocabulary

mitochondrion: cell organelle that breaks down lipids and carbohydrates and releases energy

New Vocabulary

- metabolism
- photosynthesis
- cellular respiration
- fermentation

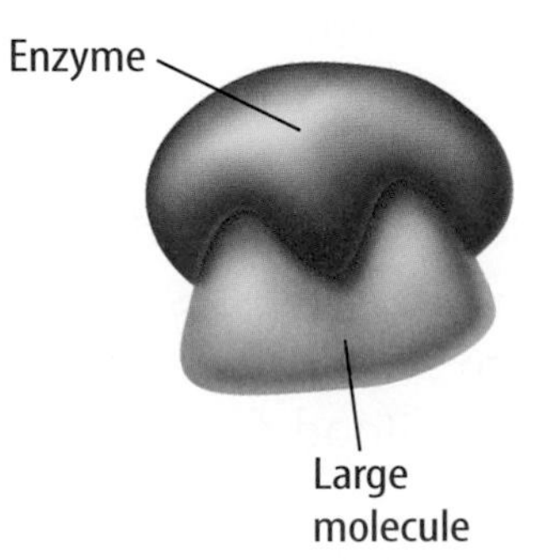

The enzyme attaches to the large molecule it will help change.

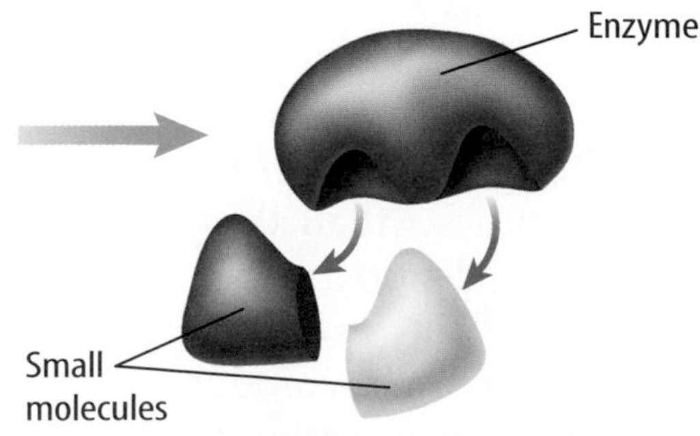

The enzyme causes the larger molecule to break down into two smaller molecules. The enzyme is not changed and can be used again.

Figure 12 Enzymes are needed for most chemical reactions that take place in cells.
Determine *What is the sum of all chemical reactions in an organism called?*

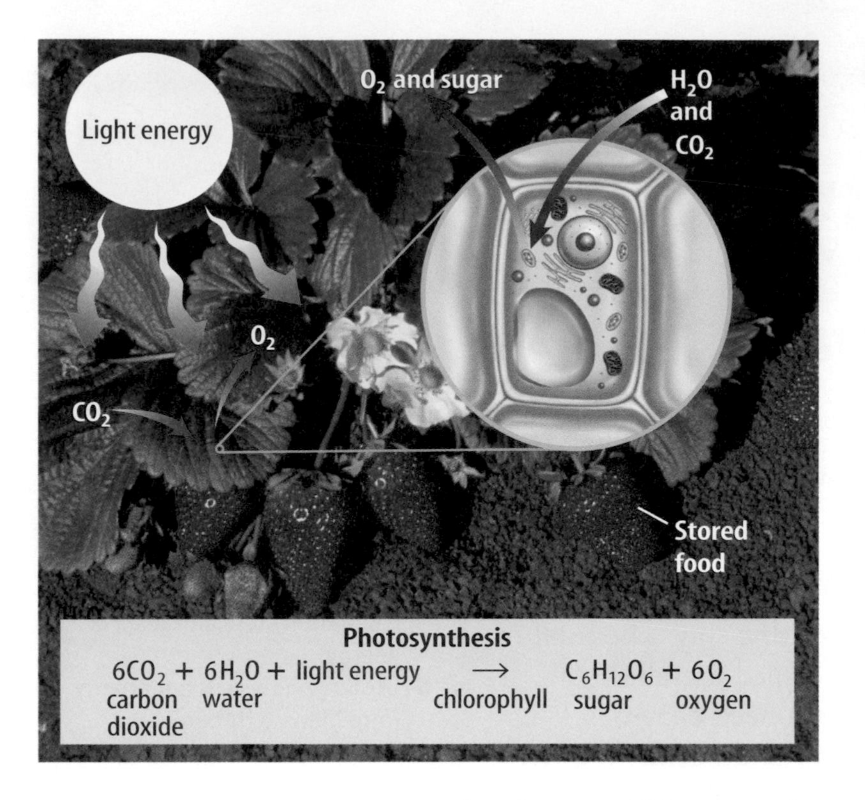

Figure 13 Plants use photosynthesis to make food.
Determine *According to the chemical equation, what raw materials would the plant in the photo need for photosynthesis?*

Photosynthesis Living things are divided into two groups—producers and consumers—based on how they obtain their food. Organisms that make their own food, such as plants, are called producers. Organisms that cannot make their own food are called consumers.

If you have ever walked barefoot across a sidewalk on a sunny summer day, you probably moved quickly because the sidewalk was hot. Sunlight energy was converted into thermal energy and heated the sidewalk. Plants and many other producers can convert light energy into another kind of energy—chemical energy. The process they use is called photosynthesis. During **photosynthesis,** producers use light energy and make sugars, which can be used as food.

Producing Carbohydrates Producers that use photosynthesis are usually green because they contain a green pigment called chlorophyll (KLOR uh fihl). Chlorophyll and other pigments are used in photosynthesis to capture light energy. In plant cells, these pigments are found in chloroplasts.

The captured light energy powers chemical reactions that produce sugar and oxygen from the raw materials, carbon dioxide and water. For plants, the raw materials come from air and soil. Some of the captured light energy is stored in the chemical bonds that hold the sugar molecules together. **Figure 13** shows what happens during photosynthesis in a plant. Enzymes also are needed before these reactions can occur.

Storing Carbohydrates Plants make more sugar during photosynthesis than they need for survival. Excess sugar is changed and stored as starches or used to make other carbohydrates. Plants use these carbohydrates as food for growth, maintenance, and reproduction.

Why is photosynthesis important to consumers? Do you eat apples? Apple trees use photosynthesis to produce apples. Do you like cheese? Some cheese comes from milk, which is produced by cows that eat plants. Consumers take in food by eating producers or other consumers. No matter what you eat, photosynthesis was involved directly or indirectly in its production.

Cellular Respiration Imagine that you get up late for school. You dress quickly, then run three blocks to school. When you get to school, you feel hot and are breathing fast. Why? Your muscle cells use a lot of energy when you run. To get this energy, muscle cells break down food. Some of the energy is used when you move and some of it becomes thermal energy, which is why you feel warm or hot. Most cells also need oxygen to break down food. You were breathing fast because your body was working to get oxygen to your muscles. Your muscle cells were using the oxygen for the process of cellular respiration. During **cellular respiration,** chemical reactions occur that break down food molecules into simpler substances and release their stored energy. Just as in photosynthesis, enzymes are needed for the chemical reactions of cellular respiration.

What must happen to food molecules for respiration to take place?

Microbiologist Dr. Harold Amos is a microbiologist who has studied cell processes in bacteria and mammals. He has a medical degree and a doctorate in bacteriology and immunology. He has also received many awards for his scientific work and his contributions to the careers of other scientists. Research microbiology careers, and write what you find in your Science Journal.

Breaking Down Carbohydrates The food molecules most easily broken down by cells are carbohydrates. Cellular respiration of carbohydrates begins in the cytoplasm of the cell. The carbohydrates are broken down into glucose molecules. Each glucose molecule is broken down further into two simpler molecules. As the glucose molecules are broken down, energy is released.

The two simpler molecules are broken down again. This breakdown occurs in the mitochondria of the cells of plants, animals, fungi, and many other organisms. This process uses oxygen, releases much more energy, and produces carbon dioxide and water as wastes. When you exhale, you breathe out carbon dioxide and some of the water.

Cellular respiration occurs in the cells of many living things. **Figure 14** shows how it occurs in one consumer. As you are reading this section of the chapter, millions of cells in your body are breaking down glucose, releasing energy, and producing carbon dioxide and water.

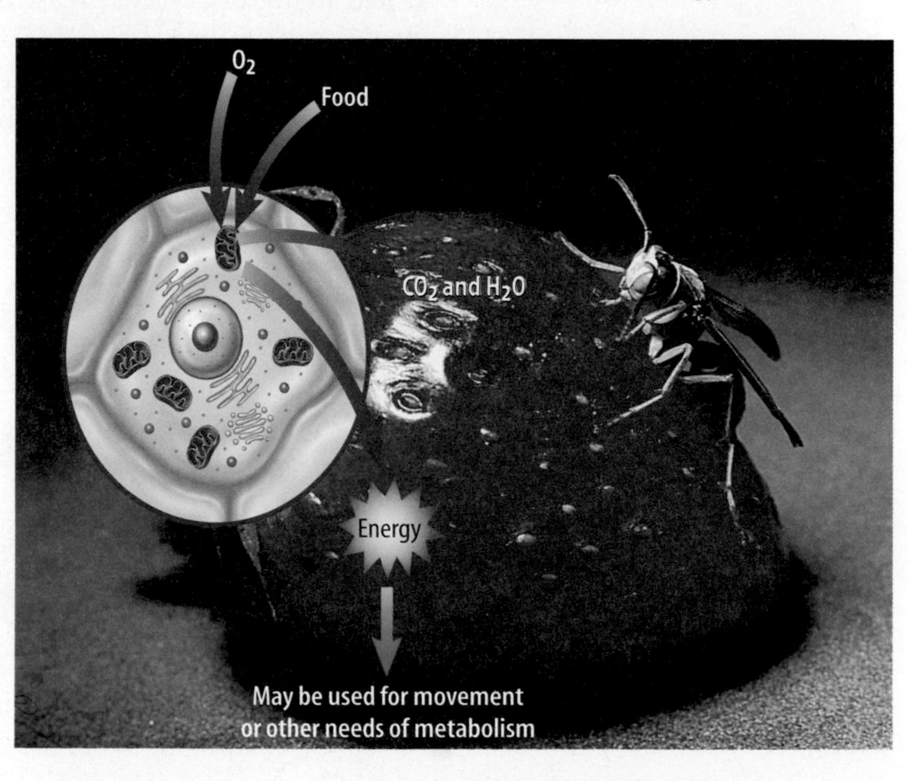

Figure 14 Producers and consumers carry on cellular respiration that releases energy from foods.

Fermentation Remember imagining you were late and had to run to school? During your run, your muscle cells might not have received enough oxygen, even though you were breathing rapidly. When cells do not have enough oxygen for cellular respiration, they use a process called **fermentation** to release some of the energy stored in glucose molecules.

Like cellular respiration, fermentation begins in the cytoplasm. Again, as the glucose molecules are broken down, energy is released. But the simple molecules from the breakdown of glucose do not move into the mitochondria. Instead, more chemical reactions occur in the cytoplasm. These reactions release some energy and produce wastes. Depending on the type of cell, the wastes may be lactic acid or alcohol and carbon dioxide, as shown in **Figure 15.** Your muscle cells can use fermentation to change the simple molecules into lactic acid while releasing energy. The presence of lactic acid is why your muscles might feel stiff and sore after exercising.

Topic: Beneficial Microorganisms
Visit booka.msscience.com for Web links to information about how microorganisms are used to produce many useful products.

Activity Find three other ways that microorganisms are beneficial.

Reading Check *Where in a cell does fermentation take place?*

Some microscopic organisms, such as bacteria, carry out fermentation and make lactic acid. Some of these organisms are used to produce yogurt and some cheeses. These organisms break down a sugar in milk and release energy. The lactic acid produced causes the milk to become more solid and gives these foods some of their flavor.

Have you ever used yeast to make bread? Yeasts are one-celled living organisms. Yeast cells use fermentation and break down sugar in bread dough. They produce alcohol and carbon dioxide as wastes. The carbon dioxide waste is a gas that makes bread dough rise before it is baked. The alcohol is lost as the bread bakes.

Figure 15 Organisms that use fermentation produce several different wastes.

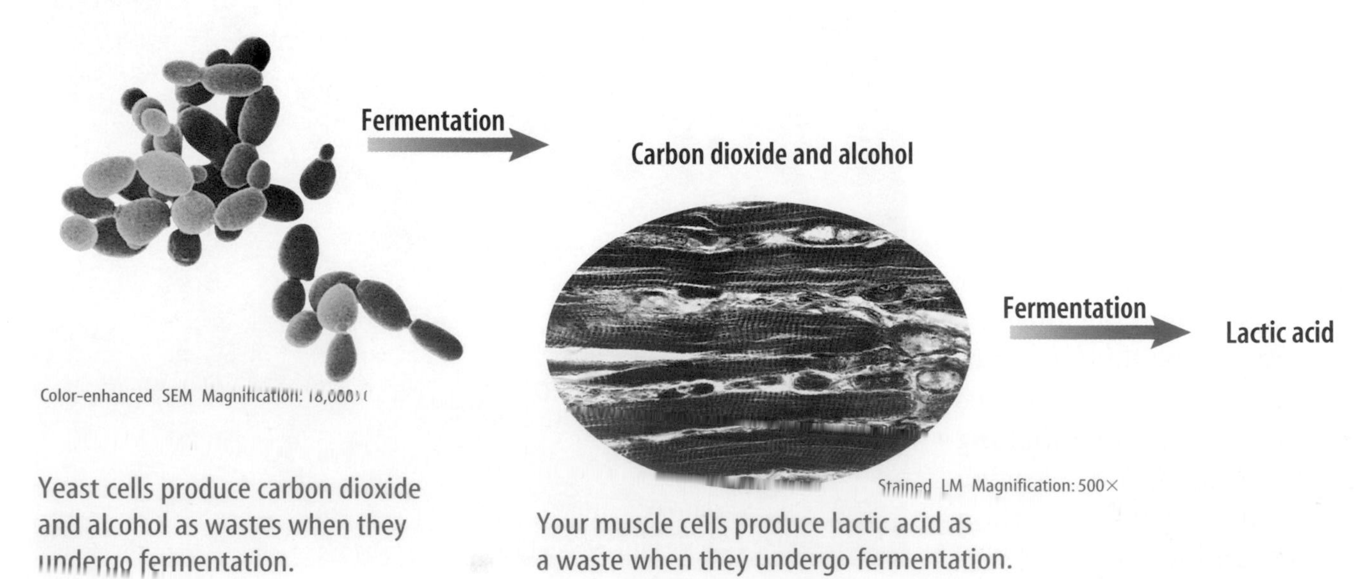

Yeast cells produce carbon dioxide and alcohol as wastes when they undergo fermentation.

Your muscle cells produce lactic acid as a waste when they undergo fermentation.

Figure 16 The chemical reactions of photosynthesis and cellular respiration could not take place without each other.

Related Processes How are photosynthesis, cellular respiration, and fermentation related? Some producers use photosynthesis to make food. All living things use respiration or fermentation to release energy stored in food. If you think carefully about what happens during photosynthesis and cellular respiration, you will see that what is produced in one is used in the other, as shown in **Figure 16.** These two processes are almost the opposite of each other. Photosynthesis produces sugars and oxygen, and cellular respiration uses these products. The carbon dioxide and water produced during cellular respiration are used during photosynthesis. Most life would not be possible without these important chemical reactions.

section 3 review

Summary

Trapping and Using Energy

- Metabolism is the total of all chemical reactions in an organism.
- During photosynthesis, light energy is transformed into chemical energy.
- Chlorophyll and other pigments capture light energy.
- Consumers take in energy by eating producers and other consumers.
- Living cells use oxygen and break down glucose that releases energy. This is called cellular respiration.
- Fermentation releases energy without oxygen.
- Without photosynthesis and cellular respiration and fermentation, most life would not be possible.

Self Check

1. **Explain** the difference between producers and consumers and give three examples of each.
2. **Infer** how the energy used by many living things on Earth can be traced back to sunlight.
3. **Compare and contrast** cellular respiration and fermentation.
4. **Think Critically** How can some indoor plants help to improve the quality of air in a room?

Applying Math

5. **Solve** Refer to the chemical equation for photosynthesis. Calculate then compare the number of carbon, hydrogen, and oxygen atoms before and after photosynthesis.

Photosynthesis and Cellular Respiration

Goals

- **Observe** green water plants in the light and dark.
- **Determine** whether plants carry on photosynthesis and cellular respiration.

Materials

16-mm test tubes (3)
150-mm test tubes with stoppers (4)
small, clear-glass baby food jars with lids (4)
test-tube rack
stirring rod
scissors
carbonated water (5 mL)
bromthymol blue solution in dropper bottle
aged tap water (20 mL)
**distilled water (20 mL)*
sprigs of *Elodea* (2)
**other water plants*
**Alternate materials*

Safety Precautions

WARNING: *Wear splash-proof safety goggles to protect eyes from hazardous chemicals.*

Real-World Question

Every living cell carries on many chemical processes. Two important chemical processes are cellular respiration and photosynthesis. All cells, including the ones in your body, carry on cellular respiration. However, some plant cells can carry on both processes. In this experiment you will investigate when these processes occur in plant cells. How could you find out when plants were using these processes? Are the products of photosynthesis and cellular respiration the same? When do plants carry on photosynthesis and cellular respiration?

Procedure

1. In your Science Journal, copy and complete the test-tube data table as you perform this lab.

Test-Tube Data

Test Tube	Color at Start	Color After 30 Minutes
1		
2	Do not write in this book.	
3		
4		

2. Label each test tube using the numbers *1, 2, 3,* and *4.* Pour 5 mL of aged tap water into each test tube.
3. Add 10 drops of carbonated water to test tubes *1* and *2.*
4. Add 10 drops of bromthymol blue to all of the test tubes. Bromthymol blue turns green to yellow in the presence of an acid.
5. Cut two 10-cm sprigs of *Elodea.* Place one sprig in test tube *1* and one sprig in test tube *3.* Stopper all test tubes.
6. Place test tubes *1* and *2* in bright light. Place tubes *3* and *4* in the dark. Observe the test tubes for 45 min or until the color changes. Record the color of each of the four test tubes.

LM Magnification: 225×

Analyze Your Data

1. **Identify** what is indicated by the color of the water in all four test tubes at the start of the activity.
2. **Infer** what process occurred in the test tube or tubes that changed color after 30 min.

Conclude and Apply

1. **Describe** the purpose of test tubes *2* and *4* in this experiment.
2. **Explain** whether or not the results of this experiment show that photosynthesis and cellular respiration occur in plants.

Communicating Your Data

Choose one of the following activities to **communicate** your data. Prepare an oral presentation that explains how the experiment showed the differences between products of photosynthesis and cellular respiration. Draw a cartoon strip to **explain** what you did in this experiment. Use each panel to show a different step. **For more help, refer to the Science Skill Handbook.**

from "Tulip"

by Penny Harter

I watched its first green push
through bare dirt, where the builders
had dropped boards, shingles,
plaster—
killing everything.
I could not recall what grew there,
what returned each spring,
but the leaves looked tulip,
and one morning it arrived,
a scarlet slash against the aluminum siding.
Mornings, on the way to my car,
I bow to the still bell
of its closed petals; evenings,
it greets me, light ringing
at the end of my driveway.
Sometimes I kneel
to stare into the yellow throat . . . .
It opens and closes my days.
It has made me weak with love . . . .

Understanding Literature

Personification Using human traits or emotions to describe an idea, animal, or inanimate object is called personification. When the poet writes that the tulip has a "yellow throat," she uses personification. Where else does the poet use personification?

Respond to the Reading

1. Why do you suppose the tulip survived the builders' abuse?
2. What is the yellow throat that the narrator is staring into?
3. **Linking Science and Writing** Keep a gardener's journal of a plant for a month, describing weekly the plant's condition, size, health, color, and other physical qualities.

Because most chemical reactions in plants take place in water, plants must have water in order to grow. The water carries nutrients and minerals from the soil into the plant. The process of active transport allows needed nutrients to enter the roots. The cell membranes of root cells contain proteins that bind with the needed nutrients. Cellular energy is used to move these nutrients through the cell membrane.

Reviewing Main Ideas

Section 1 Chemistry of Life

1. Matter is anything that has mass and takes up space.
2. Energy in matter is in the chemical bonds that hold matter together.
3. All organic compounds contain the elements hydrogen and carbon. The organic compounds in living things are carbohydrates, lipids, proteins, and nucleic acids.
4. Organic and inorganic compounds are important to living things.

Section 2 Moving Cellular Materials

1. The selectively permeable cell membrane controls which molecules can pass into and out of the cell.
2. In diffusion, molecules move from areas where there are relatively more of them to areas where there are relatively fewer of them.
3. Osmosis is the diffusion of water through a cell membrane.
4. Cells use energy to move molecules by active transport but do not use energy for passive transport.
5. Cells move large particles through cell membranes by endocytosis and exocytosis.

Section 3 Energy for Life

1. Photosynthesis is the process by which some producers change light energy into chemical energy.
2. Cellular respiration uses oxygen, releases the energy in food molecules, and produces waste carbon dioxide and water.
3. Some one-celled organisms and cells that lack oxygen use fermentation and release small amounts of energy from glucose. Wastes such as alcohol, carbon dioxide, and lactic acid are produced.

Visualizing Main Ideas

Copy and complete the following table on energy processes.

Energy Processes

	Photosynthesis	Cellular Respiration	Fermentation
Energy source		food (glucose)	food (glucose)
In plant and animal cells, occurs in		Do not write in this book.	
Reactants are			
Products are			

chapter 3 Review

Using Vocabulary

active transport p. 79	inorganic compound p. 73
cellular respiration p. 85	metabolism p. 83
diffusion p. 77	mixture p. 71
endocytosis p. 80	organic compound p. 72
enzyme p. 73	osmosis p. 78
equilibrium p. 77	passive transport p. 76
exocytosis p. 80	photosynthesis p. 85
fermentation p. 86	

Use what you know about the vocabulary words to answer the following questions.

1. What is the diffusion of water called?

2. What type of protein regulates nearly all chemical reactions in cells?

3. How do large food particles enter an amoeba?

4. What type of compound is water?

5. What process is used by some producers to convert light energy into chemical energy?

6. What type of compounds always contain carbon and hydrogen?

7. What process uses oxygen to break down glucose?

8. What is the total of all chemical reactions in an organism called?

Checking Concepts

Choose the word or phrase that best answers the question.

9. What is it called when cells use energy to move molecules?
A) diffusion **C)** active transport
B) osmosis **D)** passive transport

Use the photo below to answer question 10.

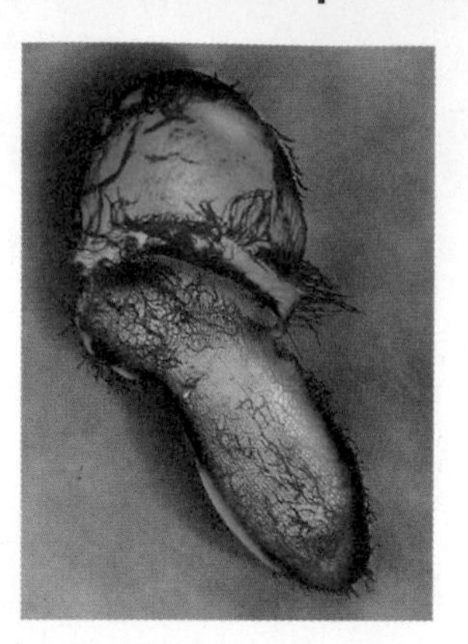

10. What cell process is occurring in the photo?
A) osmosis **C)** exocytosis
B) endocytosis **D)** diffusion

11. What occurs when the number of molecules of a substance is equal in two areas?
A) equilibrium **C)** fermentation
B) metabolism **D)** cellular respiration

12. Which substance is an example of a carbohydrate?
A) enzyme **C)** wax
B) sugar **D)** DNA

13. What is RNA an example of?
A) carbon dioxide **C)** lipid
B) water **D)** nucleic acid

14. What organic molecule stores the greatest amount of energy?
A) carbohydrate **C)** lipid
B) water **D)** nucleic acid

15. Which formula is an example of an organic compound?
A) $C_6H_{12}O_6$ **C)** H_2O
B) NO_2 **D)** O_2

16. What are organisms that cannot make their own food called?
A) biodegradables **C)** consumers
B) producers **D)** enzymes

Science Online booka.msscience.com/vocabulary_puzzlemaker

Thinking Critically

17. **Concept Map** Copy and complete the events-chain concept map to sequence the following parts of matter from smallest to largest: *atom, electron,* and *compound.*

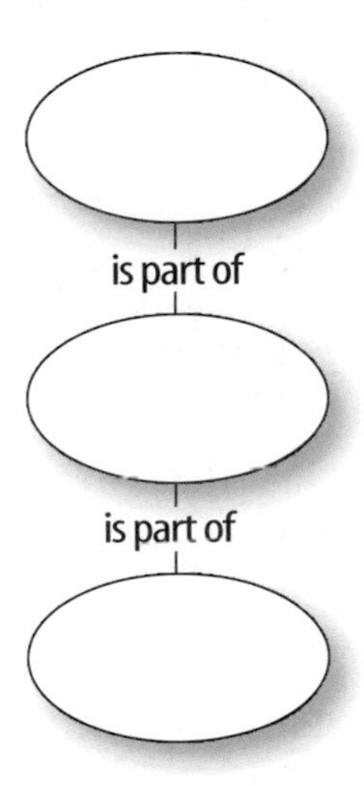

Use the table below to answer question 18.

Photosynthesis in Water Plants

Beaker Number	Distance from Light (cm)	Bubbles per Minute
1	10	45
2	30	30
3	50	19
4	70	6
5	100	1

18. **Interpret Data** Water plants were placed at different distances from a light source. Bubbles coming from the plants were counted to measure the rate of photosynthesis. What can you say about how the distance from the light affected the rate?

19. **Infer** why, in snowy places, salt is used to melt ice on the roads. Explain what could happen to many roadside plants as a result.

20. **Draw a conclusion** about why sugar dissolves faster in hot tea than in iced tea.

21. **Predict** what would happen to the consumers in a lake if all the producers died.

22. **Explain** how meat tenderizers affect meat.

23. **Form a hypothesis** about what will happen to wilted celery when placed in a glass of plain water.

Performance Activities

24. **Puzzle** Make a crossword puzzle with words describing ways substances are transported across cell membranes. Use the following words in your puzzle: *diffusion, osmosis, facilitated diffusion, active transport, endocytosis,* and *exocytosis.* Make sure your clues give good descriptions of each transport method.

Applying Math

25. **Light and Photosynthesis** Using the data from question 18, make a line graph that shows the relationship between the rate of photosynthesis and the distance from light.

26. **Importance of Water** Assume the brain is 70% water. If the average adult human brain weighs 1.4 kg, how many kilograms of water does it contain?

Use the equation below to answer question 27.

Photosynthesis

$$6CO_2 + 6H_2O + \text{light energy} \xrightarrow{\text{chlorophyll}} C_6H_{12}O_6 + 6O_2$$

carbon dioxide, water → sugar, oxygen

27. **Photosynthesis** Refer to the chemical equation above. If 18 CO_2 molecules and 18 H_2O molecules are used with light energy to make sugar, how many sugar molecules will be produced? How many oxygen molecules will be produced?

chapter 3 Standardized Test Practice

Part 1 Multiple Choice

Record your answers on the answer sheet provided by your teacher or on a sheet of paper.

1. Which describes a substance that is made up of only one kind of atom and cannot be broken down by chemical reactions?
   A. electron **C.** element
   B. carbohydrate **D.** molecule

Use the illustration below to answer questions 2 and 3.

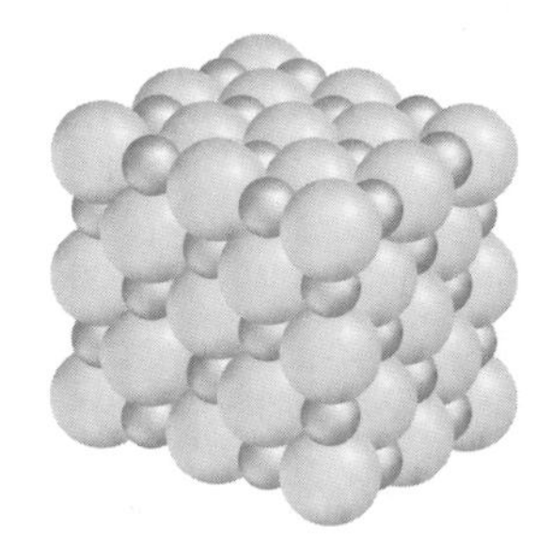

2. What kind of chemical compound do salt and water form?
   A. covalent
   B. ionic
   C. solution
   D. lipid

3. Salt is very important in the human body. What kind of compound is salt?
   A. organic
   B. carbohydrate
   C. protein
   D. inorganic

4. A cell that contains 40% water is placed in a solution that is 20% water. The cell and the solution will reach equilibrium when they both contain how much water?
   A. 30% **C.** 60%
   B. 40% **D.** 20%

5. All chemical reactions in living things take place in what kind of a solution?
   A. protein **C.** gas
   B. water **D.** solid

6. What is the sum of all the chemical reactions in an organism?
   A. cellular respiration **C.** fermentation
   B. metabolism **D.** endocytosis

7. What is needed for all chemical reactions in cells?
   A. enzymes **C.** DNA
   B. lipids **D.** cell membrane

8. What process produces the carbon dioxide that you exhale?
   A. osmosis
   B. DNA synthesis
   C. photosynthesis
   D. respiration

9. Which is needed to hold matter together or break it apart?
   A. gas
   B. liquid
   C. energy
   D. temperature

Use the table below to answer question 10.

Cell Substances		
Organic Compound	**Flexibility**	**Found in**
Keratin	Not very flexible	Hair and skin of mammals
Collagen	Not very flexible	Skin, bones, and tendons of mammals
Chitin	Very rigid	Tough outer shell of insects and crabs
Cellulose	Very flexible	Plant cell walls

10. According to this information, which organic compound is the least flexible?
    A. keratin
    B. collagen
    C. chitin
    D. cellulose

Part 2 Short Response/Grid In

Record your answers on the answer sheet provided by your teacher or on a sheet of paper.

11. Explain the structure of an atom.
12. How does chewing food affect your body's ability to release the chemical energy of the food?
13. Ice fishing is a popular sport in the winter. What properties of water is this sport based on?
14. Explain where the starch in a potato comes from.
15. Does fermentation or cellular respiration release more energy for an athlete's muscles? Which process would be responsible for making muscles sore?

Use the table below to answer question 16.

Classification of Compounds			
Compound	Organic	Inorganic	Type of organic compound
Salt			
Fat			
Skin			
DNA			
Sugar			
Water			
Potassium			

16. Copy and complete the table above. Identify each item as inorganic or organic. If the item is an organic compound further classify it as a protein, carbohydrate, lipid or nucleic acid.
17. Define selectively permeable and discuss why it is important for the cell membrane.
18. What is the source of energy for the photosynthesis reactions and where do they take place in a cell?

Part 3 Open Ended

Record your answers on a sheet of paper.

19. Give examples of each of the four types of organic molecules and why they are needed in a plant cell.
20. Trace the path of how oxygen molecules are produced in a plant cell to how they are used in human cells.
21. Describe four ways a large or small molecule can cross the cell membrane.
22. Discuss how water is bonded together and the unique properties that result from the bonds.

Use the illustration below to answer question 23.

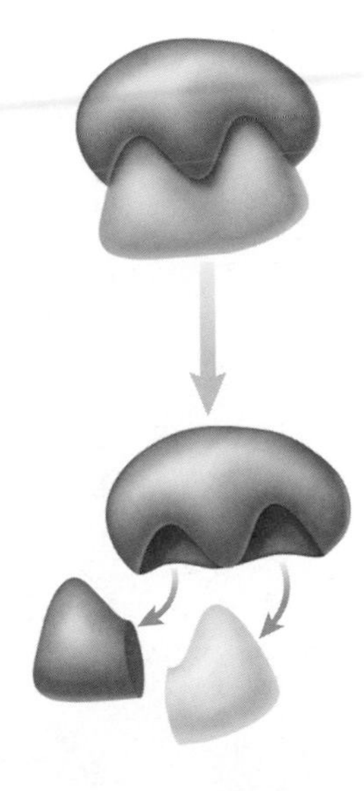

23. Describe in detail what process is taking place in this diagram and its significance for a cell.
24. How do plants use carbon dioxide? Why would plants need oxygen?

Test-Taking Tip

Diagrams Study a diagram carefully, being sure to read all labels and captions.

Cell Reproduction

The BIG Idea

Reproduction must occur for species to survive.

SECTION 1
Cell Division and Mitosis
Main Idea Different organisms can grow, repair damaged cells, and reproduce because of cell division and mitosis.

SECTION 2
Sexual Reproduction and Meiosis
Main Idea Sexual reproduction and meiosis ensure the preservation of species and diversity of life.

SECTION 3
DNA
Main Idea DNA contains the instructions for all life.

Why a turtle, not a chicken?

Several new sweet potato plants can be grown from just one potato, but turtles and most other animals need to have two parents. A cut on your finger heals. How do these things happen? In this chapter, you will find answers to these questions as you learn about cell reproduction.

Science Journal Write three things that you know about how and why cells reproduce.

Start-Up Activities

Infer About Seed Growth

Most flower and vegetable seeds sprout and grow into entire plants in just a few weeks. Although all of the cells in a seed have information and instructions to produce a new plant, only some of the cells in the seed use the information. Where are these cells in seeds? Do the following lab to find out.

1. Carefully split open two bean seeds that have soaked in water overnight.
2. Observe both halves and record your observations.
3. Wrap all four halves in a moist paper towel. Then put them into a self-sealing, plastic bag and seal the bag.
4. Make observations every day for a few days.
5. **Think Critically** Write a paragraph that describes what you observe. Hypothesize which cells in seeds use information about how plants grow.

Preview this chapter's content and activities at booka.msscience.com

FOLDABLES™ Study Organizer

How and Why Cells Divide Make the following Foldable to help you organize information from the chapter about cell reproduction.

STEP 1 **Draw** a mark at the midpoint of a vertical sheet of paper along the side edge.

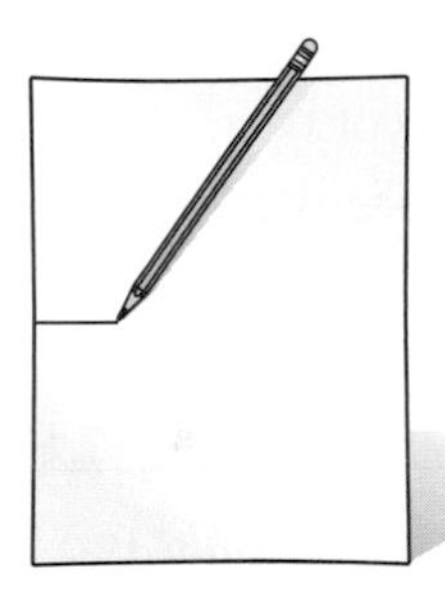

STEP 2 **Turn** the paper horizontally and **fold** the outside edges in to touch at the midpoint mark.

STEP 3 **Use** a pencil to draw a cell on the front of your Foldable as shown.

Analyze As you read the chapter, write under the flaps how cells divide. In the middle section, list why cells divide.

Get Ready to Read

Monitor

1 Learn It! An important strategy to help you improve your reading is monitoring, or finding your reading strengths and weaknesses. As you read, monitor yourself to make sure the text makes sense. Discover different monitoring techniques you can use at different times, depending on the type of test and situation.

2 Practice It! The paragraph below appears in Section 1. Read the passage and answer the questions that follow. Discuss your answers with other students to see how they monitor their reading.

> Reproduction is the process by which an organism produces others of its same kind. Among living organisms, there are two types of reproduction—sexual and asexual. Sexual reproduction usually requires two organisms. In **asexual reproduction,** a new organism (sometimes more than one) is produced from one organism. The new organism will have hereditary material identical to the hereditary material of the parent organism.
>
> —*from page 103*

- What questions do you still have after reading?
- Do you understand all of the words in the passage?
- Did you have to stop reading often? Is the reading level appropriate for you?

3 Apply It! Identify one paragraph that is difficult to understand. Discuss it with a partner to improve your understanding.

Reading Tip

Monitor your reading by slowing down or speeding up depending on your understanding of the text.

Target Your Reading

Use this to focus on the main ideas as you read the chapter.

1. **Before you read** the chapter, respond to the statements below on your worksheet or on a numbered sheet of paper.
   - Write an **A** if you **agree** with the statement.
   - Write a **D** if you **disagree** with the statement.

2. **After you read** the chapter, look back to this page to see if you've changed your mind about any of the statements.
   - If any of your answers changed, explain why.
   - Change any false statements into true statements.
   - Use your revised statements as a study guide.

Before You Read A or D		Statement	After You Read A or D
	1	All cell cycles last the same amount of time.	
	2	Interphase lasts longer than other phases of a cell's cycle.	
	3	Asexual reproduction requires two parents.	
	4	Cell division and mitosis is the same in all organisms.	
	5	Meiosis always happens before fertilization.	
	6	A zygote is the cell formed when an egg and sperm join.	
	7	Diploid cells have pairs of similar chromosomes.	
	8	The exact structure of DNA is unknown.	
	9	A gene is a section of DNA on a chromosome.	
	10	Mistakes in copying DNA result in mutations.	
	11	Budding and regeneration can occur in most organisims.	

Science Online
Print out a worksheet of this page at booka.msscience.com

section 1

Cell Division and Mitosis

as you read

What You'll Learn

- **Explain** why mitosis is important.
- **Examine** the steps of mitosis.
- **Compare** mitosis in plant and animal cells.
- **List** two examples of asexual reproduction.

Why It's Important

Your growth, like that of many organisms, depends on cell division.

Review Vocabulary

nucleus: organelle that controls all the activities of a cell and contains hereditary material made of proteins and DNA

New Vocabulary

- mitosis
- chromosome
- asexual reproduction

Why is cell division important?

What do you, an octopus, and an oak tree have in common? You share many characteristics, but an important one is that you are all made of cells—trillions of cells. Where did all of those cells come from? As amazing as it might seem, many organisms start as just one cell. That cell divides and becomes two, two become four, four become eight, and so on. Many-celled organisms, including you, grow because cell division increases the total number of cells in an organism. Even after growth stops, cell division is still important. Every day, billions of red blood cells in your body wear out and are replaced. During the few seconds it takes you to read this sentence, your bone marrow produced about six million red blood cells. Cell division is important to one-celled organisms, too—it's how they reproduce themselves, as shown in **Figure 1.** Cell division isn't as simple as just cutting the cell in half, so how do cells divide?

The Cell Cycle

A living organism has a life cycle. A life cycle begins with the organism's formation, is followed by growth and development, and finally ends in death. Right now, you are in a stage of your life cycle called adolescence, which is a period of active growth and development. Individual cells also have life cycles.

Figure 1 All organisms use cell division. Many-celled organisms, such as this octopus, grow by increasing the numbers of their cells.

Like this dividing amoeba, a one-celled organism reaches a certain size and then reproduces.

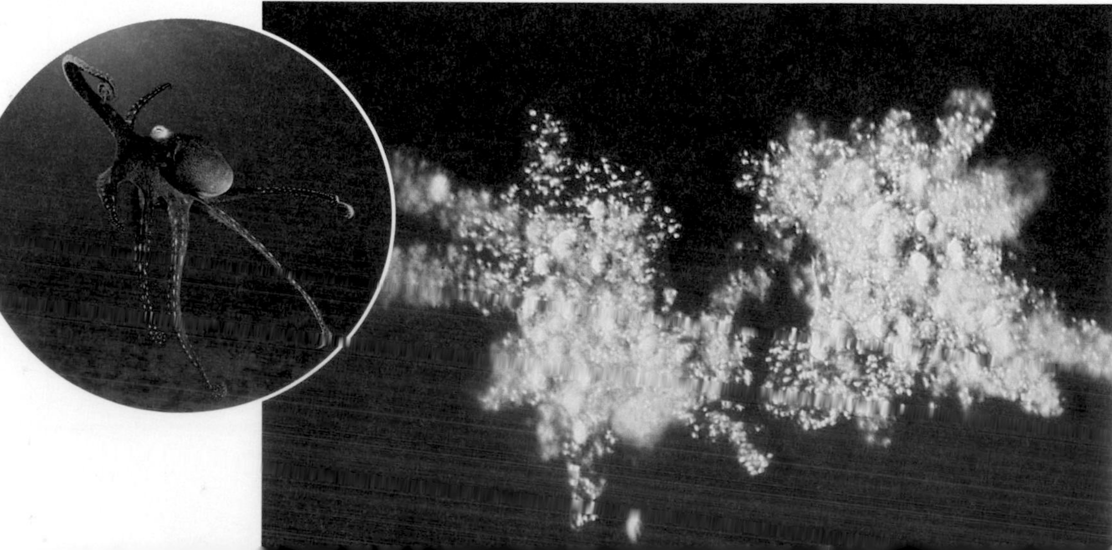

Figure 2 Interphase is the longest part of the cell cycle.
Identify *When do chromosomes duplicate?*

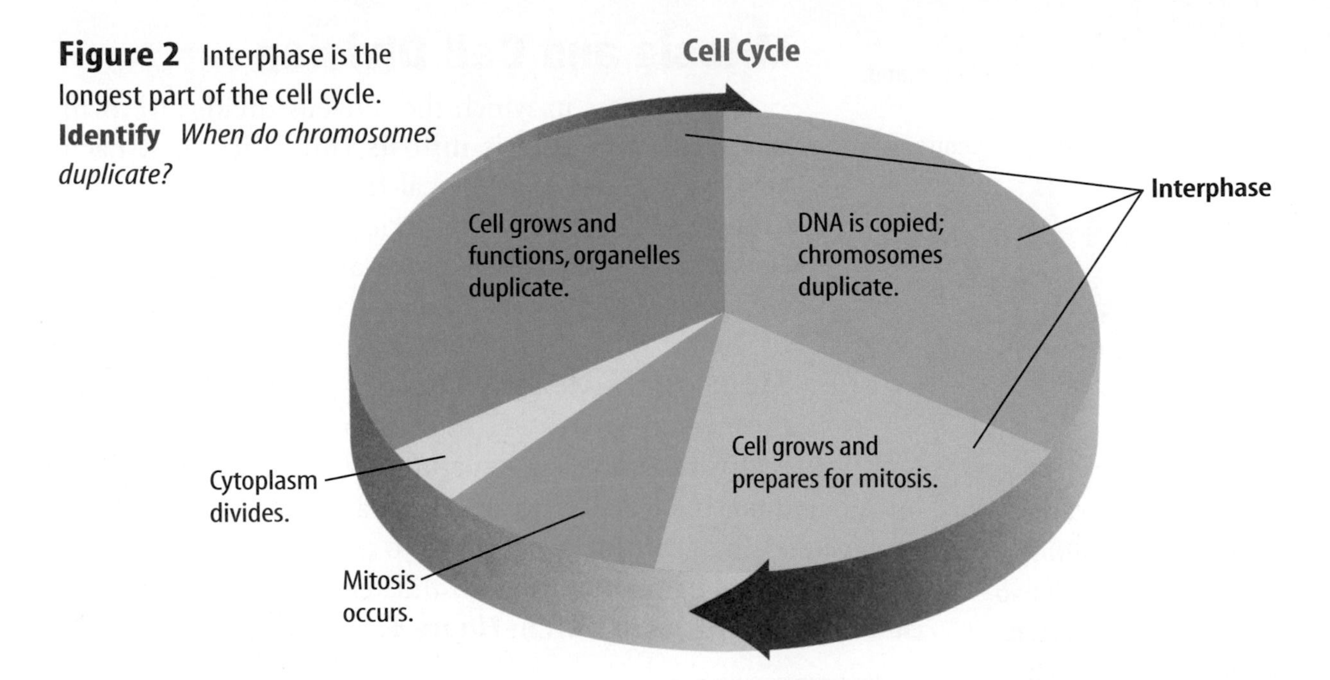

Length of Cycle The cell cycle, as shown in **Figure 2,** is a series of events that takes place from one cell division to the next. The time it takes to complete a cell cycle is not the same in all cells. For example, the cycle for cells in some bean plants takes about 19 h to complete. Cells in animal embryos divide rapidly and can complete their cycles in less than 20 min. In some human cells, the cell cycle takes about 16 h. Cells in humans that are needed for repair, growth, or replacement, like skin and bone cells, constantly repeat the cycle.

Interphase Most of the life of any eukaryotic cell—a cell with a nucleus—is spent in a period of growth and development called interphase. Cells in your body that no longer divide, such as nerve and muscle cells, are always in interphase. An actively dividing cell, such as a skin cell, copies its hereditary material and prepares for cell division during interphase.

Why is it important for a cell to copy its hereditary information before dividing? Imagine that you have a part in a play and the director has one complete copy of the script. If the director gave only one page to each person in the play, no one would have the entire script. Instead the director makes a complete, separate copy of the script for each member of the cast so that each one can learn his or her part. Before a cell divides, a copy of the hereditary material must be made so that each of the two new cells will get a complete copy. Just as the actors in the play need the entire script, each cell needs a complete set of hereditary material to carry out life functions.

After interphase, cell division begins. The nucleus divides, and then the cytoplasm separates to form two new cells.

Oncologist In most cells, the cell cycle is well controlled. Cancer cells, however, have uncontrolled cell division. Doctors who diagnose, study, and treat cancer are called oncologists. Someone wanting to become an oncologist must first complete medical school before training in oncology. Research the subspecialities of oncology. List and describe them in your Science Journal.

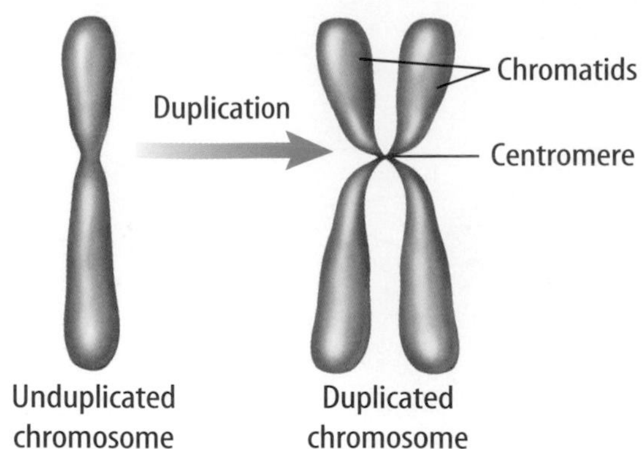

Figure 3 DNA is copied during interphase. An unduplicated chromosome has one strand of DNA. A duplicated chromosome has two identical DNA strands, called chromatids, that are held together at a region called the centromere.

Mitosis and Cell Division

The process in which the nucleus divides to form two identical nuclei is **mitosis** (mi TOH sus). Each new nucleus also is identical to the original nucleus. Mitosis is described as a series of phases, or steps. The steps of mitosis in order are named prophase, metaphase, anaphase, and telophase.

Steps of Mitosis When any nucleus divides, the chromosomes (KROH muh sohmz) play the important part. A **chromosome** is a structure in the nucleus that contains hereditary material. During interphase, each chromosome duplicates. When the nucleus is ready to divide, each duplicated chromosome coils tightly into two thickened, identical strands called chromatids, as shown in **Figure 3.**

Reading Check *How are chromosomes and chromatids related?*

During prophase, the pairs of chromatids are fully visible when viewed under a microscope. The nucleolus and the nuclear membrane disintegrate. Two small structures called centrioles (SEN tree olz) move to opposite ends of the cell. Between the centrioles, threadlike spindle fibers begin to stretch across the cell. Plant cells also form spindle fibers during mitosis but do not have centrioles.

In metaphase, the pairs of chromatids line up across the center of the cell. The centromere of each pair usually becomes attached to two spindle fibers—one from each side of the cell.

In anaphase, each centromere divides and the spindle fibers shorten. Each pair of chromatids separates, and chromatids begin to move to opposite ends of the cell. The separated chromatids are now called chromosomes. In the final step, telophase, spindle fibers start to disappear, the chromosomes start to uncoil, and two new nuclei form.

Cell Division For most cells, after the nucleus divides, the cytoplasm separates and two new cells form. In animal cells, the cell membrane pinches in the middle, like a balloon with a string tightened around it, and the cytoplasm divides. In plant cells, the appearance of a cell plate, as shown in **Figure 4,** tells you that the cytoplasm is being divided. New cell membranes form from the cell plate, and new cell walls develop from molecules released by the cell membranes. Following division of the cytoplasm, most new cells begin the period of growth, or interphase. Review cell division for an animal cell using the illustrations in **Figure 5.**

Figure 4 The cell plate shown in this plant cell appears when the cytoplasm is being divided.
Identify *what phase of mitosis will be next.*

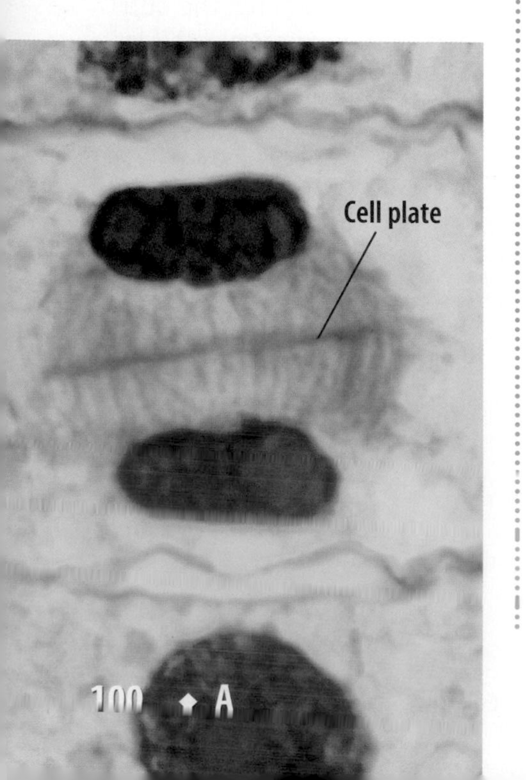

Figure 5 Cell division for an animal cell is shown here. Each micrograph shown in this figure is magnified 600 times.

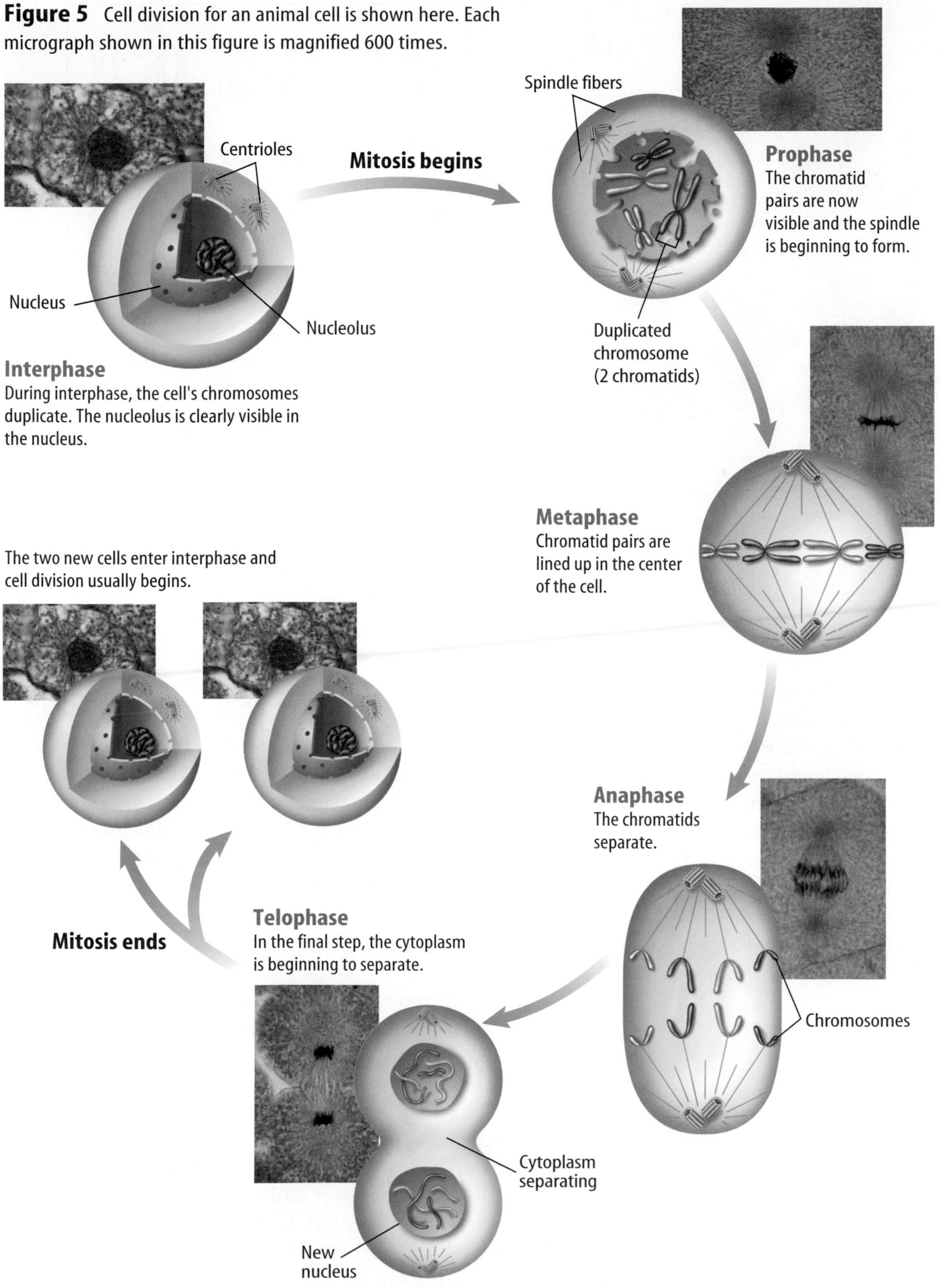

Centrioles

Nucleus

Nucleolus

Interphase
During interphase, the cell's chromosomes duplicate. The nucleolus is clearly visible in the nucleus.

Mitosis begins

Spindle fibers

Duplicated chromosome (2 chromatids)

Prophase
The chromatid pairs are now visible and the spindle is beginning to form.

Metaphase
Chromatid pairs are lined up in the center of the cell.

Anaphase
The chromatids separate.

Chromosomes

Telophase
In the final step, the cytoplasm is beginning to separate.

Cytoplasm separating

New nucleus

Mitosis ends

The two new cells enter interphase and cell division usually begins.

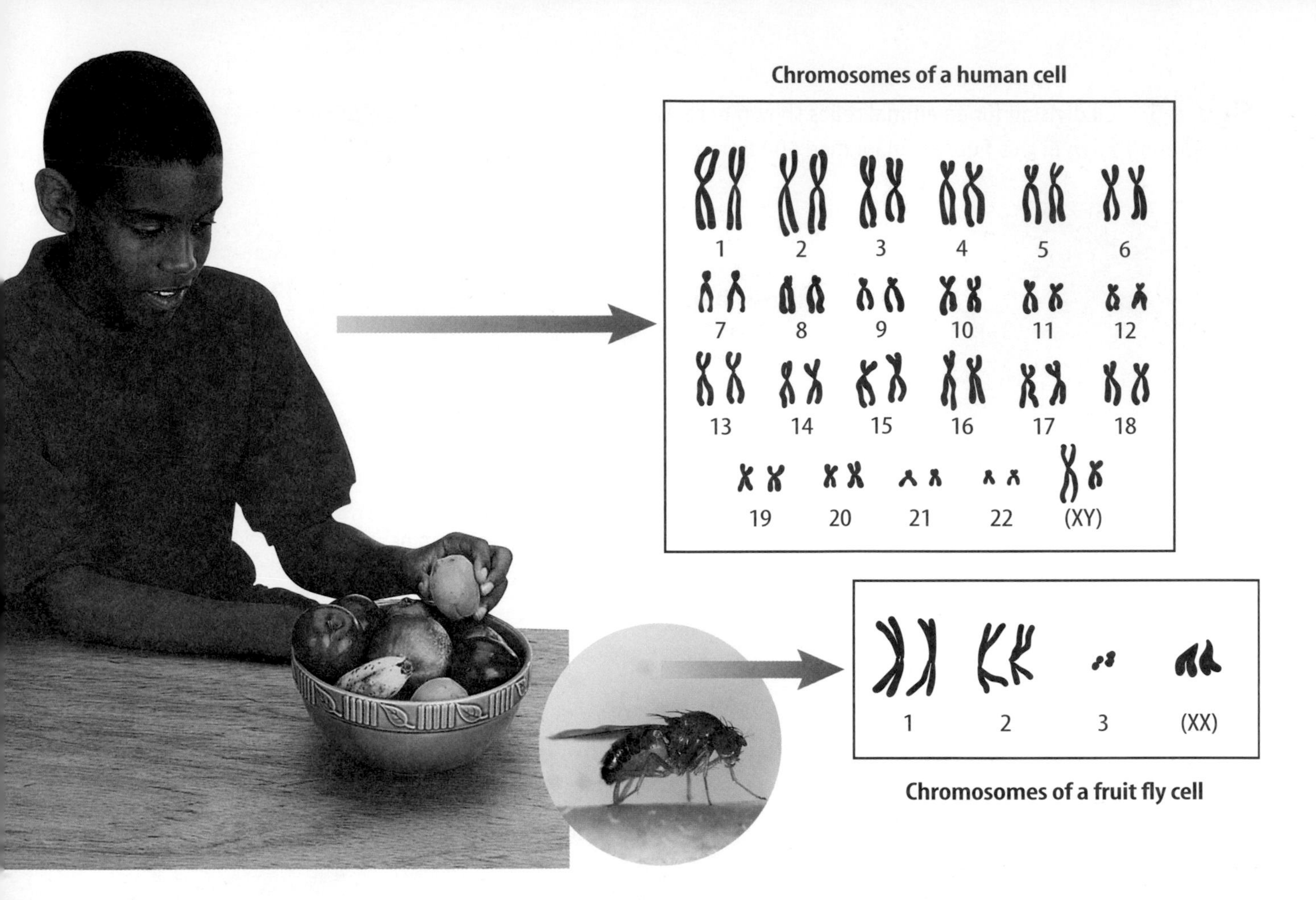

Figure 6 Pairs of chromosomes are found in the nucleus of most cells. All chromosomes shown here are in their duplicated form. Most human cells have 23 pairs of chromosomes including one pair of chromosomes that help determine a person's gender such as the XY pair above. Most fruit fly cells have four pairs of chromosomes.
Infer *What do you think the XX pair in fruit flies helps determine?*

Results of Mitosis and Cell Division There are three important things to remember about mitosis and cell division. First, mitosis is the division of a nucleus. Second, it produces two new nuclei that are identical to each other and the original nucleus. Each new nucleus has the same number and type of chromosomes. Each of the trillions of cells in your body, except sex cells, has a nucleus with a copy of the same 46 chromosomes, because you began as one cell with 46 chromosomes in its nucleus. In the same way, each cell in a fruit fly has eight chromosomes and each new cell produced by mitosis and cell division has a copy of those eight chromosomes, as shown in **Figure 6.** Third, the original cell no longer exists.

You probably know that not all of your cells are the same. Just as all actors in a play use the same script to learn different roles, cells use copies of the same hereditary material and become different cell types with specific functions. Certain cells only used that part of the heredity material with information needed to become a specific cell type.

Cell division allows growth and replaces worn out or damaged cells. You are much larger and have more cells than a baby mainly because of cell division. If you cut yourself, the wound heals because cell division replaces damaged cells. Another way some organisms use cell division is to produce new organisms.

Asexual Reproduction

Reproduction is the process by which an organism produces others of its same kind. Among living organisms, there are two types of reproduction—sexual and asexual. Sexual reproduction usually requires two organisms. In **asexual reproduction,** a new organism (sometimes more than one) is produced from one organism. The new organism will have hereditary material identical to the hereditary material of the parent organism.

How many organisms are needed for asexual reproduction?

Cellular Asexual Reproduction Organisms with eukaryotic cells asexually reproduce by mitosis and cell division. A sweet potato growing in a jar of water is an example of asexual reproduction. All the new stems, leaves, and roots that grow from the sweet potato have the same hereditary material. New strawberry plants can be reproduced asexually from horizontal stems called runners. **Figure 7** shows asexual reproduction in a potato and a strawberry plant.

Recall that mitosis is the division of a nucleus. However, a bacterium does not have a nucleus so it can't use mitosis. Instead, bacteria reproduce asexually by fission. During fission, the one-celled bacterium without a nucleus copies its genetic material and then divides into two identical organisms.

Modeling Mitosis

Procedure

1. Make models of cell division using **materials supplied by your teacher.**
2. Use four chromosomes in your model.
3. When finished, arrange the models in the order in which mitosis occurs.

Analysis

1. In which steps is the nucleus visible?
2. How many cells does a dividing cell form?

Figure 7 Many plants can reproduce asexually.

A new potato plant can grow from each sprout on this potato.

Infer *how the genetic material in the small strawberry plant above compares to the genetic material in the large strawberry plant.*

Figure 8 Some organisms use cell division for budding and regeneration.

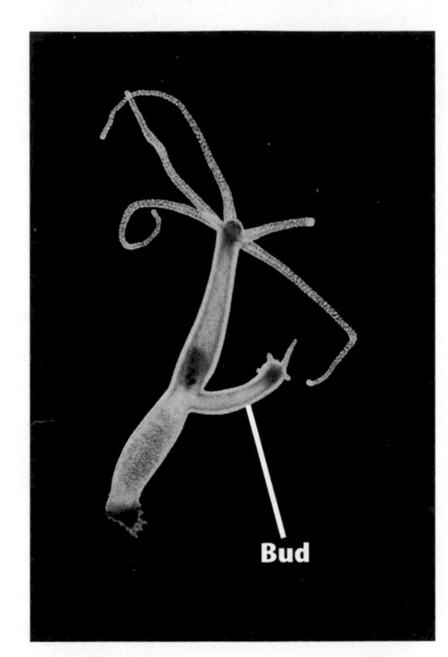

A Hydra, a freshwater animal, can reproduce asexually by budding. The bud is a small exact copy of the adult.

B This sea star is regenerating four new arms.

Budding and Regeneration Look at **Figure 8A.** A new organism is growing from the body of the parent organism. This organism, called a hydra, is reproducing by budding. Budding is a type of asexual reproduction made possible because of mitosis and cell division. When the bud on the adult becomes large enough, it breaks away to live on its own.

Some organisms can regrow damaged or lost body parts, as shown in **Figure 8B.** Regeneration is the process that uses mitosis and cell division to regrow body parts. Sponges, planaria, sea stars, and some other organisms can also use regeneration for asexual reproduction. If these organisms break into pieces, a whole new organism can grow from each piece. Because sea stars eat oysters, oyster farmers dislike them. What would happen if an oyster farmer collected sea stars, cut them into pieces, and threw them back into the ocean?

section 1 review

Summary

The Cell Cycle

- The cell cycle is a series of events from one cell division to the next.
- Most of a eukaryotic cell's life is interphase.

Mitosis

- Mitosis is a series of four phases or steps.
- Each new nucleus formed by mitosis has the same number and type of chromosomes.

Asexual Reproduction

- In asexual reproduction, a new organism is produced from one organism.
- Fission, budding, and regeneration are forms of asexual reproduction.

Self Check

1. **Define** mitosis. How does it differ in plants and animals?
2. **Identify** two examples of asexual reproduction in many-celled organisms.
3. **Describe** what happens to chromosomes before mitosis.
4. **Compare and contrast** the two new cells formed after mitosis and cell division.
5. **Think Critically** Why is it important for the nuclear membrane to disintegrate during mitosis?

Applying Math

6. **Solve One-Step Equations** If a cell undergoes cell division every 5 min, how many cells will there be after 1 h?

Science Online booka.msscience.com/self_check_quiz

Mitosis in Plant Cells

Reproduction of most cells in plants and animals uses mitosis and cell division. In this lab, you will study mitosis in plant cells by examining prepared slides of onion root-tip cells.

Real-World Question

How can plant cells in different stages of mitosis be distinguished from each other?

Goals

- **Compare** cells in different stages of mitosis and observe the location of their chromosomes.
- **Observe** what stage of mitosis is most common in onion root tips.

Materials

prepared slide of an onion root tip
microscope

Safety Precautions

Procedure

1. Copy the data table in your Science Journal.

Number of Root-Tip Cells Observed		
Stage of Mitosis	**Number of Cells Observed**	**Percent of Cells Observed**
Prophase		
Metaphase		
Anaphase	Do not write in this book.	
Telophase		
Total		

2. **Obtain** a prepared slide of cells from an onion root tip.

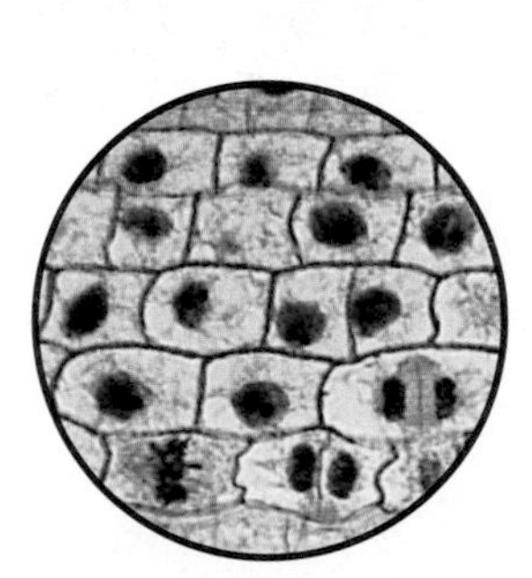
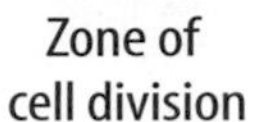
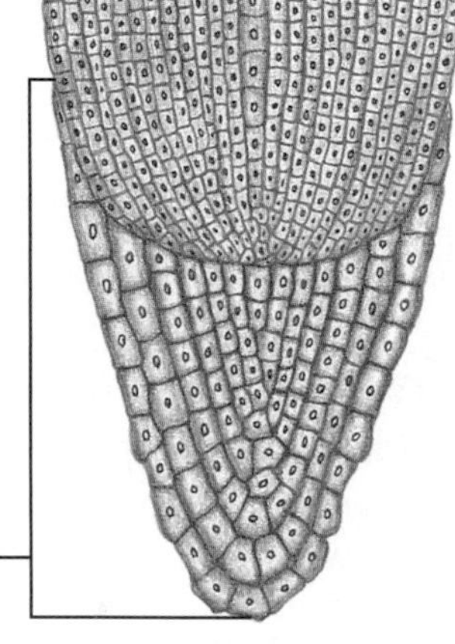

3. Set your microscope on low power and examine the slide. The large, round cells at the root tip are called the root cap. Move the slide until you see the cells just behind the root cap. Turn to the high-power objective.
4. Find an area where you can see the most stages of mitosis. Count and record how many cells you see in each stage.
5. Return the nosepiece to low power. Remove the onion root-tip slide.

Conclude and Apply

1. **Compare** the cells in the region behind the root cap to those in the root cap.
2. **Calculate** the percent of cells found in each stage of mitosis. Infer which stage of mitosis takes the longest period of time.

Communicating Your Data

Write and illustrate a story as if you were a cell undergoing mitosis. Share your story with your class. **For more help, refer to the Science Skill Handbook.**

section 2

Sexual Reproduction and Meiosis

as you read

What You'll Learn

- **Describe** the stages of meiosis and how sex cells are produced.
- **Explain** why meiosis is needed for sexual reproduction.
- **Name** the cells that are involved in fertilization.
- **Explain** how fertilization occurs in sexual reproduction.

Why It's Important

Meiosis and sexual reproduction are the reasons why no one else is exactly like you.

Review Vocabulary
organism: any living thing; uses energy, is made of cells, reproduces, responds, grows, and develops

New Vocabulary
- sexual reproduction
- sperm
- egg
- fertilization
- zygote
- diploid
- haploid
- meiosis

Sexual Reproduction

Another way that a new organism can be produced is by sexual reproduction. During **sexual reproduction,** two sex cells, sometimes called an egg and a sperm, come together. Sex cells, like those in **Figure 9,** are formed from cells in reproductive organs. **Sperm** are formed in the male reproductive organs. **Eggs** are formed in the female reproductive organs. The joining of an egg and a sperm is called **fertilization,** and the cell that forms is called a **zygote** (ZI goht). Generally, the egg and the sperm come from two different organisms of the same species. Following fertilization, mitosis and cell division begins. A new organism with a unique identity develops.

Diploid Cells Your body forms two types of cells—body cells and sex cells. Body cells far outnumber sex cells. Your brain, skin, bones, and other tissues and organs are formed from body cells. Recall that a typical human body cell has 46 chromosomes. Each chromosome has a mate that is similar to it in size and shape and has similar DNA. Human body cells have 23 pairs of chromosomes. When cells have pairs of similar chromosomes, they are said to be **diploid** (DIH ployd).

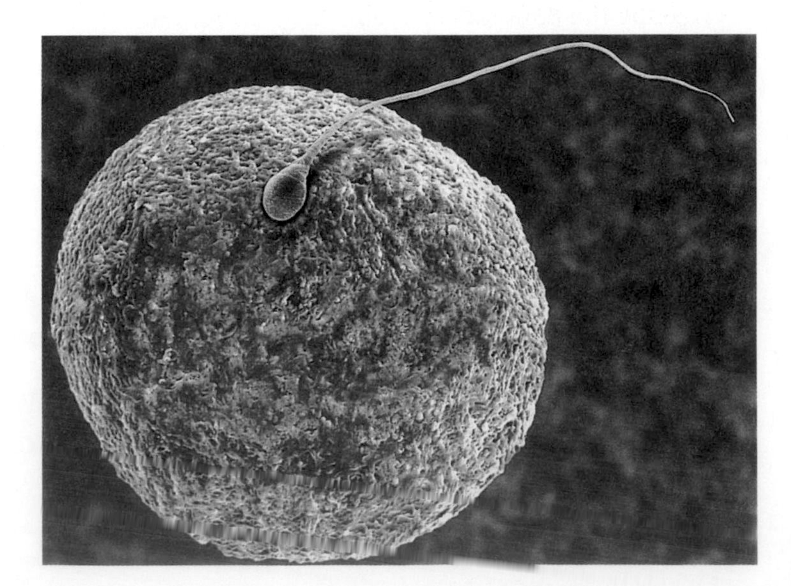

Figure 9 A human egg and a human sperm at fertilization.

Haploid Cells Because sex cells do not have pairs of chromosomes, they are said to be **haploid** (HA ployd). They have only half the number of chromosomes as body cells. *Haploid* means "single form." Human sex cells have only 23 chromosomes—one from each of the 23 pairs of similar chromosomes. Compare the number of chromosomes found in a human sex cell to the full set of human chromosomes seen in **Figure 6.**

Reading Check *How many chromosomes are usually in each human sperm?*

Meiosis and Sex Cells

A process called **meiosis** (mi OH sus) produces haploid sex cells. What would happen in sexual reproduction if two diploid cells combined? The offspring would have twice as many chromosomes as its parent. Although plants with twice the number of chromosomes as the parent plants are often produced, most animals do not survive with a double number of chromosomes. Meiosis ensures that the offspring will have the same diploid number as its parent, as shown in **Figure 10.** After two haploid sex cells combine, a diploid zygote is produced that develops into a new diploid organism.

During meiosis, two divisions of the nucleus occur. These divisions are called meiosis I and meiosis II. The steps of each division have names like those in mitosis and are numbered for the division in which they occur.

Diploid Zygote The human egg releases a chemical into the surrounding fluid that attracts sperm. Usually, only one sperm fertilizes the egg. After the sperm nucleus enters the egg, the cell membrane of the egg changes in a way that prevents other sperm from entering. What adaptation in this process guarantees that the zygote will be diploid? Write a paragraph describing your ideas in your Science Journal.

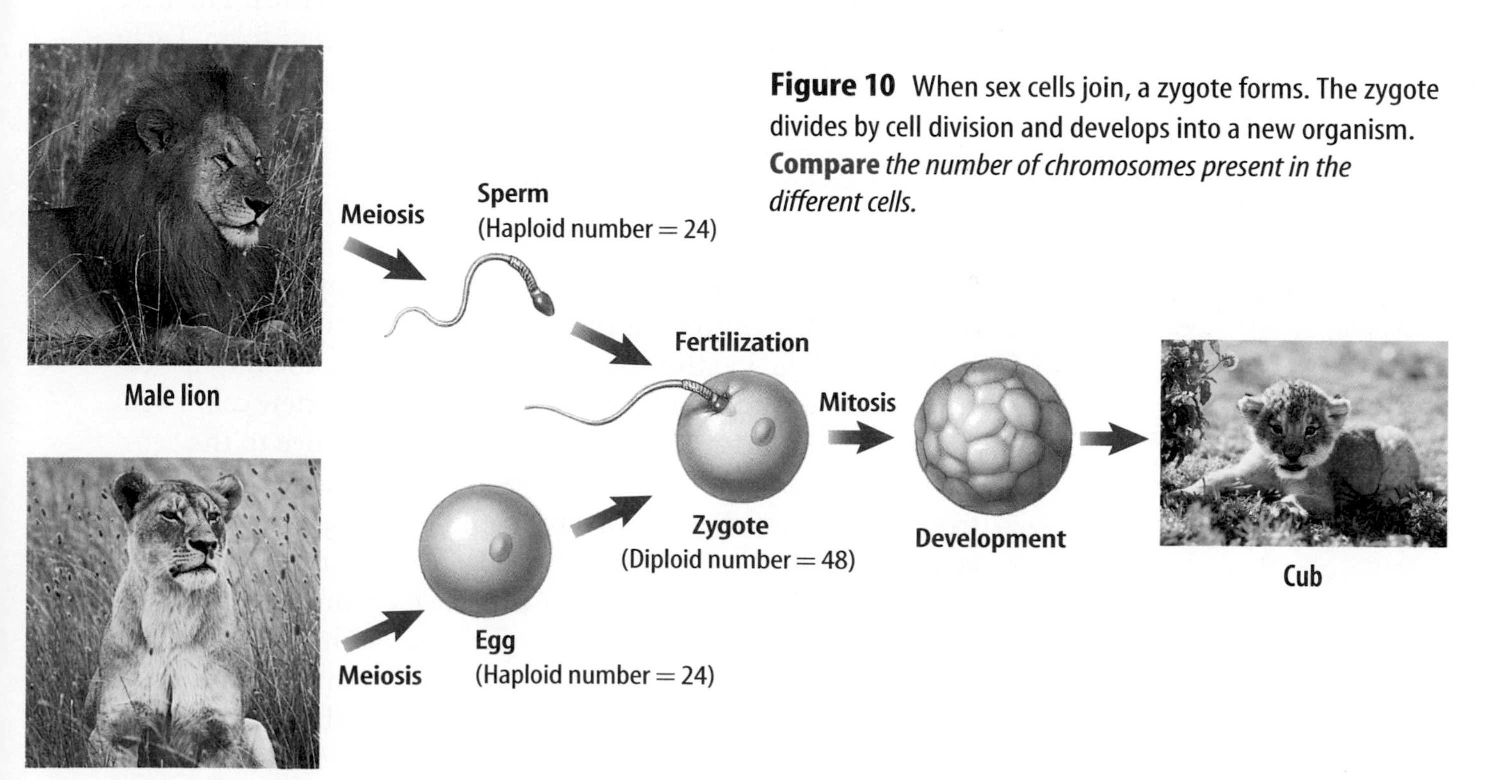

Figure 10 When sex cells join, a zygote forms. The zygote divides by cell division and develops into a new organism.
Compare *the number of chromosomes present in the different cells.*

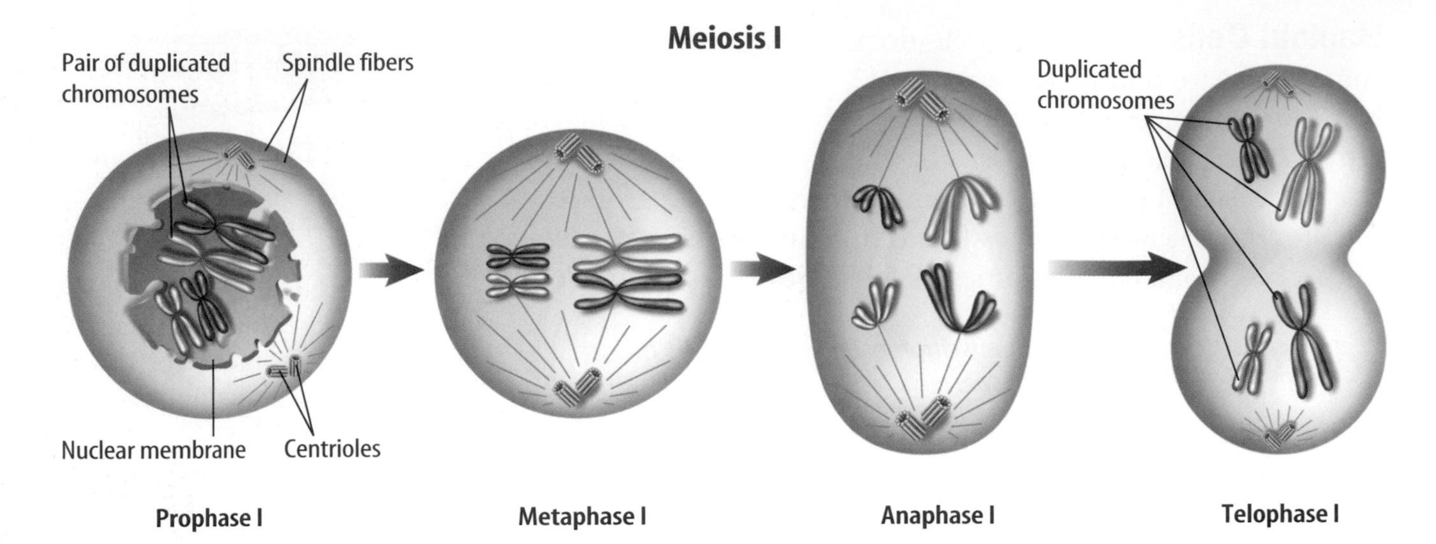

Figure 11 Meiosis has two divisions of the nucleus—meiosis I and meiosis II.
Determine *how many sex cells are finally formed after both divisions are completed.*

Meiosis I Before meiosis begins, each chromosome is duplicated, just as in mitosis. When the cell is ready for meiosis, each duplicated chromosome is visible under the microscope as two chromatids. As shown in **Figure 11,** the events of prophase I are similar to those of prophase in mitosis. In meiosis, each duplicated chromosome comes near its similar duplicated mate. In mitosis they do not come near each other.

In metaphase I, the pairs of duplicated chromosomes line up in the center of the cell. The centromere of each chromatid pair becomes attached to one spindle fiber, so the chromatids do not separate in anaphase I. The two pairs of chromatids of each similar pair move away from each other to opposite ends of the cell. Each duplicated chromosome still has two chromatids. Then, in telophase I, the cytoplasm divides, and two new cells form. Each new cell has one duplicated chromosome from each similar pair.

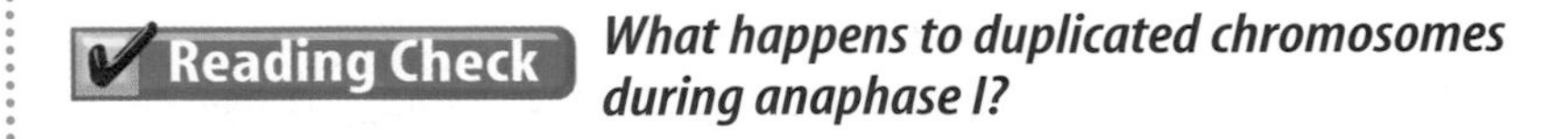

What happens to duplicated chromosomes during anaphase I?

Meiosis II The two cells formed during meiosis I now begin meiosis II. The chromatids of each duplicated chromosome will be separated during this division. In prophase II, the duplicated chromosomes and spindle fibers reappear in each new cell. Then in metaphase II, the duplicated chromosomes move to the center of the cell. Unlike what occurs in metaphase I, each centromere now attaches to two spindle fibers instead of one. The centromere divides during anaphase II. The chromatids separate and move to opposite ends of the cell. Each chromatid now is an individual chromosome. As telophase II begins, the spindle fibers disappear, and a nuclear membrane forms around each set of chromosomes. When meiosis II is finished, the cytoplasm divides.

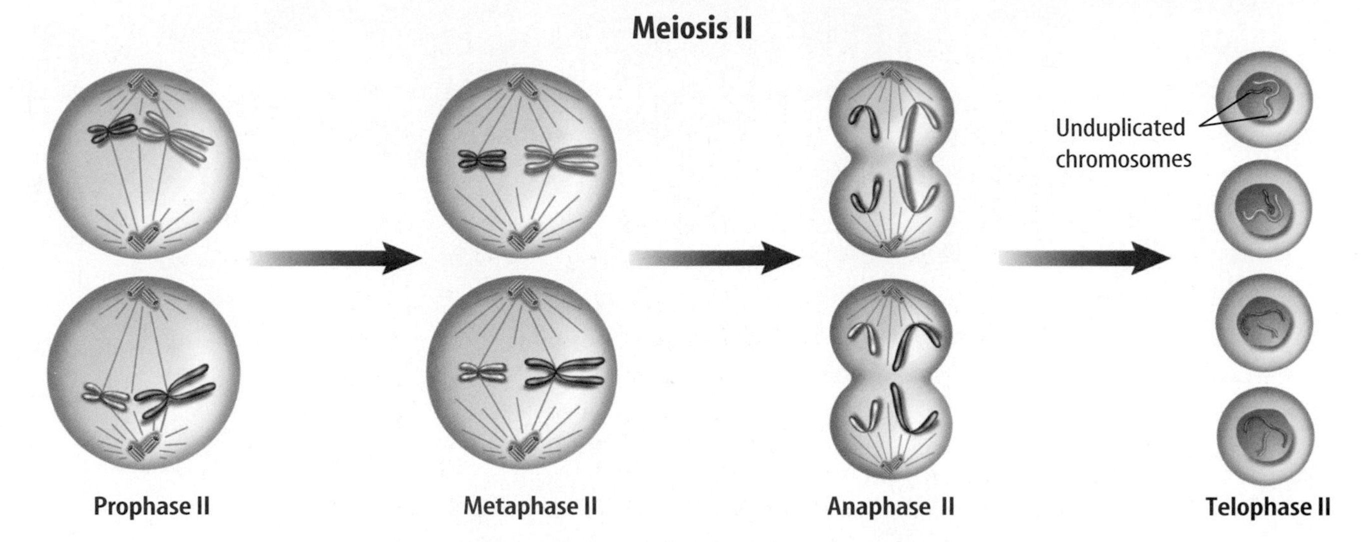

Summary of Meiosis Two cells form during meiosis I. In meiosis II, both of these cells form two cells. The two divisions of the nucleus result in four sex cells. Each has one-half the number of chromosomes in its nucleus that was in the original nucleus. From a human cell with 46 paired chromosomes, meiosis produces four sex cells each with 23 unpaired chromosomes.

Applying Science

How can chromosome numbers be predicted?

Offspring get half of their chromosomes from one parent and half from the other. What happens if each parent has a different diploid number of chromosomes?

Identifying the Problem

A Grevy's zebra and a donkey can mate to produce a zonkey, as shown below.

Solving the Problem

1. How many chromosomes would the zonkey receive from each parent?
2. What is the chromosome number of the zonkey?
3. What would happen when meiosis occurs in the zonkey's reproductive organs?
4. Predict why zonkeys are usually sterile.

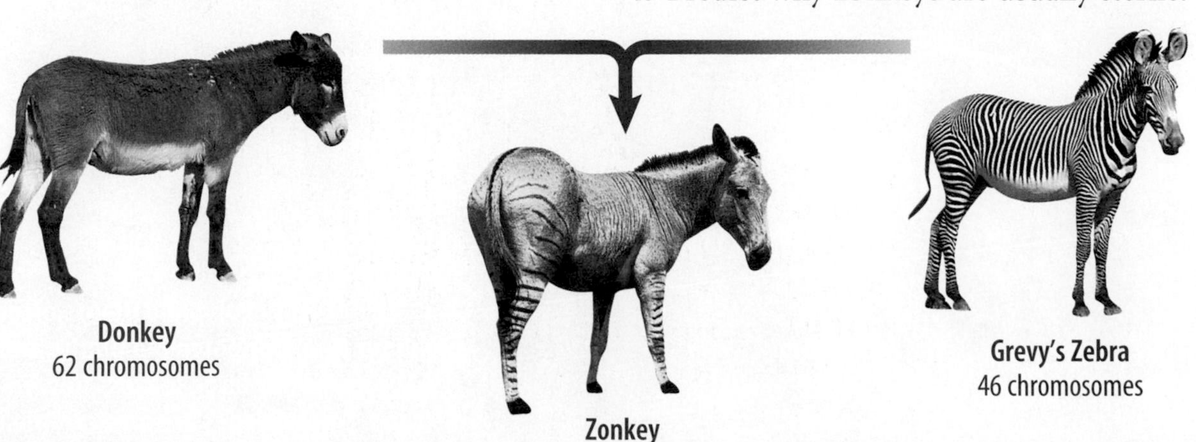

NATIONAL GEOGRAPHIC

VISUALIZING POLYPLOIDY IN PLANTS

Figure 12

You received a haploid (n) set of chromosomes from each of your parents, making you a diploid (2n) organism. In nature, however, many plants are polyploid—they have three (3n), four (4n), or more sets of chromosomes. We depend on some of these plants for food.

▲ TRIPLOID Bright yellow bananas typically come from triploid (3n) banana plants. Plants with an odd number of chromosome sets usually cannot reproduce sexually and have very small seeds or none at all.

▲ TETRAPLOID Polyploidy occurs naturally in many plants—including peanuts and daylilies—due to mistakes in mitosis or meiosis.

▼ HEXAPLOID Modern cultivated strains of oats have six sets of chromosomes, making them hexaploid (6n) plants.

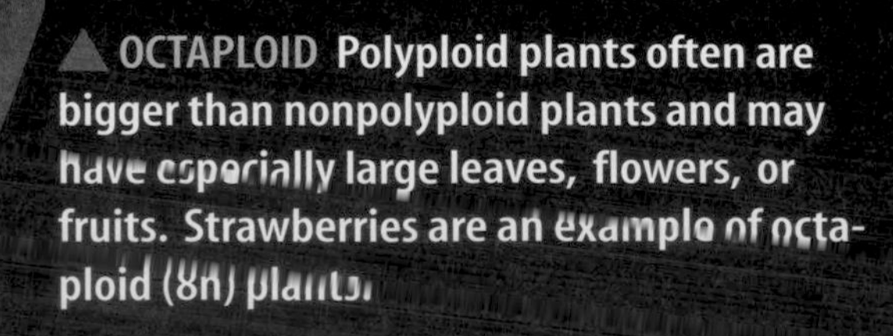

▲ OCTAPLOID Polyploid plants often are bigger than nonpolyploid plants and may have especially large leaves, flowers, or fruits. Strawberries are an example of octaploid (8n) plants.

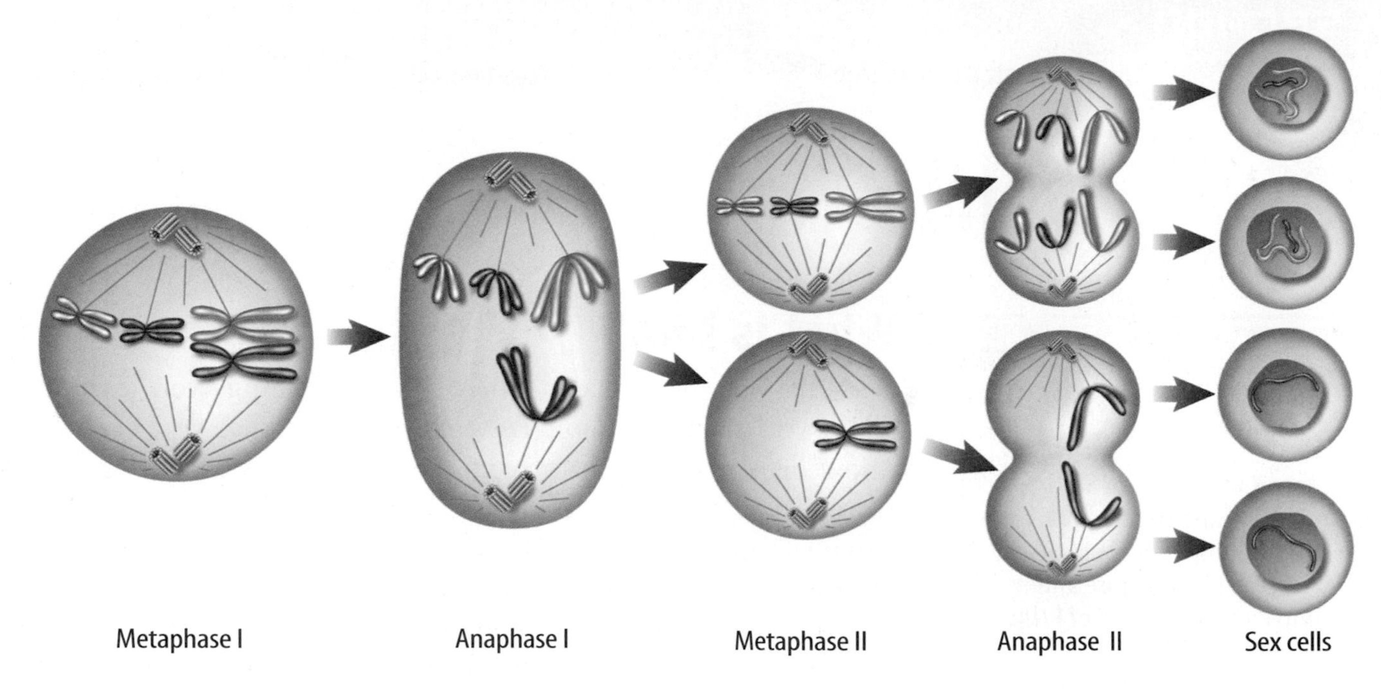

Mistakes in Meiosis Meiosis occurs many times in reproductive organs. Although mistakes in plants, as shown in **Figure 12,** are common, mistakes are less common in animals. These mistakes can produce sex cells with too many or too few chromosomes, as shown in **Figure 13.** Sometimes, zygotes produced from these sex cells die. If the zygote lives, every cell in the organism that grows from that zygote usually will have the wrong number of chromosomes. Organisms with the wrong number of chromosomes may not grow normally.

Figure 13 This diploid cell has four chromosomes. During anaphase I, one pair of duplicated chromosomes did not separate. **Infer** *how many chromosomes each sex cell usually has.*

section 2 review

Summary

Sexual Reproduction

- During sexual reproduction, two sex cells come together.
- Mitosis and cell division begin after fertilization.
- A typical human body cell has 46 chromosomes, and a human sex cell has 23 chromosomes.

Meiosis and Sex Cells

- Each chromosome is duplicated before meiosis, then two divisions of the nucleus occur.
- During meiosis I, duplicated pairs of chromosomes are separated into new cells.
- Chromatids separate during meiosis II.
- Meiosis I and meiosis II result in four sex cells.

Self Check

1. **Describe** a zygote and how it is formed.
2. **Explain** where sex cells form.
3. **Compare** what happens to chromosomes during anaphase I and anaphase II.
4. **Think Critically** Plants grown from runners and leaf cuttings have the same traits as the parent plant. Plants grown from seeds can vary from the parent plants in many ways. Why can this happen?

Applying Skills

5. **Make and use a table** to compare mitosis and meiosis in humans. Vertical headings should include: *What Type of Cell (Body or Sex), Beginning Cell (Haploid or Diploid), Number of Cells Produced, End-Product Cell (Haploid or Diploid),* and *Number of Chromosomes in New Cells.*

section 3

DNA

as you read

What You'll Learn

- **Identify** the parts of a DNA molecule and its structure.
- **Explain** how DNA copies itself.
- **Describe** the structure and function of each kind of RNA.

Why It's Important

DNA helps determine nearly everything your body is and does.

Review Vocabulary
protein: large organic molecule made of amino acid bases

New Vocabulary
- DNA
- RNA
- gene
- mutation

What is DNA?

Why was the alphabet one of the first things you learned when you started school? Letters are a code that you need to know before you learn to read. A cell also uses a code that is stored in its hereditary material. The code is a chemical called deoxyribonucleic (dee AHK sih ri boh noo klay ihk) acid, or **DNA.** It contains information for an organism's growth and function. **Figure 14** shows how DNA is stored in cells that have a nucleus. When a cell divides, the DNA code is copied and passed to the new cells. In this way, new cells receive the same coded information that was in the original cell. Every cell that has ever been formed in your body or in any other organism contains DNA.

INTEGRATE Chemistry

Discovering DNA Since the mid-1800s, scientists have known that the nuclei of cells contain large molecules called nucleic acids. By 1950, chemists had learned what the nucleic acid DNA was made of, but they didn't understand how the parts of DNA were arranged.

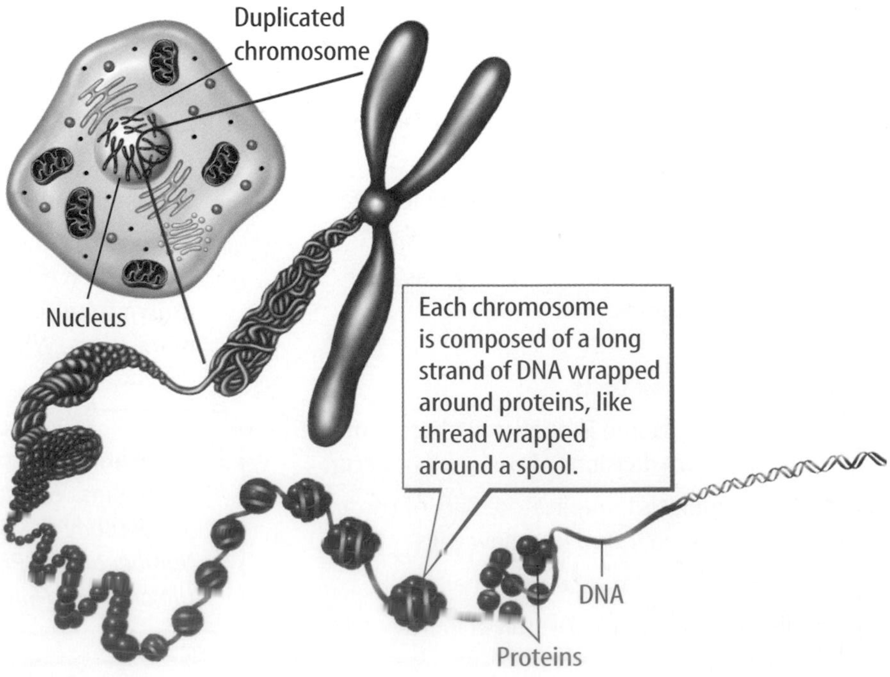

Figure 14 DNA is part of the chromosomes found in a cell's nucleus.

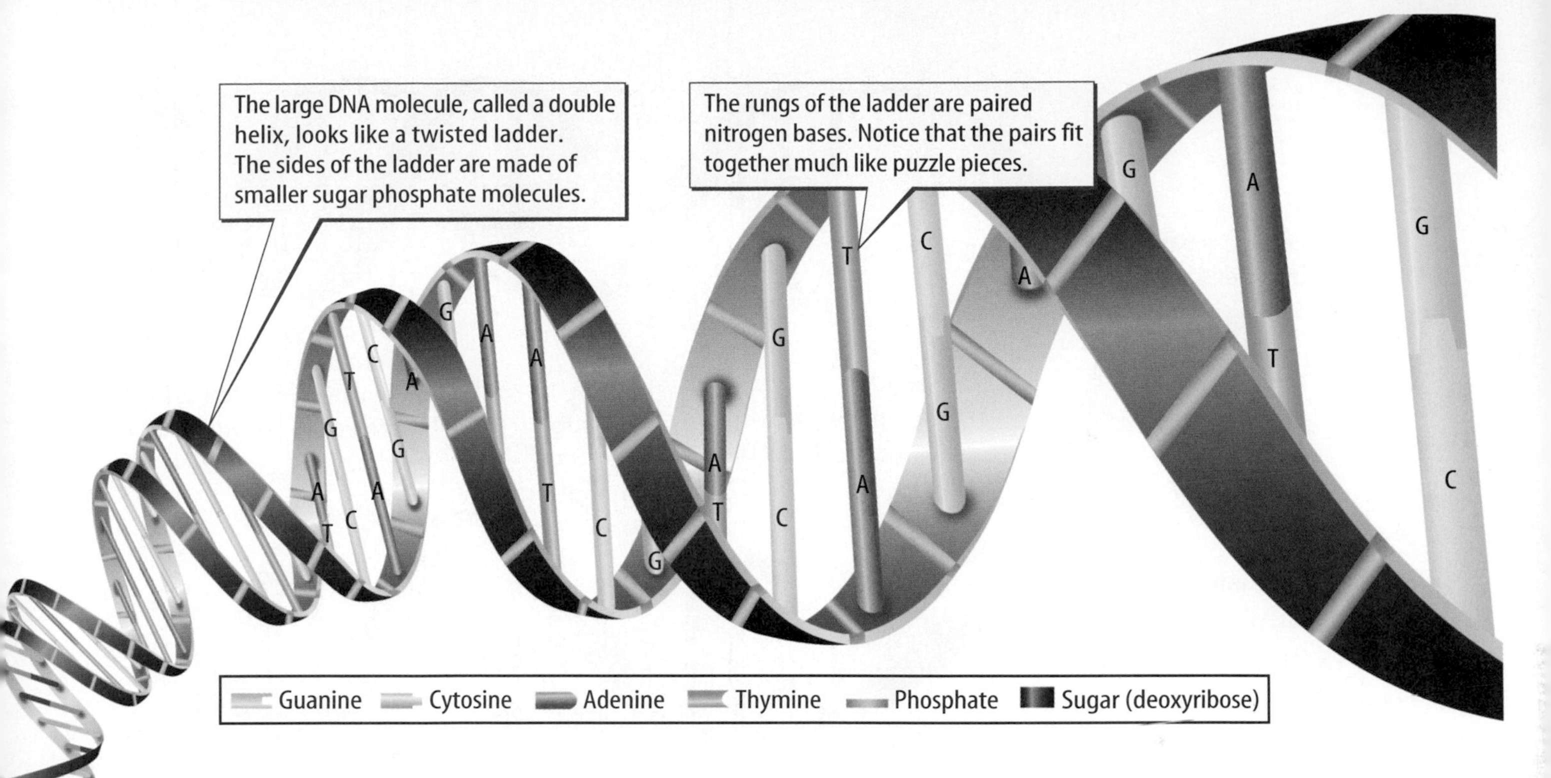

DNA's Structure In 1952, scientist Rosalind Franklin discovered that DNA is two chains of molecules in a spiral form. By using an X-ray technique, Dr. Franklin showed that the large spiral was probably made up of two spirals. As it turned out, the structure of DNA is similar to a twisted ladder. In 1953, using the work of Franklin and others, scientists James Watson and Francis Crick made a model of a DNA molecule.

A DNA Model What does DNA look like? According to the Watson and Crick DNA model, each side of the ladder is made up of sugar-phosphate molecules. Each molecule consists of the sugar called deoxyribose (dee AHK sih ri bohs) and a phosphate group. The rungs of the ladder are made up of other molecules called nitrogen bases. DNA has four kinds of nitrogen bases—adenine (A duh neen), guanine (GWAH neen), cytosine (SI tuh seen), and thymine (THI meen). The nitrogen bases are represented by the letters A, G, C, and T. The amount of cytosine in cells always equals the amount of guanine, and the amount of adenine always equals the amount of thymine. This led to the hypothesis that the nitrogen bases occur as pairs in DNA. **Figure 14** shows that adenine always pairs with thymine, and guanine always pairs with cytosine. Like interlocking pieces of a puzzle, each base bonds only with its correct partner.

Reading Check *What are the nitrogen base pairs in a DNA molecule?*

Modeling DNA Replication

Procedure

1. Suppose you have a segment of DNA that is six nitrogen base pairs in length. On **paper,** using the letters A, T, C, and G, write a combination of six pairs, remembering that A and T are always a pair and C and G are always a pair.
2. Duplicate your segment of DNA. On paper, diagram how this happens and show the new DNA segments.

Analysis

Compare the order of bases of the original DNA to the new DNA molecules.

Try at Home

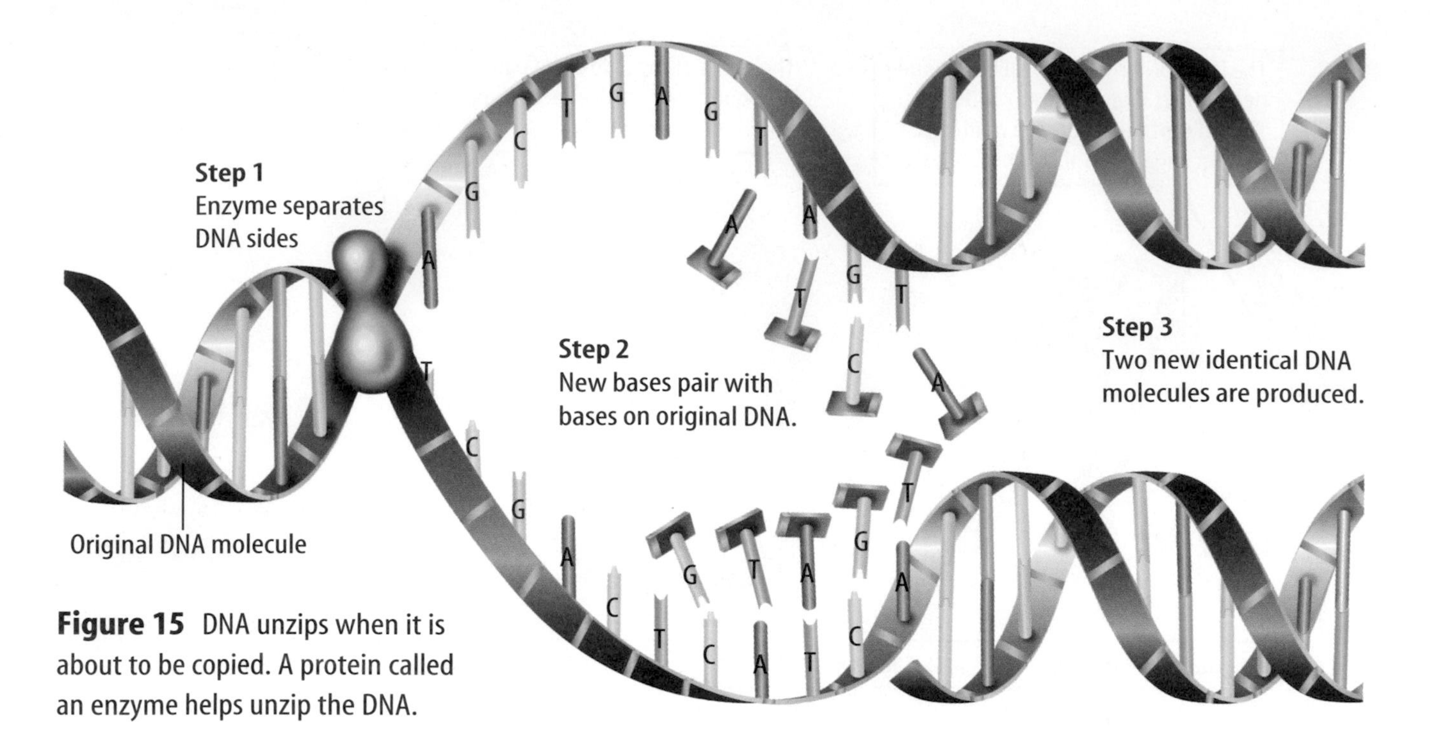

Figure 15 DNA unzips when it is about to be copied. A protein called an enzyme helps unzip the DNA.

Copying DNA When chromosomes are duplicated before mitosis or meiosis, the amount of DNA in the nucleus is doubled. The Watson and Crick model shows how this takes place. The two sides of DNA unwind and separate. Each side then becomes a pattern on which a new side forms, as shown in **Figure 15.** The new DNA has bases that are identical to those of the original DNA and are in the same order.

Genes

Most of your characteristics, such as the color of your hair, your height, and even how things taste to you, depend on the kinds of proteins your cells make. DNA in your cells stores the instructions for making these proteins.

Proteins build cells and tissues or work as enzymes. The instructions for making a specific protein are found in a **gene** which is a section of DNA on a chromosome. As shown in **Figure 16,** each chromosome contains hundreds of genes. Proteins are made of chains of hundreds or thousands of amino acids. The gene determines the order of amino acids in a protein. Changing the order of the amino acids makes a different protein. What might occur if an important protein couldn't be made or if the wrong protein was made in your cells?

Making Proteins Genes are found in the nucleus, but proteins are made on ribosomes in cytoplasm. The codes for making proteins are carried from the nucleus to the ribosomes by another type of nucleic acid called ribonucleic acid, or **RNA.**

Figure 16 This diagram shows just a few of the genes that have been identified on human chromosome 7. The bold print is the name that has been given to each gene.

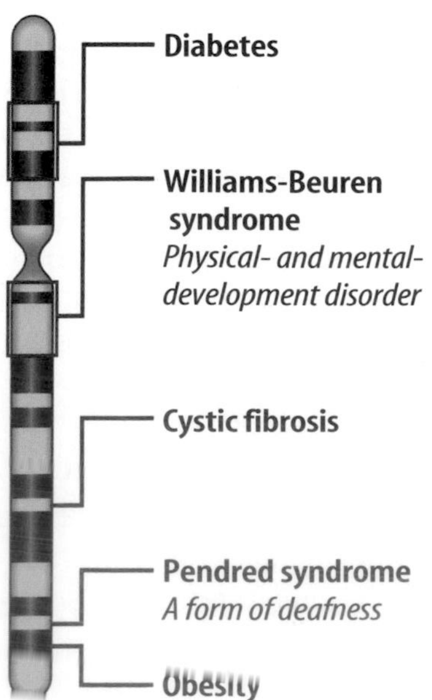

Ribonucleic Acid RNA is made in the nucleus on a DNA pattern but is different from DNA. If DNA is like a ladder, RNA is like a ladder that has all its rungs cut in half. Compare the DNA molecule in **Figure 14** to the RNA molecule in **Figure 17.** RNA has the nitrogen bases A, G, and C like DNA but has the nitrogen base uracil (U) instead of thymine (T). The sugar-phosphate molecules in RNA contain the sugar ribose, not deoxyribose.

The three main kinds of RNA made from DNA in a cell's nucleus are messenger RNA (mRNA), ribosomal RNA (rRNA), and transfer RNA (tRNA). Protein production begins when mRNA moves into the cytoplasm. There, ribosomes attach to it. Ribosomes are made of rRNA. Transfer RNA molecules in the cytoplasm bring amino acids to these ribosomes. Inside the ribosomes, three nitrogen bases on the mRNA temporarily match with three nitrogen bases on the tRNA. The same thing happens for the mRNA and another tRNA molecule, as shown in **Figure 17.** The amino acids that are attached to the two tRNA molecules bond. This is the beginning of a protein.

The code carried on the mRNA directs the order in which the amino acids bond. After a tRNA molecule has lost its amino acid, it can move about the cytoplasm and pick up another amino acid just like the first one. The ribosome moves along the mRNA. New tRNA molecules with amino acids match up and add amino acids to the protein molecule.

Topic: The Human Genome Project

Visit booka.msscience.com for Web links to information about the Human Genome Project.

Activity Find out when chromosomes 5, 16, 29, 21, and 22 were completely sequenced. Write about what scientists learned about each of these chromosomes.

Figure 17 Cells need DNA, RNA, and amino acids to make proteins.

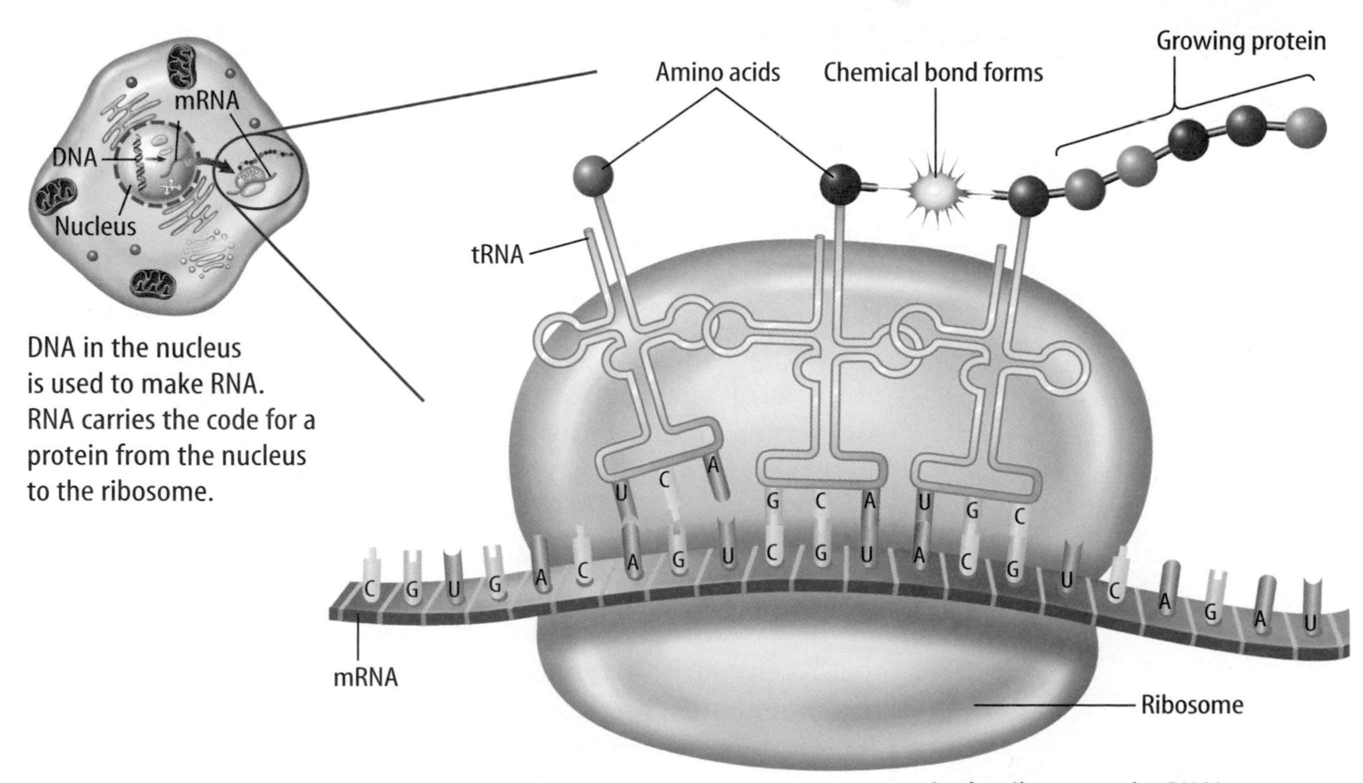

DNA in the nucleus is used to make RNA. RNA carries the code for a protein from the nucleus to the ribosome.

At the ribosome, the RNA's message is translated into a specific protein.

Figure 18 Each cell in the body produces only the proteins that are necessary to do its job.

Controlling Genes You might think that because most cells in an organism have exactly the same chromosomes and the same genes, they would make the same proteins, but they don't. In many-celled organisms like you, each cell uses only some of the thousands of genes that it has to make proteins. Just as each actor uses only the lines from the script for his or her role, each cell uses only the genes that direct the making of proteins that it needs. For example, muscle proteins are made in muscle cells, as represented in **Figure 18,** but not in nerve cells.

Cells must be able to control genes by turning some genes off and turning other genes on. They do this in many different ways. Sometimes the DNA is twisted so tightly that no RNA can be made. Other times, chemicals bind to the DNA so that it cannot be used. If the incorrect proteins are produced, the organism cannot function properly.

Mutations

Sometimes mistakes happen when DNA is being copied. Imagine that the copy of the script the director gave you was missing three pages. You use your copy to learn your lines. When you begin rehearsing for the play, everyone is ready for one of the scenes except for you. What happened? You check your copy of the script against the original and find that three of the pages are missing. Because your script is different from the others, you cannot perform your part correctly.

If DNA is not copied exactly, the proteins made from the instructions might not be made correctly. These mistakes, called **mutations,** are any permanent change in the DNA sequence of a gene or chromosome of a cell. Some mutations include cells that receive an entire extra chromosome or are missing a chromosome. Outside factors such as X rays, sunlight, and some chemicals have been known to cause mutations.

Reading Check *When are mutations likely to occur?*

Figure 19 Because of a defect on chromosome 2, the mutant fruit fly has short wings and cannot fly.
Predict *Could this defect be transferred to the mutant's offspring? Explain.*

Results of a Mutation Genes control the traits you inherit. Without correctly coded proteins, an organism can't grow, repair, or maintain itself. A change in a gene or chromosome can change the traits of an organism, as illustrated in **Figure 19.**

If the mutation occurs in a body cell, it might or might not be life threatening to the organism. However, if a mutation occurs in a sex cell, then all the cells that are formed from that sex cell will have that mutation. Mutations add variety to a species when the organism reproduces. Many mutations are harmful to organisms, often causing their death. Some mutations do not appear to have any effect on the organism, and some can even be beneficial. For example, a mutation to a plant might cause it to produce a chemical that certain insects avoid. If these insects normally eat the plant, the mutation will help the plant survive.

Topic: Fruit Fly Genes
Visit booka.msscience.com for Web links to information about what genes are present on the chromosomes of a fruit fly.

Activity Draw a picture of one of the chromosomes of a fruit fly and label some of its genes.

section 3 review

Summary

What is DNA?

- Each side of the DNA ladder is made up of sugar-phosphate molecules, and the rungs of the ladder are made up of nitrogen bases.
- When DNA is copied, the new DNA has bases that are identical to those of the original DNA.

Genes

- The instructions for making a specific protein are found in genes in the cell nucleus. Proteins are made on ribosomes in the cytoplasm.
- There are three main kinds of RNA—mRNA, rRNA, and tRNA.

Mutations

- If DNA is not copied exactly, the resulting mutations may cause proteins to be made incorrectly.

Self Check

1. **Describe** how DNA makes a copy of itself.
2. **Explain** how the codes for proteins are carried from the nucleus to the ribosomes.
3. **Apply** A strand of DNA has the bases AGTAAC. Using letters, show a matching DNA strand.
4. **Determine** how tRNA is used when cells build proteins.
5. **Think Critically** You begin as one cell. Compare the DNA in your brain cells to the DNA in your heart cells.

Applying Skills

6. **Concept Map** Using a Venn diagram, compare and contrast DNA and RNA.
7. **Use a word processor** to make an outline of the events that led up to the discovery of DNA. Use library resources to find this information.

Use the Internet

Mutations

Fantail pigeon

Real-World Question

Mutations can result in dominant or recessive genes. A recessive characteristic can appear only if an organism has two recessive genes for that characteristic. However, a dominant characteristic can appear if an organism has one or two dominant genes for that characteristic. Why do some mutations result in more common traits while others do not? Form a hypothesis about how a mutation can become a common trait.

Make a Plan

1. **Observe** common traits in various animals, such as household pets or animals you might see in a zoo.
2. **Learn** what genes carry these traits in each animal.
3. **Research** the traits to discover which ones are results of mutations. Are all mutations dominant? Are any of these mutations beneficial?

Goals

- **Observe** traits of various animals.
- **Research** how mutations become traits.
- Gather data about mutations.
- Make a frequency table of your findings and communicate them to other students.

Data Source

Visit **booka.msscience.com/internet_lab** for more information on common genetic traits in different animals, recessive and dominant genes, and data from other students.

White tiger

Follow Your Plan

1. Make sure your teacher approves your plan before you start.
2. Visit the link shown below to access different Web sites for information about mutations and genetics.
3. **Decide** if a mutation is beneficial, harmful, or neither. Record your data in your Science Journal.

Analyze Your Data

1. **Record** in your Science Journal a list of traits that are results of mutations.
2. **Describe** an animal, such as a pet or an animal you've seen in the zoo. Point out which traits are known to be the result of a mutation.
3. **Make** a chart that compares recessive mutations to dominant mutations. Which are more common?
4. **Share** your data with other students by posting it at the link shown below.

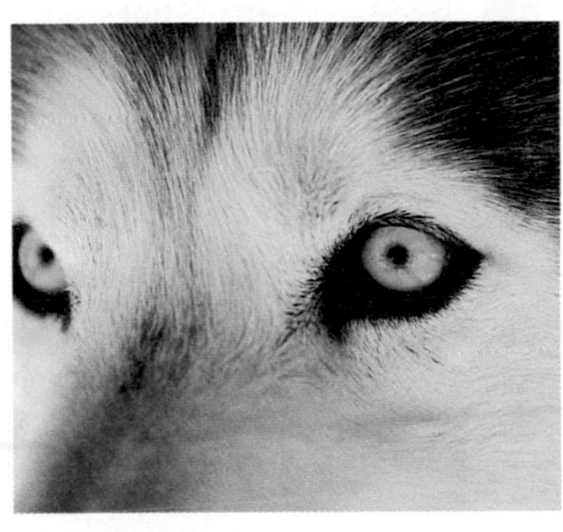

Siberian Husky's eyes

Conclude and Apply

1. **Compare** your findings to those of your classmates and other data at the link shown below. What were some of the traits your classmates found that you did not? Which were the most common?
2. Look at your chart of mutations. Are all mutations beneficial? When might a mutation be harmful to an organism?
3. **Predict** how your data would be affected if you had performed this lab when one of these common mutations first appeared. Do you think you would see more or less animals with this trait?
4. Mutations occur every day but we only see a few of them. Infer how many mutations over millions of years can lead to a new species.

Find this lab using the link below. **Post** your data in the table provided. Combine your data with that of other students and make a chart that shows all of the data.

booka.msscience.com/internet_lab

A Tangled Tale

How did a scientist get chromosomes to separate?

Thanks to chromosomes, each of us is unique!

Viewed under the microscope, chromosomes in cells sometimes look a lot like spaghetti. That's why scientists had such a hard time figuring out how many chromosomes are in each human cell. Imagine, then, how Dr. Tao-Chiuh Hsu (dow shew•SEW) must have felt when he looked into a microscope and saw "beautifully scattered chromosomes." The problem was, Hsu didn't know what he had done to separate the chromosomes into countable strands.

"I tried to study those slides and set up some more cultures to repeat the miracle," Hsu explained. "But nothing happened."

For three months Hsu tried changing every variable he could think of to make the chromosomes separate again. In April 1952, his efforts were finally rewarded. Hsu quickly realized that the chromosomes separated because of osmosis.

Osmosis is the movement of water molecules through cell membranes. This movement occurs in predictable ways. The water molecules move from areas with higher concentrations of water to areas with lower concentrations of water. In Hsu's case, the solution he used to prepare the cells had a higher concentration of water then the cell did. So water moved from the solution into the cell and the cell swelled until it finally exploded. The chromosomes suddenly were visible as separate strands.

What made the cells swell the first time? Apparently a technician had mixed the solution incorrectly. "Since nearly four months had elapsed, there was no way to trace who actually had prepared that particular [solution]," Hsu noted. "Therefore, this heroine must remain anonymous."

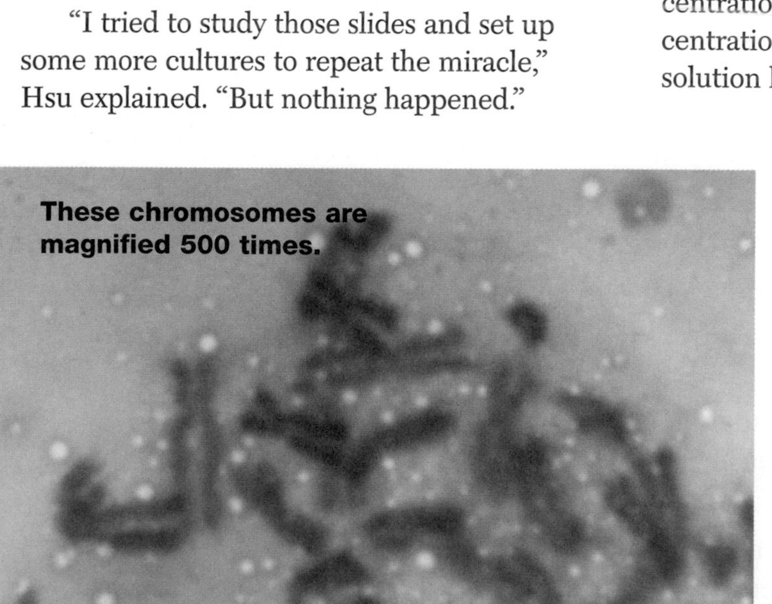

These chromosomes are magnified 500 times.

Research What developments led scientists to conclude that the human cell has 46 chromosomes? Visit the link shown to the right to get started.

For more information, visit booka.msscience.com/oops

Reviewing Main Ideas

Section 1 Cell Division and Mitosis

1. The life cycle of a cell has two parts—growth and development, and cell division.
2. In mitosis, the nucleus divides to form two identical nuclei. Mitosis occurs in four continuous steps, or phases—prophase, metaphase, anaphase, and telophase.
3. Cell division in animal cells and plant cells is similar, but plant cells do not have centrioles and animal cells do not form cell walls.
4. Organisms use mitosis and cell division to grow, to replace cells, and for asexual reproduction. Asexual reproduction produces organisms with DNA identical to the parent's DNA. Fission, budding, and regeneration can be used for asexual reproduction.

Section 2 Sexual Reproduction and Meiosis

1. Sexual reproduction results when an egg and sperm join. This event is called fertilization, and the cell that forms is called the zygote.
2. Meiosis occurs in the reproductive organs, producing four haploid sex cells.
3. During meiosis, two divisions of the nucleus occur.
4. Meiosis ensures that offspring produced by fertilization have the same number of chromosomes as their parents.

Section 3 DNA

1. DNA is a large molecule made up of two twisted strands of sugar-phosphate molecules and nitrogen bases.
2. All cells contain DNA. The section of DNA on a chromosome that directs the making of a specific protein is a gene.
3. DNA can copy itself and is the pattern from which RNA is made. Messenger RNA, ribosomal RNA, and transfer RNA are used to make proteins.
4. Permanent changes in DNA are called mutations.

Visualizing Main Ideas

Copy and complete the spider diagram below about how organisms can use mitosis and cell division.

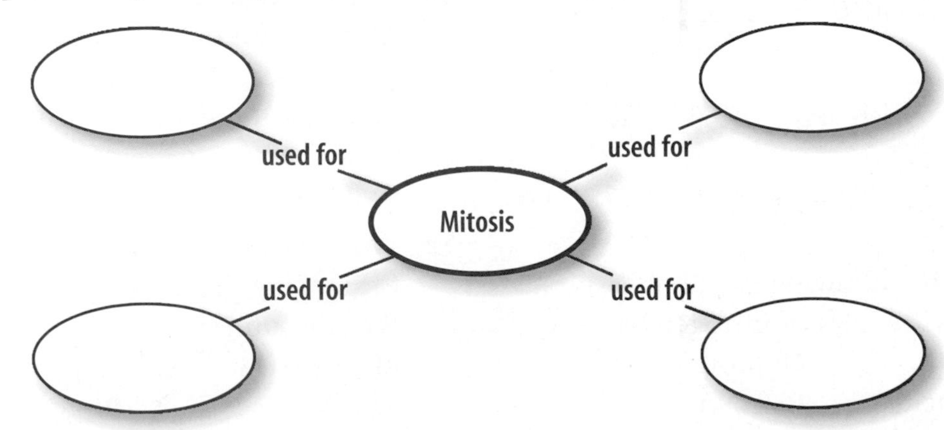

Using Vocabulary

asexual reproduction p. 103
chromosome p. 100
diploid p. 106
DNA p. 112
egg p. 106
fertilization p. 106
gene p. 114
haploid p. 107
meiosis p. 107
mitosis p. 100
mutation p. 116
RNA p. 114
sexual reproduction p. 106
sperm p. 106
zygote p. 106

Fill in the blanks with the correct vocabulary word or words.

1. _________ and _________ cells are sex cells.
2. _________ produces two identical cells.
3. An example of a nucleic acid is _________.
4. A(n) _________ is the code for a protein.
5. A(n) _________ sperm is formed during meiosis.
6. Budding is a type of _________.
7. A(n) _________ is a structure in the nucleus that contains hereditary material.
8. _________ produces four sex cells.
9. As a result of _________, a new organism develops that has its own unique identity.
10. An error made during the copying of DNA is called a(n) _________.

Checking Concepts

Choose the word or phrase that best answers the question.

11. Which of the following is a double spiral molecule with pairs of nitrogen bases?
    A) RNA **C)** protein
    B) amino acid **D)** DNA
12. What is in RNA but not in DNA?
    A) thymine **C)** adenine
    B) thyroid **D)** uracil
13. If a diploid tomato cell has 24 chromosomes, how many chromosomes will the tomato's sex cells have?
    A) 6 **C)** 24
    B) 12 **D)** 48
14. During a cell's life cycle, when do chromosomes duplicate?
    A) anaphase **C)** interphase
    B) metaphase **D)** telophase
15. When do chromatids separate during mitosis?
    A) anaphase **C)** metaphase
    B) prophase **D)** telophase
16. How is the hydra shown in the picture reproducing?
    A) asexually, by budding
    B) sexually, by budding
    C) asexually, by fission
    D) sexually, by fission

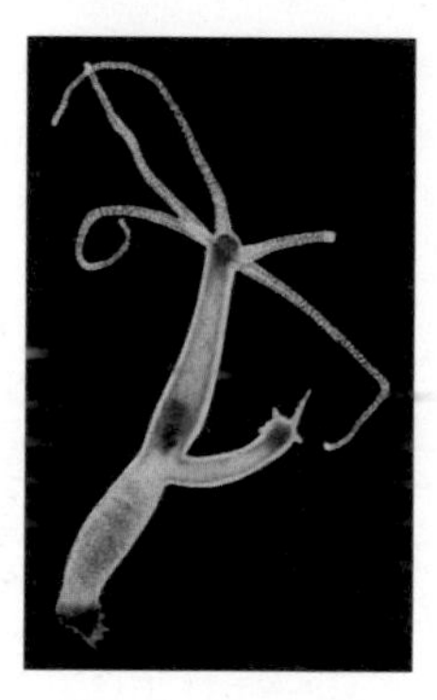

17. What is any permanent change in a gene or a chromosome called?
    A) fission **C)** replication
    B) reproduction **D)** mutation
18. What does meiosis produce?
    A) cells with the diploid chromosome number
    B) cells with identical chromosomes
    C) sex cells
    D) a zygote
19. What type of nucleic acid carries the codes for making proteins from the nucleus to the ribosome?
    A) DNA **C)** protein
    B) RNA **D)** genes

 Science Online booka.msscience.com/vocabulary_puzzlemaker

Thinking Critically

20. **List** the base sequence of a strand of RNA made using the DNA pattern ATCCGTC. Look at **Figure 14** for a hint.

21. **Predict** whether a mutation in a human skin cell can be passed on to the person's offspring. Explain.

22. **Explain** how a zygote could end up with an extra chromosome.

23. **Classify** Copy and complete this table about DNA and RNA.

DNA and RNA

	DNA	RNA
Number of strands		
Type of sugar	Do not write in this book.	
Letter names of bases		
Where found		

24. **Concept Map** Make an events-chain concept map of what occurs from interphase in the parent cell to the formation of the zygote. Tell whether the chromosome's number at each stage is haploid or diploid.

25. **Concept Map** Copy and complete the events-chain concept map of DNA synthesis.

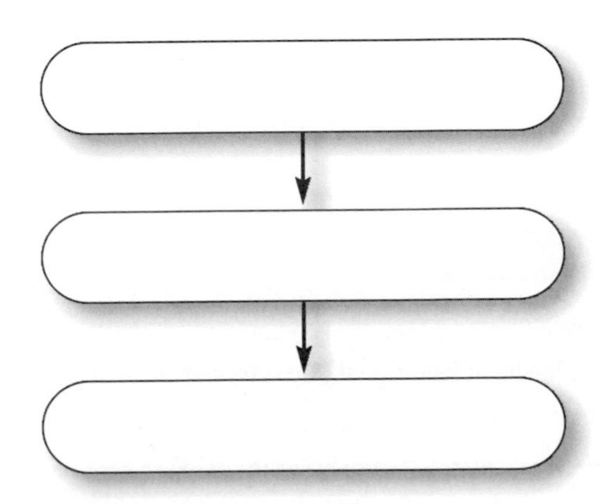

26. **Compare and Contrast** Meiosis is two divisions of a reproductive cell's nucleus. It occurs in a continuous series of steps. Compare and contrast the steps of meiosis I to the steps of meiosis II.

27. **Describe** what occurs in mitosis that gives the new cells identical DNA.

28. **Form a hypothesis** about the effect of an incorrect mitotic division on the new cells produced.

29. **Determine** how many chromosomes are in the original cell compared to those in the new cells formed by cell division. Explain.

Performance Activities

30. **Flash Cards** Make a set of 11 flash cards with drawings of a cell that show the different stages of meiosis. Shuffle your cards and then put them in the correct order. Give them to another student in the class to try.

Applying Math

31. **Cell Cycle** Assume an average human cell has a cell cycle of 20 hours. Calculate how many cells there would be after 80 hours.

Use the diagram below to answer question 32.

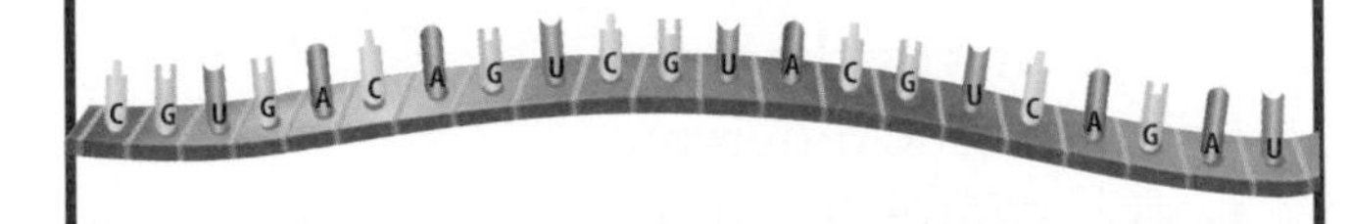

32. **Amino Acids** Sets of three nitrogen bases code for an amino acid. How many amino acids will make up the protein molecule that is coded for by the mRNA molecule above?

chapter 4 Standardized Test Practice

Part 1 Multiple Choice

Record your answers on the answer sheet provided by your teacher or on a sheet of paper.

1. What stage of the cell cycle involves growth and function?

A. prophase
B. interphase
C. mitosis
D. cytoplasmic division

2. During interphase, which structure of a cell is duplicated?

A. cell plate
B. mitochondrion
C. chromosome
D. chloroplast

Use the figure below to answer questions 3 and 4.

3. What form of asexual reproduction is shown here?

A. regeneration
B. cell division
C. sprouting
D. meiosis

4. How does the genetic material of the new organism above compare to that of the parent organism?

A. It is exactly the same.
B. It is a little different.
C. It is completely different.
D. It is haploid.

5. Organisms with three or more sets of chromosomes are called

A. monoploid.
B. diploid.
C. haploid.
D. polyploid.

6. If a sex cell has eight chromosomes, how many chromosomes will there be after fertilization?

A. 8
B. 16
C. 32
D. 64

Use the diagram below to answer questions 7 and 8.

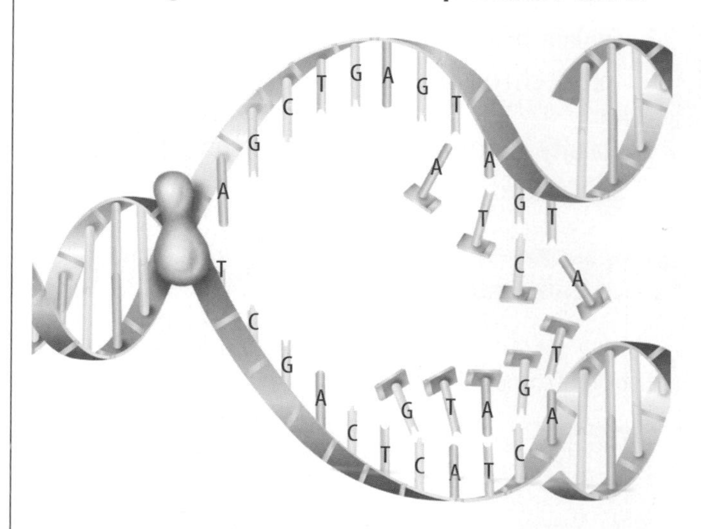

7. What does this diagram illustrate?

A. DNA duplication
B. RNA
C. cell reproduction
D. RNA synthesis

8. When does the process shown occur in the cell cycle?

A. prophase
B. metaphase
C. interphase
D. anaphase

9. Proteins are made of

A. genes
B. bases
C. amino acids
D. chromosomes

Test-Taking Tip

Prepare Avoid rushing on test day. Prepare your clothes and test supplies the night before. Wake up early and arrive at school on time on test day.

Part 2 | Short Response/Grid In

Record your answers on the answer sheet provided by your teacher or on a sheet of paper.

10. In the human body, which cells are constantly dividing? Why is this important? How can this be potentially harmful?

11. Arrange the following terms in the correct order: *fertilization, sex cells, meiosis, zygote, mitosis.*

12. What are the three types of RNA used during protein synthesis? What is the function of each type of RNA?

13. Describe the relationship between gene, protein, DNA and chromosome.

Use the table below to answer question 14.

Phase of Cell Cycle	Action within the Cell
	Chromosomes duplicate
Prophase	
Metaphase	
	Chromosomes have separated
Telophase	

14. Fill in the blanks in the table with the appropriate term or definition.

15. What types of cells would constantly be in interphase?

16. Why is regeneration important for some organisms? In what way could regeneration of nerve cells be beneficial for humans?

17. What types of organisms are polyploidy? Why are they important?

18. What happens to chromosomes in meiosis I and meiosis II?

19. Describe several different ways that organisms can reproduce.

Part 3 | Open Ended

Record your answers on a sheet of paper.

Use the photo below to answer question 20.

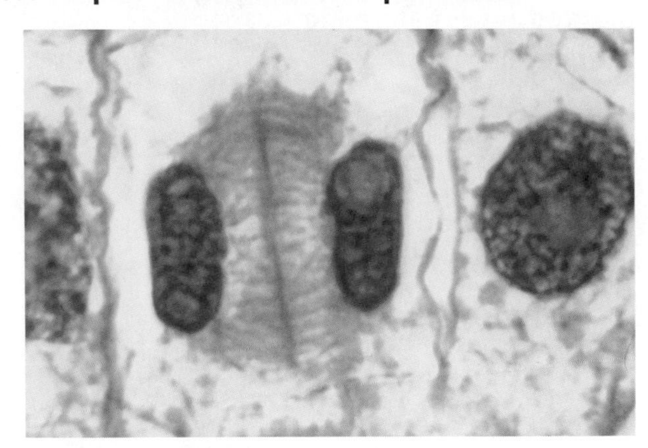

20. Is this a plant or an animal cell? Compare and contrast animal and plant cell division.

21. Describe in detail the structure of DNA.

Use the diagram below to answer question 22.

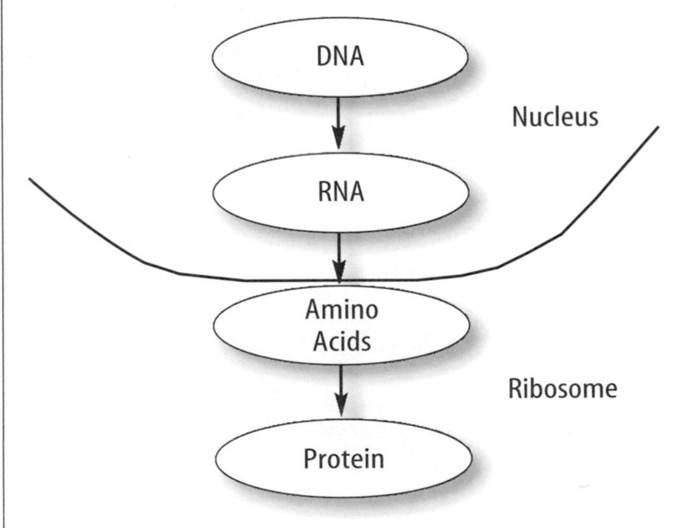

22. Discuss in detail what is taking place at each step of protein synthesis diagrammed above.

23. If a skin cell and a stomach cell have the same DNA then why are they so different?

24. What is mutation? Give examples where mutations could be harmful, beneficial or neutral.

chapter 5

Heredity

The BIG Idea

Inherited genes determine an organism's traits.

Section 1
Genetics
Main Idea Using scientific methods, Gregor Mendel discovered the basic principles of genetics.

Section 2
Genetics Since Mendel
Main Idea It is now known that interactions among alleles, genes, and the environment determine an organism's traits.

Section 3
Advances in Genetics
Main Idea Through genetic engineering, scientists can change the DNA of organisms to improve them, increase resistance to insects and diseases, or produce medicines.

Why do people look different?

People have different skin colors, different kinds of hair, and different heights. Knowing how these differences are determined will help you predict when certain traits might appear. This will help you understand what causes hereditary disorders and how these are passed from generation to generation.

Science Journal Write three traits that you have and how you would determine how those traits were passed to you.

Start-Up Activities

Who around you has dimples?

You and your best friend enjoy the same sports, like the same food, and even have similar haircuts. But, there are noticeable differences between your appearances. Most of these differences are controlled by the genes you inherited from your parents. In the following lab, you will observe one of these differences.

1. Notice the two students in the photographs. One student has dimples when she smiles, and the other student doesn't have dimples.
2. Ask your classmates to smile naturally. In your Science Journal, record the name of each classmate and whether each one has dimples.
3. **Think Critically** In your Science Journal, calculate the percentage of students who have dimples. Are facial dimples a common feature among your classmates?

Classify Characteristics As you read this chapter about heredity, you can use the following Foldable to help you classify characteristics as inherited or not inherited.

STEP 1 **Fold** the top of a vertical piece of paper down and the bottom up to divide the paper into thirds.

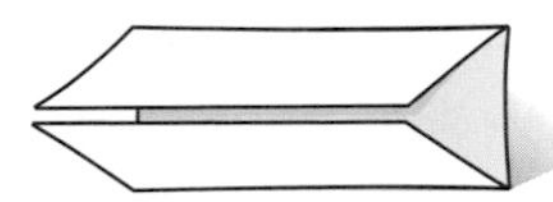

STEP 2 **Turn** the paper horizontally; **unfold and label** the three columns as shown.

Personal Characteristics | Inherited | Not Inherited
eyes
hair
dimples

Read for Main Ideas Before you read the chapter, list personal characteristics and predict which are inherited or not inherited. As you read the chapter, check and change your list.

Preview this chapter's content and activities at booka.msscience.com

Get Ready to Read

Visualize

1 Learn It! Visualize by forming mental images of the text as you read. Imagine how the text descriptions look, sound, feel, smell, or taste. Look for any pictures or diagrams on the page that may help you add to your understanding.

2 Practice It! Read the following paragraph. As you read, use the underlined details to form a picture in your mind.

> In a Punnett square for predicting one trait, the letters representing the two alleles from one parent are written along the top of the grid, one letter per section. Those of the second parent are placed down the side of the grid, one letter per section. Each square of the grid is filled in with one allele donated by each parent. The letters that you use to fill in each of the squares represent the genotypes of possible offspring that the parents could produce.
>
> *—from page 133*

Based on the description above, try to visualize a Punnett square. Now look at the *Applying Math* feature on page 133.

- How closely do these Punnett squares match your mental picture?
- Reread the passage and look at the picture again. Did your ideas change?
- Compare your image with what others in your class visualized.

3 Apply It! Read the chapter and list three subjects you were able to visualize. Make a rough sketch showing what you visualized.

Target Your Reading

Use this to focus on the main ideas as you read the chapter.

1. **Before you read** the chapter, respond to the statements below on your worksheet or on a numbered sheet of paper.
   - Write an **A** if you **agree** with the statement.
   - Write a **D** if you **disagree** with the statement.

2. **After you read** the chapter, look back to this page to see if you've changed your mind about any of the statements.
   - If any of your answers changed, explain why.
   - Change any false statements into true statements.
   - Use your revised statements as a study guide.

Reading Tip

Forming your own mental images will help you remember what you read.

Before You Read A or D	Statement	After You Read A or D
	1 The two alleles of a gene can be the same or different.	
	2 Alleles are either dominant or recessive.	
	3 An organism's phenotype determines its genotype.	
	4 A Punnett square shows the actual genetics of offspring from two parents.	
	5 Traits are determined by more than one gene.	
	6 Some organisms inherit extra chromosomes.	
	7 A pedigrees chart can show the inheritance of a trait within a family.	
	8 The female parent determines whether an offspring will be male or female.	
	9 Genetically engineered organisms can produce medicines.	
	10 Sex-linked disorders are more common in females than in males.	

Science Online

Print out a worksheet of this page at booka.msscience.com

section 1

Genetics

as you read

What You'll Learn

- **Explain** how traits are inherited.
- **Identify** Mendel's role in the history of genetics.
- **Use** a Punnett square to predict the results of crosses.
- **Compare and contrast** the difference between an individual's genotype and phenotype.

Why It's Important

Heredity and genetics help explain why people are different.

Review Vocabulary
meiosis: reproductive process that produces four haploid sex cells from one diploid cell

New Vocabulary

- heredity
- allele
- genetics
- hybrid
- dominant
- recessive
- Punnett square
- genotype
- phenotype
- homozygous
- heterozygous

Inheriting Traits

Do you look more like one parent or grandparent? Do you have your father's eyes? What about Aunt Isabella's cheekbones? Eye color, nose shape, and many other physical features are some of the traits that are inherited from parents, as **Figure 1** shows. An organism is a collection of traits, all inherited from its parents. **Heredity** (huh REH duh tee) is the passing of traits from parent to offspring. What controls these traits?

What is genetics? Generally, genes on chromosomes control an organism's form and function or traits. The different forms of a trait that make up a gene pair are called **alleles** (uh LEELZ). When a pair of chromosomes separates during meiosis (mi OH sus), alleles for each trait also separate into different sex cells. As a result, every sex cell has one allele for each trait. In **Figure 2,** the allele in one sex cell controls one form of the trait for having facial dimples. The allele in the other sex cell controls a different form of the trait—not having dimples. The study of how traits are inherited through the interactions of alleles is the science of **genetics** (juh NE tihks).

Figure 1 Note the strong family resemblance among these four generations.

Figure 2 An allele is one form of a gene. Alleles separate into separate sex cells during meiosis. In this example, the alleles that control the trait for dimples include *D,* the presence of dimples, and *d,* the absence of dimples.

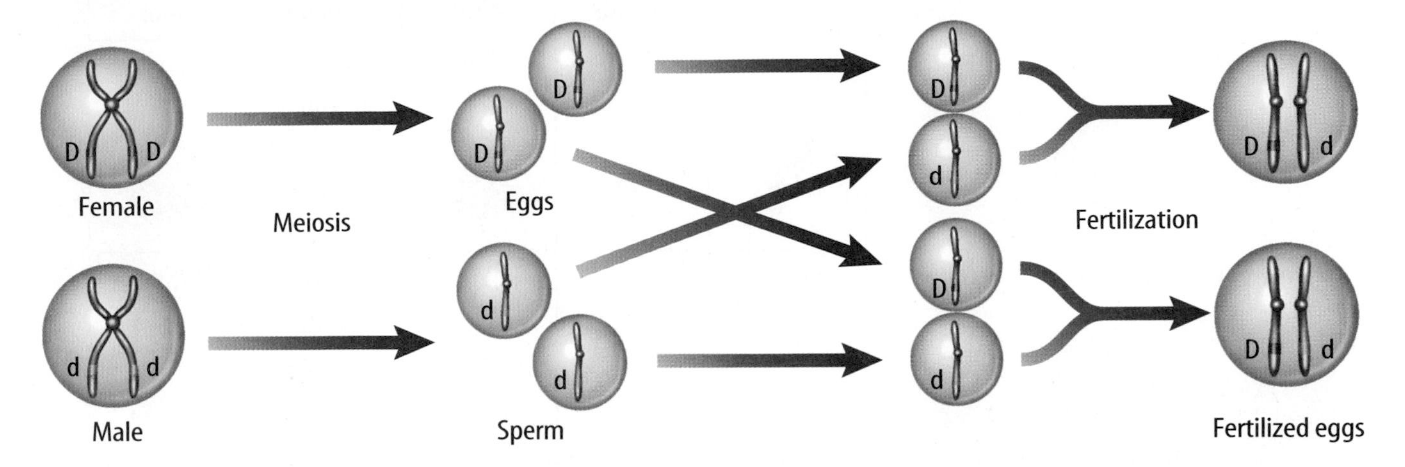

The alleles that control a trait are located on each duplicated chromosome.

During meiosis, duplicated chromosomes separate.

During fertilization, each parent donates one chromosome. This results in two alleles for the trait of dimples in the new individual formed.

Mendel—The Father of Genetics

Did you know that an experiment with pea plants helped scientists understand why your eyes are the color that they are? Gregor Mendel was an Austrian monk who studied mathematics and science but became a gardener in a monastery. His interest in plants began as a boy in his father's orchard where he could predict the possible types of flowers and fruits that would result from crossbreeding two plants. Curiosity about the connection between the color of a pea flower and the type of seed that same plant produced inspired him to begin experimenting with garden peas in 1856. Mendel made careful use of scientific methods, which resulted in the first recorded study of how traits pass from one generation to the next. After eight years, Mendel presented his results with pea plants to scientists.

Before Mendel, scientists mostly relied on observation and description, and often studied many traits at one time. Mendel was the first to trace one trait through several generations. He was also the first to use the mathematics of probability to explain heredity. The use of math in plant science was a new concept and not widely accepted then. Mendel's work was forgotten for a long time. In 1900, three plant scientists, working separately, reached the same conclusions as Mendel. Each plant scientist had discovered Mendel's writings while doing his own research. Since then, Mendel has been known as the father of genetics.

Topic: Genetics
Visit booka.msscience.com for Web links to information about early genetics experiments.

Activity List two other scientists who studied genetics, and what organism they used in their research.

Table 1 Traits Compared by Mendel

Traits	Shape of Seeds	Color of Seeds	Color of Pods	Shape of Pods	Plant Height	Position of Flowers	Flower Color
Dominant trait	Round	Yellow	Green	Full	Tall	At leaf junctions	Purple
Recessive trait	Wrinkled	Green	Yellow	Flat, constricted	short	At tips of branches	White

Genetics in a Garden

Each time Mendel studied a trait, he crossed two plants with different expressions of the trait and found that the new plants all looked like one of the two parents. He called these new plants **hybrids** (HI brudz) because they received different genetic information, or different alleles, for a trait from each parent. The results of these studies made Mendel even more curious about how traits are inherited.

Garden peas are easy to breed for pure traits. An organism that always produces the same traits generation after generation is called a purebred. For example, tall plants that always produce seeds that produce tall plants are purebred for the trait of tall height. **Table 1** shows other pea plant traits that Mendel studied.

Reading Check *Why might farmers plant purebred crop seeds?*

Dominant and Recessive Factors In nature, insects randomly pollinate plants as they move from flower to flower. In his experiments, Mendel used pollen from the flowers of purebred tall plants to pollinate by hand the flowers of purebred short plants. This process is called cross-pollination. He found that tall plants crossed with short plants produced seeds that produced all tall plants. Whatever caused the plants to be short had disappeared. Mendel called the tall form the **dominant** (DAH muh nunt) factor because it dominated, or covered up, the short form. He called the form that seemed to disappear the **recessive** (rih SE sihv) factor. Today, these are called dominant alleles and recessive alleles. What happened to the recessive form? **Figure 3** answers this question.

Comparing Common Traits

Procedure

1. Safely survey as many **dogs** in your neighborhood as you can for the presence of a solid color or spotted coat, short or long hair, and floppy or upright ears.
2. Make a data table that lists each of the traits. Record your data in the data table.

Analysis

1. Compare the number of dogs that have one form of a trait with those that have the other form.
2. What can you conclude about the variations you noticed in the dogs?

Try at Home

VISUALIZING MENDEL'S EXPERIMENTS

NATIONAL GEOGRAPHIC

Figure 3

Gregor Mendel discovered that the experiments he carried out on garden plants provided an understanding of heredity. For eight years he crossed plants that had different characteristics and recorded how those characteristics were passed from generation to generation. One such characteristic, or trait, was the color of pea pods. The results of Mendel's experiment on pea pod color are shown below.

Parents

1st Generation

Ⓐ One of the so-called "parent plants" in Mendel's experiment had pods that were green, a dominant trait. The other parent plant had pods that were yellow, a recessive trait.

Ⓑ Mendel discovered that the two "parents" produced a generation of plants with green pods. The recessive color—yellow—did not appear in any of the pods.

2nd Generation

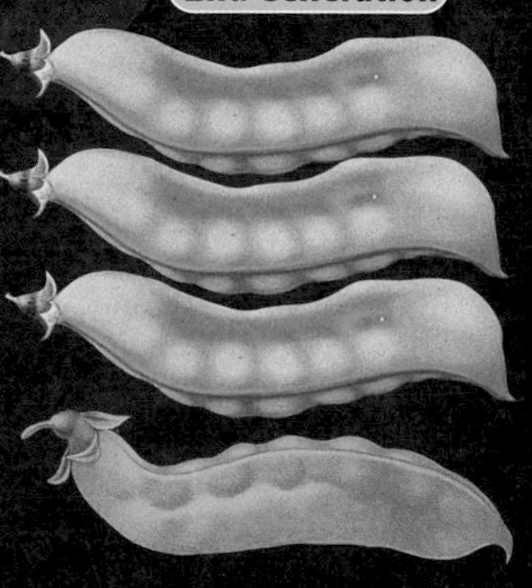

Ⓒ Next, Mendel collected seeds from the first-generation plants and raised a second generation. He discovered that these second-generation plants produced plants with either green or yellow pods in a ratio of about three plants with green pods for every one plant with yellow pods. The recessive trait had reappeared. This 3:1 ratio proved remarkably consistent in hundreds of similar crosses, allowing Mendel to accurately predict the ratio of pod color in second-generation plants.

Using Probability to Make Predictions If you and your sister can't agree on what movie to see, you could solve the problem by tossing a coin. When you toss a coin, you're dealing with probabilities. Probability is a branch of mathematics that helps you predict the chance that something will happen. If your sister chooses tails while the coin is in the air, what is the probability that the coin will land tail-side up? Because a coin has two sides, there are two possible outcomes, heads or tails. Therefore, the probability of tails is one out of two, or 50 percent.

Mendel also dealt with probabilities. One of the things that made his predictions accurate was that he worked with large numbers of plants. He studied almost 30,000 pea plants over a period of eight years. By doing so, Mendel increased his chances of seeing a repeatable pattern. Valid scientific conclusions need to be based on results that can be duplicated.

Figure 4 This snapdragon's phenotype is red.
Determine *Can you tell what the flower's genotype for color is? Explain your answer.*

Punnett Squares Suppose you wanted to know what colors of pea plant flowers you would get if you pollinated white flowers on one pea plant with pollen from purple flowers on a different plant. How could you predict what the offspring would look like without making the cross? A handy tool used to predict results in Mendelian genetics is the **Punnett** (PUH nut) **square.** In a Punnett square, letters represent dominant and recessive alleles. An uppercase letter stands for a dominant allele. A lowercase letter stands for a recessive allele. The letters are a form of code. They show the **genotype** (JEE nuh tipe), or genetic makeup, of an organism. Once you understand what the letters mean, you can tell a lot about the inheritance of a trait in an organism.

The way an organism looks and behaves as a result of its genotype is its **phenotype** (FEE nuh tipe), as shown in **Figure 4.** If you have brown hair, then the phenotype for your hair color is brown.

Alleles Determine Traits Most cells in your body have at least two alleles for every trait. These alleles are located on similar pairs of chromosomes within the nucleus of cells. An organism with two alleles that are the same is called **homozygous** (hoh muh ZI gus) for that trait. For Mendel's peas, this would be written as *TT* (homozygous for the tall-dominant trait) or *tt* (homozygous for the short-recessive trait). An organism that has two different alleles is called **heterozygous** (he tuh roh ZI gus) for that trait. The hybrid plants Mendel produced were all heterozygous for height, *Tt*.

Reading Check *What is the difference between homozygous and heterozygous organisms?*

Making a Punnett Square In a Punnett square for predicting one trait, the letters representing the two alleles from one parent are written along the top of the grid, one letter per section. Those of the second parent are placed down the side of the grid, one letter per section. Each square of the grid is filled in with one allele donated by each parent. The letters that you use to fill in each of the squares represent the genotypes of possible offspring that the parents could produce.

Applying Math — Calculate Percentages

PUNNET SQUARE One dog carries heterozygous, black-fur traits (*Bb*), and its mate carries homogeneous, blond-fur traits (*bb*). Use a Punnett square to determine the probability of one of their puppies having black fur.

Solution

1 *This is what you know:*

- dominant allele is represented by *B*
- recessive allele is represented by *b*

2 *This is what you need to find out:*

What is the probability of a puppy's fur color being black?

3 *This is the procedure you need to use:*

- Complete the Punnett square.
- There are two *Bb* genotypes and four possible outcomes.
- %(black fur) = number of ways to get black fur

$$= \frac{2}{4} = \frac{1}{2} = 50\%$$

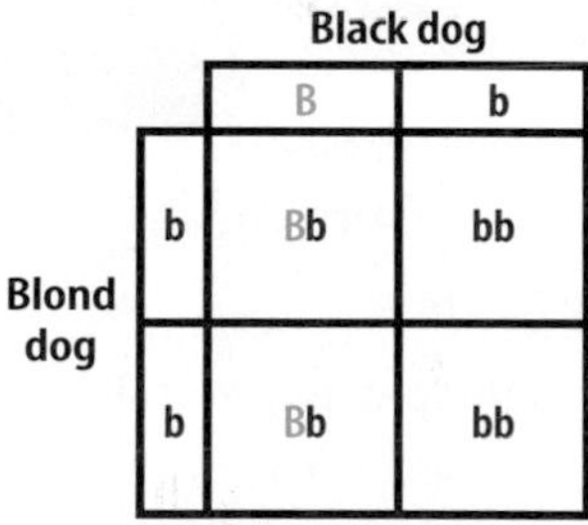

Genotypes of offspring: 2Bb, 2bb
Phenotypes of offspring: 2 black, 2 blond

4 *Check your answer:*

$\frac{1}{2}$ of 4 is 2, which is the number of black dogs.

Practice Problems

1. In peas, the color yellow (*Y*) is dominant to the color green (*y*). According to the Punnett square, what is the probability of an offspring being yellow?
2. What is the probability of an offspring having the *yy* genotype?

Parent (Yy) across the top; Parent (Yy) down the side:

	Y	y
Y	YY	Yy
y	Yy	yy

Science Online — For more practice, visit booka.msscience.com/math_practice

Principles of Heredity Even though Gregor Mendel didn't know anything about DNA, genes, or chromosomes, he succeeded in beginning to describe and mathematically represent how inherited traits are passed from parents to offspring. He realized that some factor in the pea plant produced certain traits. Mendel also concluded that these factors separated when the pea plant reproduced. Mendel arrived at his conclusions after years of detailed observation, careful analysis, and repeated experimentation. **Table 2** summarizes Mendel's principles of heredity.

Table 2 Principles of Heredity

1	Traits are controlled by alleles on chromosomes.
2	An allele's effect is dominant or recessive.
3	When a pair of chromosomes separates during meiosis, the different alleles for a trait move into separate sex cells.

section 1 review

Summary

Inheriting Traits

- Heredity is the passing of traits from parent to offspring.
- Genetics is the study of how traits are inherited through the interactions of alleles.

Mendel—The Father of Genetics

- In 1856, Mendel began experimenting with garden peas, using careful scientific methods.
- Mendel was the first to trace one trait through several generations.
- In 1900, three plant scientists separately reached the same conclusions as Mendel.

Genetics in a Garden

- Hybrids receive different genetic information for a trait from each parent.
- Genetics involves dominant and recessive factors.
- Punnett squares can be used to predict the results of a cross.
- Mendel's conclusions led to the principles of heredity.

Self Check

1. **Contrast** Alleles are described as being dominant or recessive. What is the difference between a dominant and a recessive allele?
2. **Describe** how dominant and recessive alleles are represented in a Punnett square.
3. **Explain** the difference between genotype and phenotype. Give examples.
4. **Infer** Gregor Mendel, an Austrian monk who lived in the 1800s, is known as the father of genetics. Explain why Mendel has been given this title.
5. **Think Critically** If an organism expresses a recessive phenotype, can you tell the genotype? Explain your answer by giving an example.

Applying Math

6. **Use Percentages** One fruit fly is heterozygous for long wings, and another fruit fly is homozygous for short wings. Long wings are dominant to short wings. Use a Punnett square to find the expected percent of offspring with short wings.

Science Online booka.msscience.com/self_check_quiz

Predicting Results

Could you predict how many brown rabbits would result from crossing two heterozygous black rabbits? Try this investigation to find out. Brown color is a recessive trait for hair color in rabbits.

Real-World Question

How does chance affect combinations of genes?

Goals

- **Model** chance events in heredity.
- **Compare and contrast** predicted and actual results.

Materials

paper bags (2)
red beans (100)
white beans (100)

Safety Precautions

WARNING: *Do not taste, eat, or drink any materials used in the lab.*

Procedure

1. Make a Punnett square for a cross between two heterozygous black rabbits, (*Bb* ⊆ *Bb*). *B* represents the black allele and b represents the brown allele.
2. Model the above cross by placing 50 red beans and 50 white beans in one paper bag and 50 red beans and 50 white beans in a second bag. Red beans represent black alleles and white beans represent brown alleles.
3. Label one of the bags *Female* for the female parent. Label the other bag *Male* for the male parent.
4. Use a data table to record the combination each time you remove two beans. Your table will need to accommodate 50 picks.
5. Without looking, remove one bean from each bag and record the results on your data table. Return the beans to their bags.
6. Repeat step five 49 more times.
7. **Count and record** the total numbers for each of the three combinations in your data table.
8. **Compile and record** the class totals.

Conclude and Apply

1. **Name** the combination that occurred most often.
2. **Calculate** the ratio of red/red to red/white to white/white. What hair color in rabbits do these combinations represent?
3. **Compare** your predicted (expected) results with your observed (actual) results.
4. **Hypothesize** how you could get predicted results to be closer to actual results.

Gene Combinations

Rabbits	Red/ Red	Red/ White	White/ White
Your total	Do not write in this book.		
Class total			

Communicating Your Data

Write a paragraph that clearly describes your results. Have another student read your paragraph. Ask if he or she could understand what happened. If not, rewrite your paragraph and have the other student read it again. **For more help, refer to the Science Skill Handbook.**

section 2

Genetics Since Mendel

as you read

What You'll Learn

- **Explain** how traits are inherited by incomplete dominance.
- **Compare** multiple alleles and polygenic inheritance, and give examples of each.
- **Describe** two human genetic disorders and how they are inherited.
- **Explain** how sex-linked traits are passed to offspring.

Why It's Important

Most of your inherited traits involve more complex patterns of inheritance than Mendel discovered.

Review Vocabulary
gene: section of DNA on a chromosome that contains instructions for making specific proteins

New Vocabulary
- incomplete dominance
- polygenic inheritance
- sex-linked gene

Incomplete Dominance

Not even in science do things remain the same. After Mendel's work was rediscovered in 1900, scientists repeated his experiments. For some plants, such as peas, Mendel's results proved true. However, when different plants were crossed, the results were sometimes different. One scientist crossed purebred red-flowered four-o'clock plants with purebred white-flowered four-o'clock plants. He expected to get all red flowers, but they were pink. Neither allele for flower color seemed dominant. Had the colors become blended like paint colors? He crossed the pink-flowered plants with each other, and red, pink, and white flowers were produced. The red and white alleles had not become blended. Instead, when the allele for white flowers and the allele for red flowers combined, the result was an intermediate phenotype—a pink flower.

When the offspring of two homozygous parents show an intermediate phenotype, this inheritance is called **incomplete dominance.** Other examples of incomplete dominance include the flower color of some plant breeds and the coat color of some horse breeds, as shown in **Figure 5.**

Chestnut horse

Cremello horse

Figure 5 When a chestnut horse is bred with a cremello horse, all offspring will be palomino. The Punnett square shown on the opposite page can be used to predict this result. **Explain** *how the color of the palomino horse shows that the coat color of horses may be inherited by incomplete dominance.*

Multiple Alleles Mendel studied traits in peas that were controlled by just two alleles. However, many traits are controlled by more than two alleles. A trait that is controlled by more than two alleles is said to be controlled by multiple alleles. Traits controlled by multiple alleles produce more than three phenotypes of that trait.

Imagine that only three types of coins are made—nickels, dimes, and quarters. If every person can have only two coins, six different combinations are possible. In this problem, the coins represent alleles of a trait. The sum of each two-coin combination represents the phenotype. Can you name the six different phenotypes possible with two coins?

Blood type in humans is an example of multiple alleles that produce only four phenotypes. The alleles for blood types are called A, B, and O. The O allele is recessive to both the A and B alleles. When a person inherits one A allele and one B allele for blood type, both are expressed—phenotype AB. A person with phenotype A blood has the genetic makeup, or genotype—AA or AO. Someone with phenotype B blood has the genotype BB or BO. Finally, a person with phenotype O blood has the genotype OO.

What are the six different blood type genotypes?

Topic: Blood Types
Visit booka.msscience.com for Web links to information about the importance of blood types in blood transfusions.

Activity Make a chart showing which blood types can be used for transfusions into people with A, B, AB, or O blood phenotypes.

Palomino horse

Punnett square

Cremello horse (C'C') \ Chestnut horse (CC)	C	C
C'	CC'	CC'
C'	CC'	CC'

Genotypes: All CC'
Phenotypes: All palomino horses

Interpreting Polygenic Inheritance

Procedure

1. Measure the hand spans of your classmates.
2. Using a **ruler,** measure from the tip of the thumb to the tip of the little finger when the hand is stretched out. Read the measurement to the nearest centimeter.
3. Record the name and hand-span measurement of each person in a data table.

Analysis

1. What range of hand spans did you find?
2. What are the mean, median, and mode for your class's data?
3. Are hand spans inherited as a simple Mendelian pattern or as a polygenic or incomplete dominance pattern? Explain.

Polygenic Inheritance

Eye color is an example of a trait that is produced by a combination of many genes. **Polygenic** (pah lih JEH nihk) **inheritance** occurs when a group of gene pairs acts together to produce a trait. The effects of many alleles produces a wide variety of phenotypes. For this reason, it may be hard to classify all the different shades of eye color.

Your height and the color of your eyes and skin are just some of the many human traits controlled by polygenic inheritance. It is estimated that three to six gene pairs control your skin color. Even more gene pairs might control the color of your hair and eyes. The environment also plays an important role in the expression of traits controlled by polygenic inheritance. Polygenic inheritance is common and includes such traits as grain color in wheat and milk production in cows. Egg production in chickens is also a polygenic trait.

Impact of the Environment Your environment plays a role in how some of your genes are expressed or whether they are expressed at all, as shown in **Figure 6.** Environmental influences can be internal or external. For example, most male birds are more brightly colored than females. Chemicals in their bodies determine whether the gene for brightly colored feathers is expressed.

Although genes determine many of your traits, you might be able to influence their expression by the decisions you make. Some people have genes that make them at risk for developing certain cancers. Whether they get cancer might depend on external environmental factors. For instance, if some people at risk for skin cancer limit their exposure to the Sun and take care of their skin, they might never develop cancer.

Reading Check *What environmental factors might affect the size of leaves on a tree?*

Figure 6 Himalayan rabbits have alleles for dark-colored fur. However, this allele is able to express itself only at lower temperatures. Only the areas located farthest from the rabbit's main body heat (ears, nose, feet, tail) have dark-colored fur.

Human Genes and Mutations

Sometimes a gene undergoes a change that results in a trait that is expressed differently. Occasionally errors occur in the DNA when it is copied inside of a cell. Such changes and errors are called mutations. Not all mutations are harmful. They might be helpful or have no effect on an organism.

Certain chemicals are known to produce mutations in plants or animals, including humans. X rays and radioactive substances are other causes of some mutations. Mutations are changes in genes.

Chromosome Disorders In addition to individual mutations, problems can occur if the incorrect number of chromosomes is inherited. Every organism has a specific number of chromosomes. However, mistakes in the process of meiosis can result in a new organism with more or fewer chromosomes than normal. A change in the total number of human chromosomes is usually fatal to the unborn embryo or fetus, or the baby may die soon after birth.

Look at the photo of human chromosome 21 in **Figure 7.** If three copies of this chromosome are in the fertilized human egg, Down syndrome results. Individuals with Down syndrome can be short, exhibit learning disabilities, and have heart problems. Such individuals can lead normal lives if they have no severe health complications.

Genetic Counselor Testing for genetic disorders may allow many affected individuals to seek treatment and cope with their diseases. Genetic counselors are trained to analyze a family's history to determine a person's health risk. Research what a genetic counselor does and how to become a genetic counselor. Record what you learn in your Science Journal.

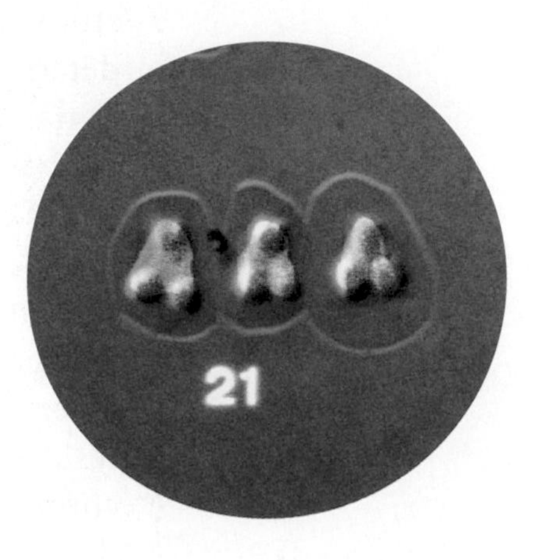

Figure 7 Humans usually have 46 chromosomes arranged as 23 pairs. If a person inherits three copies of chromosome 21 instead of the usual two, Down syndrome results. Chris Burke, a well-known actor, has Down syndrome.

Color-enhanced SEM Magnification: 16000×

Figure 8 Sex in many organisms is determined by X and Y chromosomes.
Observe *How do the X (left) and Y (right) chromosomes differ from one another in shape and size?*

Recessive Genetic Disorders

Many human genetic disorders, such as cystic fibrosis, are caused by recessive genes. Some recessive genes are the result of a mutation within the gene. Many of these alleles are rare. Such genetic disorders occur when both parents have a recessive allele for this disorder. Because the parents are heterozygous, they don't show any symptoms. However, if each parent passes the recessive allele to the child, the child inherits both recessive alleles and will have a recessive genetic disorder.

How is cystic fibrosis inherited?

Cystic fibrosis is a homozygous recessive disorder. It is the most common genetic disorder that can lead to death among Caucasian Americans. In most people, a thin fluid is produced that lubricates the lungs and intestinal tract. People with cystic fibrosis produce thick mucus instead of this thin fluid. The thick mucus builds up in the lungs and makes it hard to breathe. This buildup often results in repeated bacterial respiratory infections. The thick mucus also reduces or prevents the flow of substances necessary for digesting food. Physical therapy, special diets, and new drug therapies have increased the life spans of patients with cystic fibrosis.

Gender Determination

What determines the gender or sex of an individual? Much information on gender inheritance came from studies of fruit flies. Fruit flies have only four pairs of chromosomes. Because the chromosomes are large and few in number, they are easy to study. Scientists identified one pair that contains genes that determine the sex of the organism. They labeled the pair XX in females and XY in males. Geneticists use these labels when studying organisms, including humans. You can see human X and Y chromosomes in **Figure 8.**

Each egg produced by a female normally contains one X chromosome. Males produce sperm that normally have either an X or a Y chromosome. When a sperm with an X chromosome fertilizes an egg, the offspring is a female, XX. A male offspring, XY, is the result of a Y-containing sperm fertilizing an egg. What pair of sex chromosomes is in each of your cells? Sometimes chromosomes do not separate during meiosis. When this occurs, an individual can inherit an abnormal number of sex chromosomes.

Sex-Linked Disorders

A **sex-linked gene** is an allele on a sex chromosome. Some conditions that result from inheriting a sex-linked gene are called sex-linked disorders. Red-green color blindness in humans is a sex-linked disorder because the related genes are on the X chromosome. People who inherit this disorder have difficulty seeing the difference between green and red, and sometimes, yellow. This condition is a recessive sex-linked disorder. A female is color-blind when each of her X chromosomes has the recessive allele. A male has only one X chromosome and, if it has the recessive allele, he will be color-blind.

Dominant sex-linked disorders are rare and result when a person inherits at least one dominant sex-linked allele. Vitamin D-resistant rickets is an X-linked dominant disorder. The kidneys of an affected person cannot absorb adequate amounts of phosphorus. The person might have low blood-phosphorus levels, soft bones, and poor teeth formation.

Pedigrees Trace Traits

How can you trace a trait through a family? A pedigree is a visual tool for following a trait through generations of a family. Males are represented by squares and females by circles. A completely filled circle or square shows that the trait is seen in that person. Half-colored circles or squares indicate carriers. A carrier is heterozygous for the trait, and it is not seen. People represented by empty circles or squares do not have the trait and are not carriers. The pedigree in **Figure 9** shows how the trait for color blindness is carried through a family.

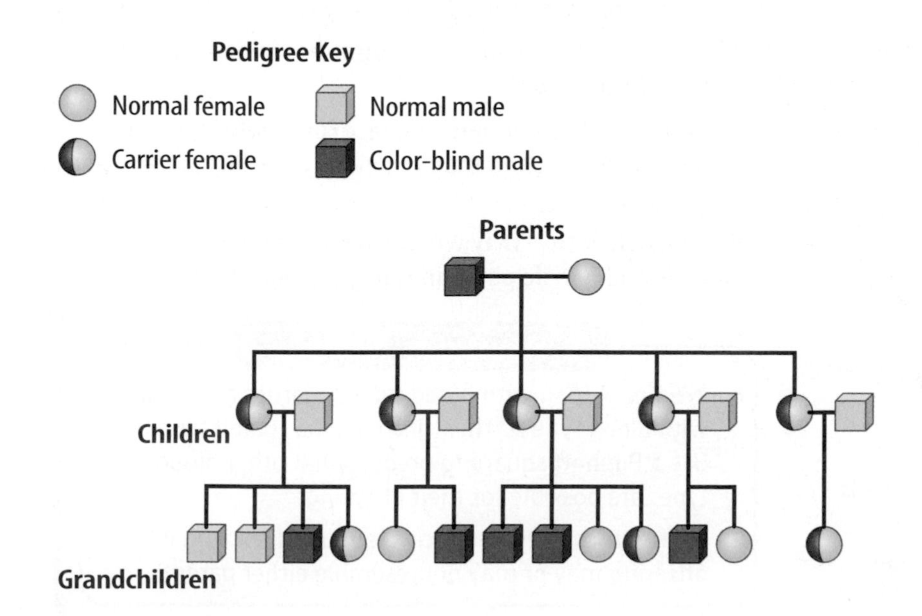

Figure 9 The symbols in this pedigree's key mean the same thing on all pedigree charts. The grandfather in this family was color-blind and married to a woman who was not a carrier of the color-blind allele.
Infer *why no women in this family are color-blind.*

Shih Tzu

Black Labrador

Figure 10
A variety of traits are considered when breeding dogs.

Using Pedigrees A pedigree is a useful tool for a geneticist. Sometimes a geneticist needs to understand who has had a trait in a family over several generations to determine its pattern of inheritance. A geneticist determines if a trait is recessive, dominant, sex-linked, or has some other pattern of inheritance. When geneticists understand how a trait is inherited, they can predict the probability that a baby will be born with a specific trait.

Pedigrees also are important in breeding animals or plants. Because livestock and plant crops are used as sources of food, these organisms are bred to increase their yield and nutritional content. Breeders of pets and show animals, like the dogs pictured in **Figure 10,** also examine pedigrees carefully for possible desirable physical and ability traits. Issues concerning health also are considered when researching pedigrees.

section 2 review

Summary

Incomplete Dominance

- Incomplete dominance is when a dominant and recessive allele for a trait show an intermediate phenotype.
- Many traits are controlled by more than two alleles.
- A wide variety of phenotypes is produced by polygenic inheritance.

Human Genes and Mutations

- Errors can occur when DNA is copied.
- Mistakes in meiosis can result in an unequal number of chromosomes in sex cells.
- Recessive genes control many human genetic disorders.

Sex Determination

- An allele inherited on a sex chromosome is called a sex-linked gene.
- Pedigrees are visual tools to trace a trait through generations of a family.

Self Check

1. **Compare** how inheritance by multiple alleles and polygenic inheritance are similar.
2. **Explain** why a trait inherited by incomplete dominance is not a blend of two alleles.
3. **Discuss** Choose two genetic disorders and discuss how they are inherited.
4. **Apply** Using a Punnett square, explain why males are affected more often than females by sex-linked genetic disorders.
5. **Think Critically** Why wouldn't a horse breeder mate male and female palominos to get palomino colts?

Applying Skills

6. **Predict** A man with blood type B marries a woman with blood type A. Their first child has blood type O. Use a Punnett square to predict what other blood types are possible for their offspring.
7. **Communicate** In your Science Journal, explain why offspring may or may not resemble either parent.

booka.msscience.com/self_check_quiz

section 3

Advances in Genetics

Why is genetics important?

If Mendel were to pick up a daily newspaper in any country today, he'd probably be surprised. News articles about developments in genetic research appear almost daily. The term *gene* has become a common word. The principles of heredity are being used to change the world.

Genetic Engineering

You might recall that chromosomes are made of DNA and are in the nucleus of a cell. Sections of DNA in chromosomes that direct cell activities are called genes. Through **genetic engineering,** scientists are experimenting with biological and chemical methods to change the arrangement of DNA that makes up a gene. Genetic engineering already is used to help produce large volumes of medicine. Genes also can be inserted into cells to change how those cells perform their normal functions, as shown in **Figure 11.** Other research is being done to find new ways to improve crop production and quality, including the development of plants that are resistant to disease.

as you read

What You'll Learn

- **Evaluate** the importance of advances in genetics.
- **Sequence** the steps in making genetically engineered organisms.

Why It's Important

Advances in genetics can affect your health, the foods that you eat, and your environment.

Review Vocabulary
DNA: deoxyribonucleic acid; the genetic material of all organisms

New Vocabulary
- genetic engineering

Figure 11 DNA from one organism is placed into another species. This method is used to produce human insulin, human growth hormone, and other chemicals by bacteria.

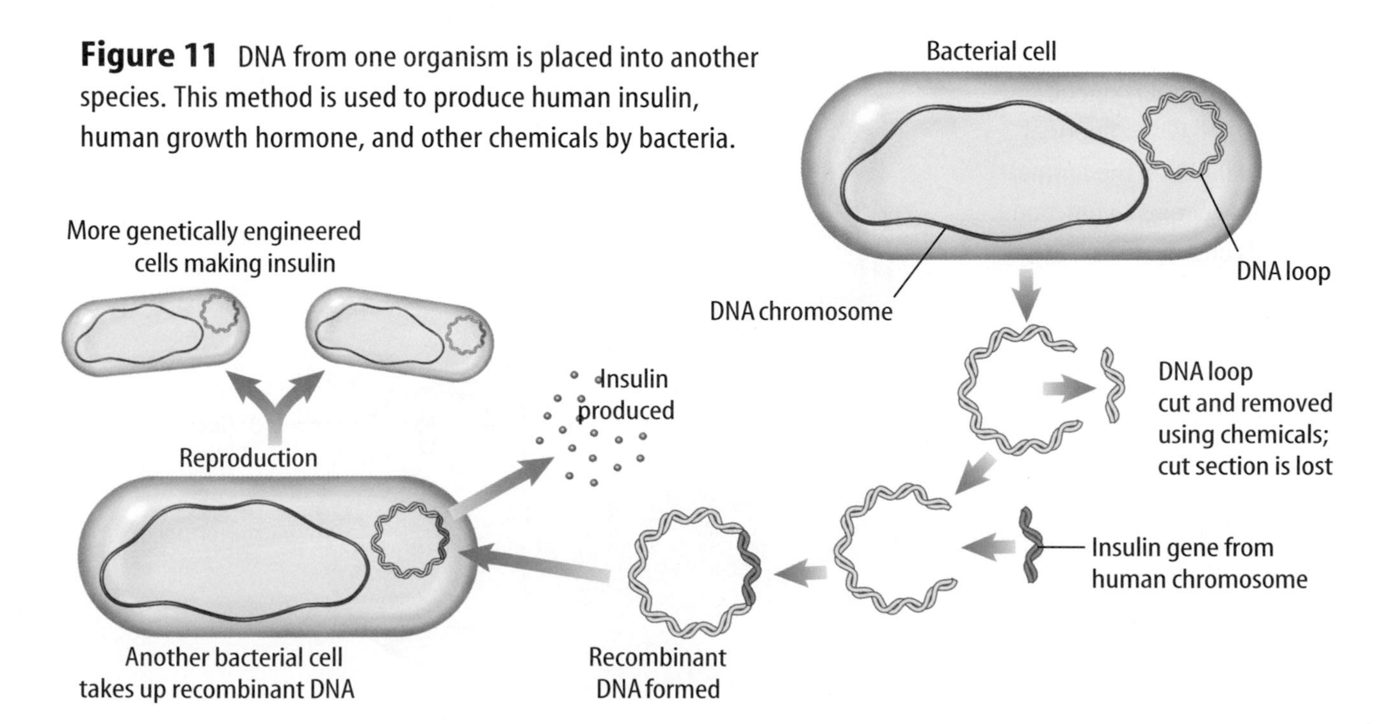

Genetically Engineered Crops Crop plants are now being genetically engineered to produce chemicals that kill specific pests that feed on them. Some of the pollen from pesticide-resistant canola crops is capable of spreading up to 8 km from the plant, while corn and potato pollen can spread up to 1 km. What might be the effects of pollen landing on other plants?

Recombinant DNA Making recombinant DNA is one method of genetic engineering. Recombinant DNA is made by inserting a useful segment of DNA from one organism into a bacterium, as illustrated in **Figure 11.** Large quantities of human insulin are made by some genetically engineered organisms. People with Type 1 diabetes need this insulin because their pancreases produce too little or no insulin. Other uses include the production of growth hormone to treat dwarfism and chemicals to treat cancer.

Gene Transfer Another application of genetic-engineering is gene transfer. A goal of this experimental procedure is to replace abnormal genetic material with normal genetic material. First, normal DNA or RNA is placed in a virus. Then the virus delivers the normal DNA or RNA to target cells, as shown in **Figure 12.** Gene transfer, also known as gene therapy, might help correct genetic disorders such as cystic fibrosis. It also is being studied as a possible treatment for cancer, heart disease, and certain infectious diseases.

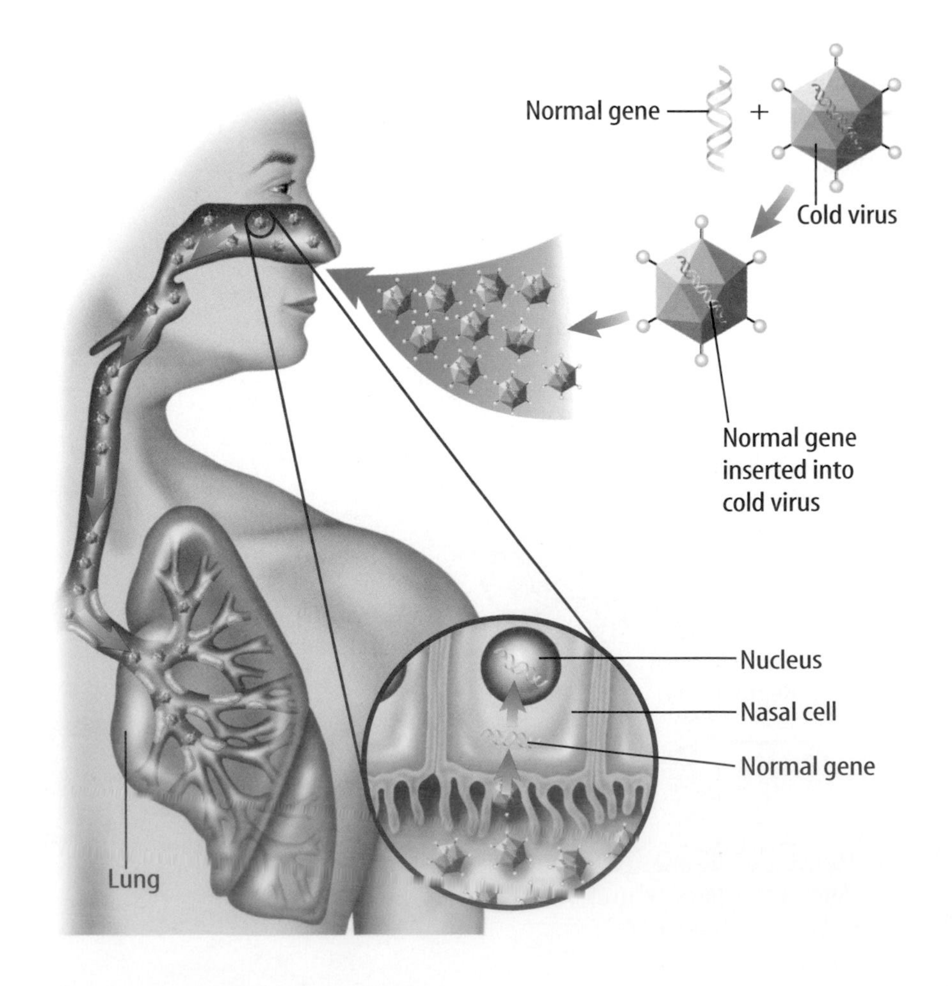

Figure 12 Gene transfer involves placing normal genetic material into a cell with abnormal genetic material. When the normal genetic material begins to function, the abnormal condition is corrected.

Genetically Engineered Plants For thousands of years people have improved the plants they use for food and clothing even without the knowledge of genotypes. Until recently, these improvements were the results of selecting plants with the most desired traits to breed for the next generation. This process is called selective breeding. Recent advances in genetics have not replaced selective breeding. Although a plant can be bred for a particular phenotype, the genotype and pedigree of the plants also are considered.

Genetic engineering can produce improvements in crop plants, such as corn, wheat, and rice. One type of genetic engineering involves finding the genes that produce desired traits in one plant and then inserting those genes into a different plant. Scientists recently have made genetically engineered tomatoes with a gene that allows tomatoes to be picked green and transported great distances before they ripen completely. Ripe, firm tomatoes are then available in the local market. In the future, additional food crops may be genetically engineered so that they are not desirable food for insects.

What other types of traits would be considered desirable in plants?

Because some people might prefer foods that are not changed genetically, some stores label such produce, as shown in **Figure 13.** The long-term effects of consuming genetically engineered plants are unknown.

Figure 13 Genetically engineered produce is sometimes labeled. This allows consumers to make informed choices about their foods.

section 3 review

Summary

Why is genetics important?

- Developments in genetic research appear in newspapers almost daily.
- The world is being changed by the principles of heredity.

Genetic Engineering

- Scientists work with biological and chemical methods to change the arrangement of DNA that makes up a gene.
- One method of genetic engineering is making recombinant DNA.
- A normal allele is replaced in a virus and then delivers the normal allele when it infects its target cell.

Self Check

1. **Apply** Give examples of areas in which advances in genetics are important.
2. **Compare and contrast** the technologies of using recombinant DNA and gene therapy.
3. **Infer** What are some benefits of genetically engineered crops?
4. **Describe** how selective breeding differs from genetic engineering.
5. **Think Critically** Why might some people be opposed to genetically engineered plants?

Applying Skills

6. **Concept Map** Make an events-chain concept map of the steps used in making recombinant DNA.

Design Your Own

Tests for Color Blindness

Goals

- **Design** an experiment that tests for a specific type of color blindness in males and females.
- **Calculate** the percentage of males and females with the disorder.

Possible Materials

white paper or poster board
colored markers: red, orange, yellow, bright green, dark green, blue
**computer and color printer*
**Alternate materials*

Real-World Question

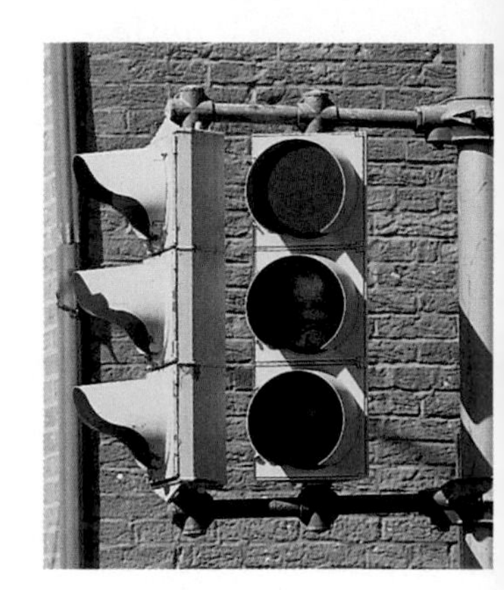

What colors do color-blind people see? That depends on the type of color blindness that they inherit. The most common type is red-green color blindness in which people have difficulty seeing any difference between red and green. People with another inherited type cannot distinguish between blue and yellow. In rare instances, a person can inherit a type of color blindness where the only colors seen are shades of gray. What percentages of males and females in your school are color-blind?

Form a Hypothesis

Based on your reading and your own experiences, form a hypothesis about how common color blindness is among males and females.

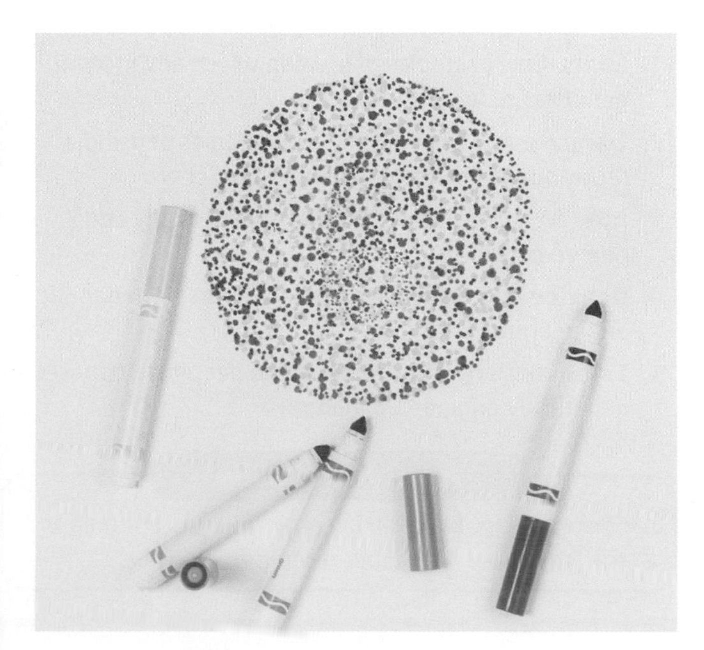

Test Your Hypothesis

Make a Plan

1. Decide what type of color blindness you will test for—the common green-red color blindness or the more rare green-blue color blindness.
2. **List** the materials you will need and describe how you will create test pictures. Tests for color blindness use many circles of red, orange, and yellow as a background, with circles of dark and light green to make a picture or number. List the steps you will take to test your hypothesis.
3. Prepare a data table in your Science Journal to record your test results.

4. **Examine** your experiment to make sure all steps are in logical order.
5. **Identify** which pictures you will use as a control and which pictures you will use as variables.

Follow Your Plan

1. Make sure your teacher approves your plan before you start.
2. **Draw** the pictures that you will use to test for color blindness.
3. Carry out your experiment as planned and record your results in your data table.

Analyze Your Data

1. **Calculate** the percentage of males and females that tested positive for color blindness.
2. **Compare** the frequency of color blindness in males with the frequency of color blindness in females.

Conclude and Apply

1. **Explain** whether or not the results supported your hypothesis.
2. **Explain** why color blindness is called a sex-linked disorder.
3. **Infer** how common the color-blind disorder is in the general population.
4. **Predict** your results if you were to test a larger number of people.

Communicating Your Data

Using a word processor, write a short article for the advice column of a fashion magazine about how a color-blind person can avoid wearing outfits with clashing colors. **For more help, refer to the Science Skill Handbook.**

Science Stats

The Human Genome

Did you know...

. . . The biggest advance in genetics in years took place in February 2001. Scientists successfully mapped the human genome. There are 30,000 to 40,000 genes in the human genome. Genes are in the nucleus of each of the several trillion cells in your body.

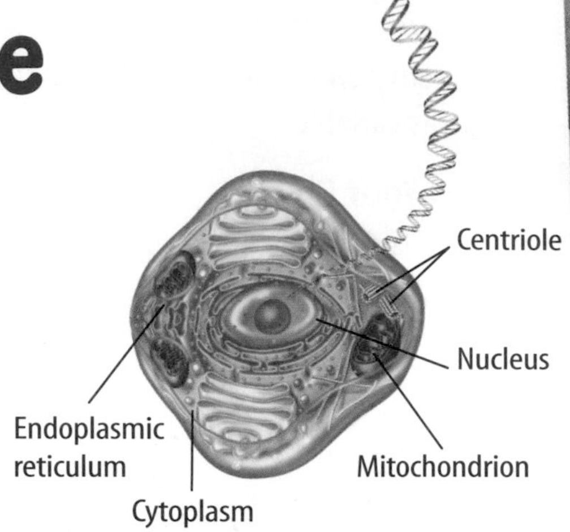

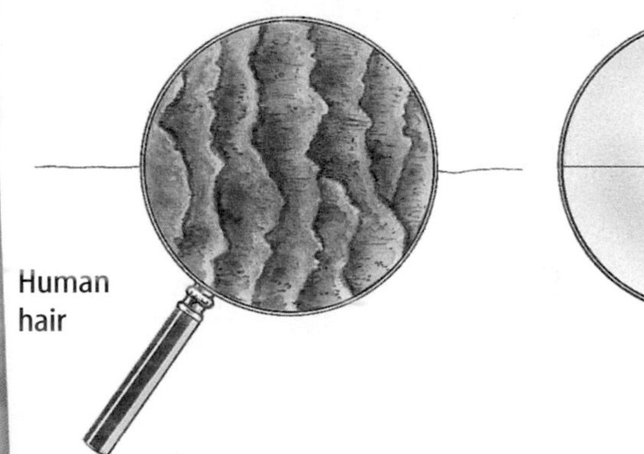

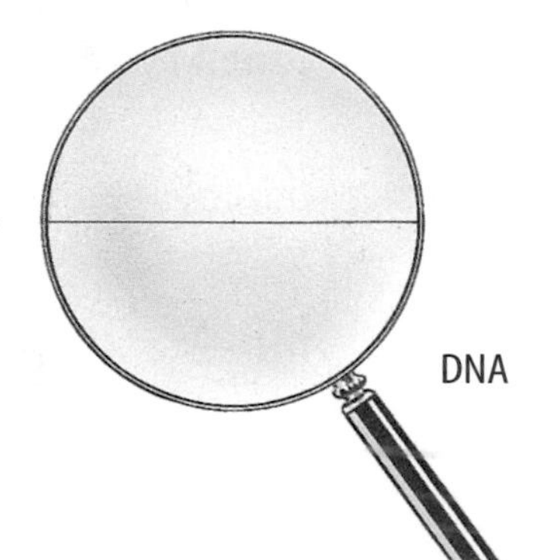

. . . The strands of DNA in the human genome, if unwound and connected end to end, would be more than 1.5 m long—but only about 130 trillionths of a centimeter wide. Even an average human hair is as much as 200,000 times wider than that.

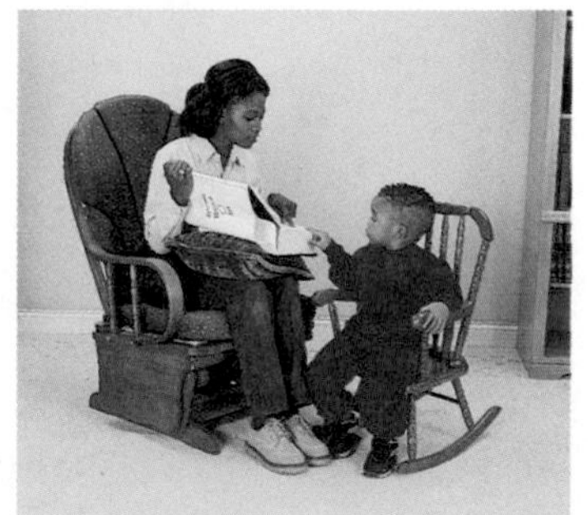

. . . It would take about nine and one-half years to read aloud without stopping the 3 billion bits of instructions (called base pairs) in your genome.

Applying Math If one million base pairs of DNA take up 1 megabyte of storage space on a computer, how many gigabytes (1,024 megabytes) would the whole genome fill?

Find Out About It

Human genome scientists hope to identify the location of disease-causing genes. Visit booka.msscience.com/science_stats to research a genetic disease and share your results with your class.

Reviewing Main Ideas

Section 1 Genetics

1. Genetics is the study of how traits are inherited. Gregor Mendel determined the basic laws of genetics.
2. Traits are controlled by alleles on chromosomes.
3. Some alleles can be dominant or recessive.
4. When a pair of chromosomes separates during meiosis, the different alleles move into separate sex cells. Mendel found that he could predict the outcome of genetic crosses.

Section 2 Genetics Since Mendel

1. Inheritance patterns studied since Mendel include incomplete dominance, multiple alleles, and polygenic inheritance.
2. These inheritance patterns allow a variety of phenotypes to be produced.
3. Some disorders are the results of inheritance and can be harmful and even deadly.
4. Pedigree charts help reveal patterns of the inheritance of a trait in a family. Pedigrees show that sex-linked traits are expressed more often in males than in females.

Section 3 Advances in Genetics

1. Genetic engineering uses biological and chemical methods to change genes.
2. Recombinant DNA is one method of genetic engineering to make useful chemicals, including hormones.
3. Gene transfer shows promise for correcting many human genetic disorders, cancer, and other diseases.
4. Breakthroughs in the field of genetic engineering are allowing scientists to do many things, such as producing plants that are resistant to disease.

Visualizing Main Ideas

Examine the following pedigree for diabetes and explain the inheritance pattern.

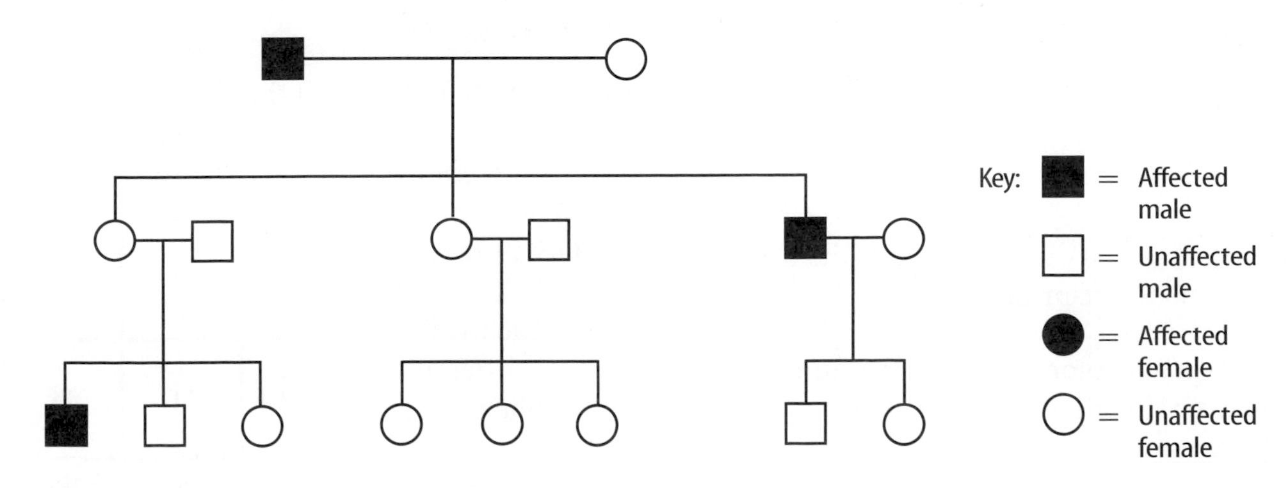

chapter 5 Review

Using Vocabulary

allele p. 128	hybrid p. 130
dominant p. 130	incomplete dominance p. 136
genetic engineering p. 143	phenotype p. 132
genetics p. 128	polygenic inheritance p. 138
genotype p. 132	Punnett square p. 132
heredity p. 128	recessive p. 130
heterozygous p. 132	sex-linked gene p. 141
homozygous p. 132	

Fill in the blanks with the correct word.

1. Alternate forms of a gene are called _________.
2. The outward appearance of a trait is a(n) _________.
3. Human height, eye color, and skin color are all traits controlled by _________.
4. An allele that produces a trait in the heterozygous condition is _________.
5. _________ is the science that deals with the study of heredity.
6. The actual combination of alleles of an organism is its _________.
7. _________ is moving fragments of DNA from one organism and inserting them into another organism.
8. A(n) _________ is a helpful device for predicting the probabilities of possible genotypes.
9. _________ is the passing of traits from parents to offspring.
10. Red-green color blindness is a human genetic disorder caused by a(n) _________.

Checking Concepts

Choose the word or phrase that best answers the question.

11. Which describes the allele that causes color blindness?
    A) dominant
    B) carried on the Y chromosome
    C) carried on the X chromosome
    D) present only in males

12. What is it called when the presence of two different alleles results in an intermediate phenotype?
    A) incomplete dominance
    B) polygenic inheritance
    C) multiple alleles
    D) sex-linked genes

13. What separates during meiosis?
    A) proteins
    B) phenotypes
    C) alleles
    D) pedigrees

14. What controls traits in organisms?
    A) cell membrane
    B) cell wall
    C) genes
    D) Punnett squares

15. What term describes the inheritance of cystic fibrosis?
    A) polygenic inheritance
    B) multiple alleles
    C) incomplete dominance
    D) recessive genes

16. What phenotype will the offspring represented in the Punnett square have?
    A) all recessive
    B) all dominant
    C) half recessive, half dominant
    D) Each will have a different phenotype.

	F	f
F	FF	Ff
F	FF	Ff

Thinking Critically

17. **Explain** the relationship between DNA, genes, alleles, and chromosomes.

18. **Classify** these inheritance patterns:
    a. many different phenotypes produced by one pair of alleles
    b. many phenotypes produced by more than one pair of alleles; two phenotypes from two alleles; three phenotypes from two alleles.

Use the illustration below to answer question 19.

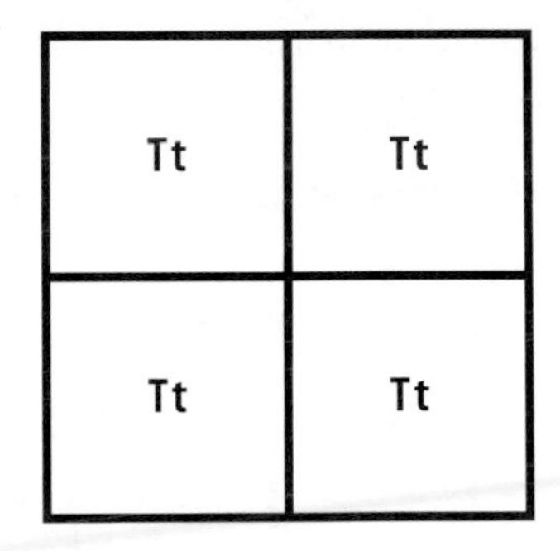

19. **Interpret Scientific Illustrations** What were the genotypes of the parents that produced the Punnett Square shown above?

20. **Explain** why two rabbits with the same genes might not be colored the same if one is raised in northern Maine and one is raised in southern Texas.

21. **Apply** Can a person with a genetic disorder that has been corrected by gene transfer pass the corrected condition to his or her children? Explain.

22. **Predict** Two organisms were found to have different genotypes but the same phenotype. Predict what these phenotypes might be. Explain.

23. **Compare and contrast** Mendelian inheritance with incomplete dominance.

Performance Activities

24. **Newspaper Article** Write a newspaper article to announce a new, genetically engineered plant. Include the method of developing the plant, the characteristic changed, and the terms that you would expect to see. Read your article to the class.

25. **Predict** In humans, the widow's peak allele is dominant, and the straight hairline allele is recessive. Predict how both parents with widow's peaks could have a child without a widow's peak hairline.

26. **Use a word processor** or program to write predictions about how advances in genetics might affect your life in the next ten years.

Applying Math

27. **Human Genome** If you wrote the genetic information for each gene in the human genome on a separate sheet of 0.2-mm-thick paper and stacked the sheets, how tall would the stack be?

Use the table below to answer question 28.

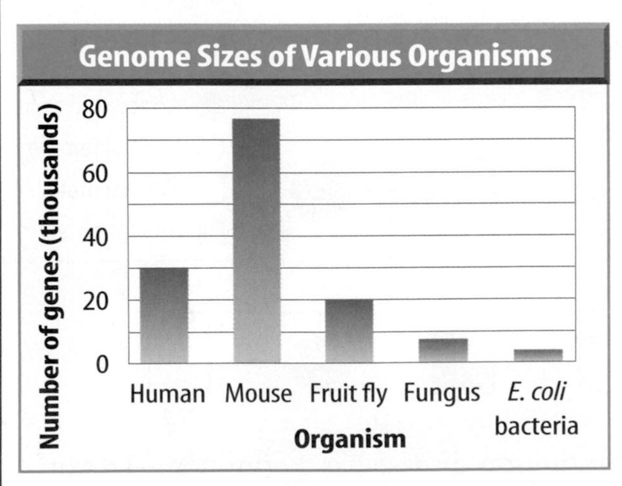

28. **Genes** Consult the graph above. How many more genes are in the human genome than the genome of the fruit fly?

Part 1 Multiple Choice

Record your answers on the answer sheet provided by your teacher or on a sheet of paper.

1. Heredity includes all of the following except
   - **A.** traits.
   - **B.** chromosomes.
   - **C.** nutrients.
   - **D.** phenotype.

2. What is a mutation?
   - **A.** A change in a gene which is harmful, beneficial, or has no effect at all.
   - **B.** A change in a gene which is only beneficial.
   - **C.** A change in a gene which is only harmful.
   - **D.** No change in a gene.

3. Sex of the offspring is determined by
   - **A.** only the mother, because she has two X chromosomes.
   - **B.** only the father, because he has one X and one Y chromosome.
   - **C.** an X chromosome from the mother and either an X or Y chromosome from the father.
   - **D.** mutations.

Use the pedigree below to answer questions 4–6.

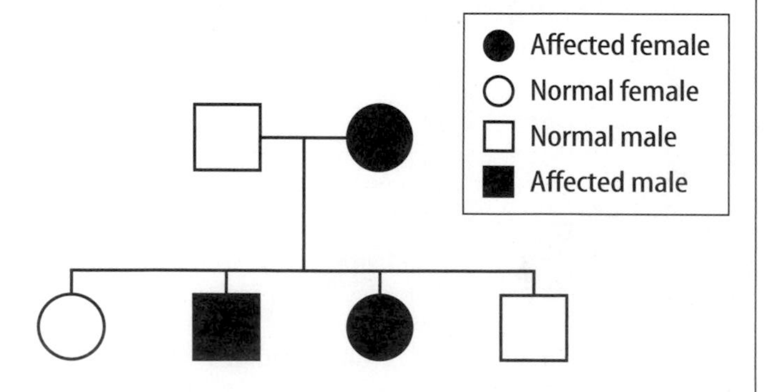

Huntington disease has a dominant (DD or Dd) inheritance pattern

4. What is the genotype of the father?
   - **A.** DD
   - **B.** Dd
   - **C.** dd
   - **D.** D

5. What is the genotype of the mother?
   - **A.** DD
   - **B.** Dd
   - **C.** dd
   - **D.** D

6. The genotype of the unaffected children is
   - **A.** DD.
   - **B.** Dd.
   - **C.** dd.
   - **D.** D.

7. Manipulating the arrangement of DNA that makes up a gene is called
   - **A.** genetic engineering.
   - **B.** chromosomal migration.
   - **C.** viral reproduction.
   - **D.** cross breeding.

Use the Punnett square below to answer question 8.

	A	O
A	AA	AO
B	AB	BO

8. How many phenotypes would result from the following Punnett square?
   - **A.** 1
   - **B.** 2
   - **C.** 3
   - **D.** 4

9. Down syndrome is an example of
   - **A.** incomplete dominance.
   - **B.** genetic engineering.
   - **C.** a chromosome disorder.
   - **D.** a sex linked disorder.

Test-Taking Tip

Complete Charts Write directly on complex charts such as a Punnett square.

Question 10 Draw a Punnett square to answer all parts of the question.

Part 2 | Short Response/Grid In

Record your answers on the answer sheet provided by your teacher or on a sheet of paper.

Use the table below to answer questions 10–11.

Some Traits Compared by Mendel			
Traits	Shape of Seeds	Shape of Pods	Flower Color
Dominant Trait	Round	Full	Purple
Recessive Trait	Wrinkled	Flat, constricted	White

10. Create a Punnett square using the *Shape of Pods* trait crossing heterozygous parents. What percentage of the offspring will be heterozygous? What percentage of the offspring will be homozygous? What percentage of the offspring will have the same phenotype as the parents?

11. Gregor Mendel studied traits in pea plants that were controlled by single genes. Explain what would have happened if the alleles for flower color were an example of incomplete dominance. What phenotypes would he have observed?

12. Why are heterozygous individuals called carriers for non-sex-linked and X-linked recessive patterns of inheritance?

13. How many alleles does a body cell have for each trait? What happens to the alleles during meiosis?

Part 3 | Open Ended

Record your answers on a sheet of paper.

14. Genetic counseling helps individuals determine the genetic risk or probability a disorder will be passed to offspring. Why would a pedigree be a very important tool for the counselors? Which patterns of inheritance (dominant, recessive, x-linked) would be the easiest to detect?

15. Explain the process of gene transfer. What types of disorders would this be best suited? How might gene transfer help patients with cystic fibrosis?

Refer to the figure below to answer question 16.

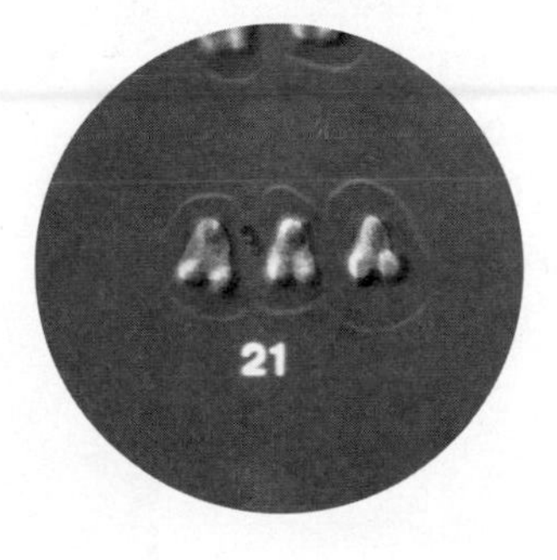

16. What is the disorder associated with the karyotype shown above? How does this condition occur? What are the characteristics of someone with this disorder?

17. Explain why the parents of someone with cystic fibrosis do not show any symptoms. How are the alleles for cystic fibrosis passed from parents to offspring?

18. What is recombinant DNA and how is it used to help someone with Type I diabetes?

19. If each kernel on an ear of corn represents a separate genetic cross, would corn be a good plant to use to study genetics? Why or why not? What process could be used to control pollination?

chapter 6

Adaptations over Time

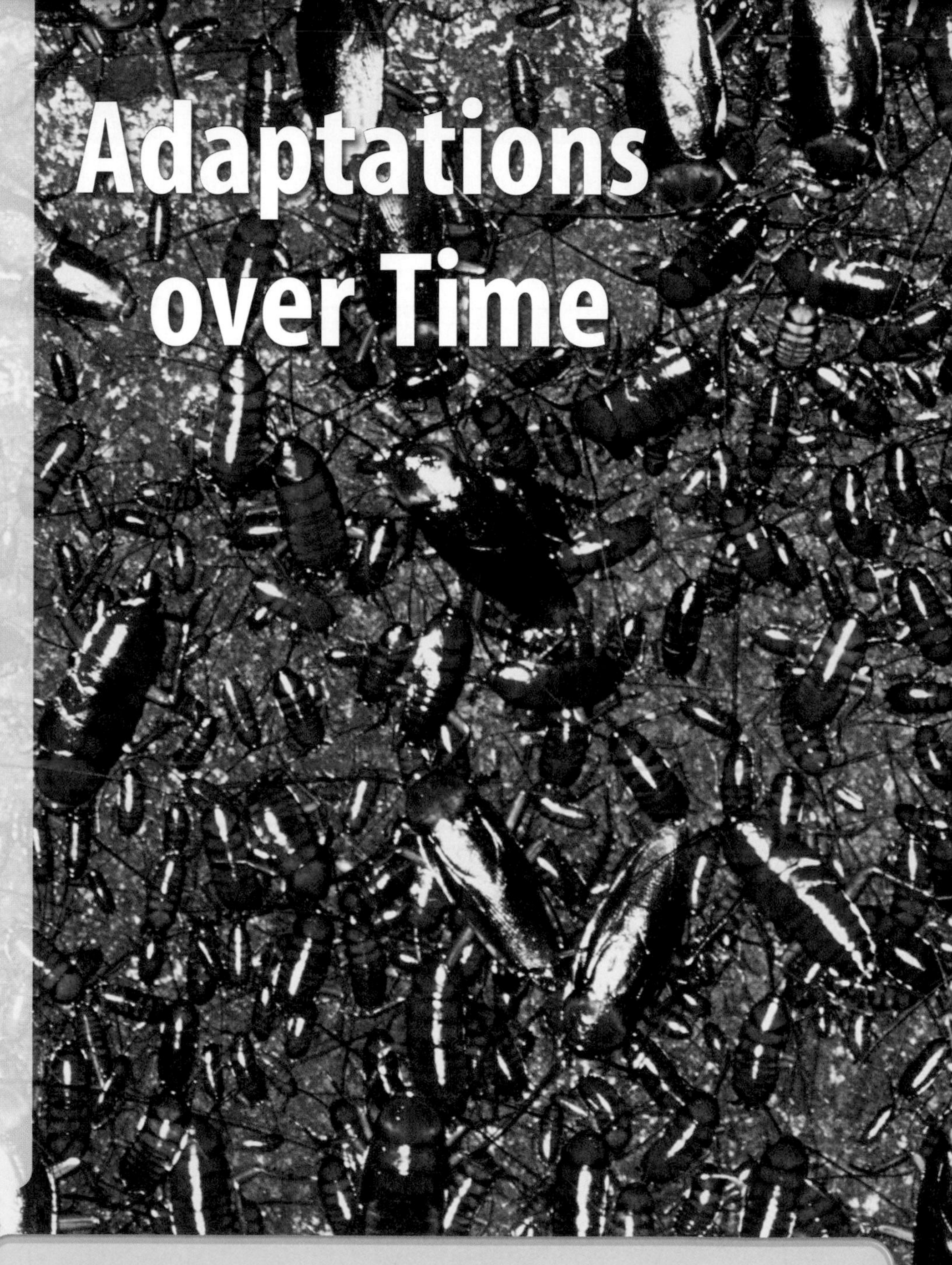

The BIG Idea

Life-forms have changed over time.

SECTION 1
Ideas About Evolution
Main Idea Charles Darwin and other scientists observed that species change over time by different methods.

SECTION 2
Clues About Evolution
Main Idea Scientists find clues about evolution by studying fossils, development of embryos, structures of organisms, and DNA.

SECTION 3
The Evolution of Primates
Main Idea Evidence indicates that the ancient ancestor of present-day humans appeared on Earth for 4–6 million years ago.

Adaptation? No problem.

Cockroaches have existed for millions of years, yet they are still adapted to their environment. Since they first appeared, many species have disappeared, and other well-adapted species have evolved.

Science Journal Pick a favorite plant or animal and list in your Science Journal all the ways it is well-suited to its environment.

Start-Up Activities

Adaptation for a Hunter

The cheetah is nature's fastest hunter, but it can run swiftly for only short distances. Its fur blends in with tall grass, making it almost invisible as it hides and waits for prey. Then the cheetah pounces, capturing the prey before it can run away.

1. Spread a sheet of newspaper classified ads on the floor.
2. Using a hole puncher, make 100 circles from each of the following types of paper: white paper, black paper, and classified ads.
3. Scatter all the circles on the newspaper on the floor. For 10 s, pick up as many circles as possible, one at a time. Have a partner time you.
4. Count the number of each kind of paper circle that you picked up. Record your results in your Science Journal.
5. **Think Critically** Which paper circles were most difficult to find? What can you infer about a cheetah's coloring from this activity? Enter your responses to these questions in your Science Journal.

FOLDABLES™ Study Organizer

Principles of Natural Selection Make the following Foldable to help you understand the process of natural selection.

STEP 1 **Fold** a sheet of paper in half lengthwise.

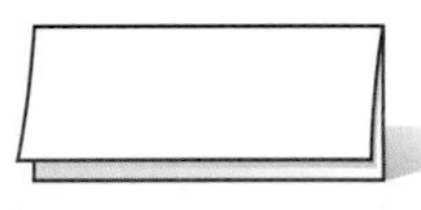

STEP 2 **Fold** paper down 2.5 cm from the top. (Hint: From the tip of your index finger to your middle knuckle is about 2.5 cm.)

STEP 3 **Open and draw** lines along the 2.5-cm fold and the center fold. **Label** as shown.

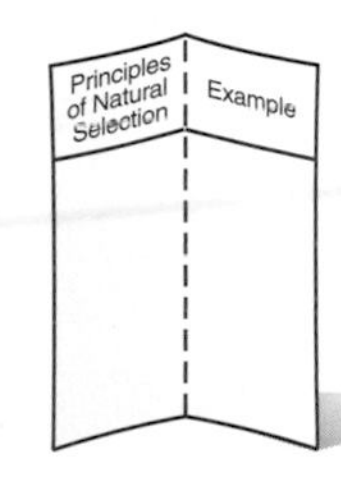

Summarize in a Table As you read, list the five principles of natural selection in the left-hand column. In the right-hand column, briefly write an example for each principle.

Preview this chapter's content and activities at booka.msscience.com

Get Ready to Read

Questioning

1 Learn It! Asking questions helps you to understand what you read. As you read, think about the questions you'd like answered. Often you can find the answer in the next paragraph or lesson. Learn to ask good questions by asking who, what, when, where, why, and how.

2 Practice It! Read the following passage from Section 2.

> One way to find the approximate age of fossils within a rock layer is relative dating. Relative dating is based on the idea that, in undisturbed areas, younger rock layers are deposited on top of older rock layers, as shown in **Figure 10.** Relative dating provides only an estimate of a fossil's age. The estimate is made by comparing the ages of rock layers found above and below the fossil layer. For example, suppose a 50 million-year-old rock layer lies below a fossil, and a 35-million-year-old layer lies above it. According to relative dating, the fossil is probably between 35 million and 50 million years old.
>
> *—from page 167*

Here are some questions you might ask about this paragraph:

- How do the ages of rock layers help determine the age of a fossil?
- What must be true of the area where rock layers are used for relative dating?
- Does relative dating determine the actual fossil age or an estimate of a fossil's age?

3 Apply It! As you read the chapter, look for answers to questions that are part of the text.

Test yourself. Create questions and then read to find answers to your own questions.

Target Your Reading

Use this to focus on the main ideas as you read the chapter.

1. **Before you read** the chapter, respond to the statements below on your worksheet or on a numbered sheet of paper.
   - Write an **A** if you **agree** with the statement.
   - Write a **D** if you **disagree** with the statement.

2. **After you read** the chapter, look back to this page to see if you've changed your mind about any of the statements.
   - If any of your answers changed, explain why.
   - Change any false statements into true statements.
   - Use your revised statements as a study guide.

Before You Read A or D	Statement		After You Read A or D
	1	Darwin's observations in the Galápagos Islands helped him develop his theory of evolution by natural selection.	
	2	When some geographic barrier, such as mountains, separate members of a species, each group remains unchanged over time.	
	3	One principle of natural selection is that organisms best able to survive in an environment are more likely to reproduce and pass their traits to future generations.	
	4	Variation makes one member of a species different from other members of the same species.	
	5	Fossils can be the actual remains of an organism.	
	6	Evolution only happens slowly over time.	
	7	Present-day organisms can provide clues about evolution.	
	8	Plants were the first forms of life to evolve.	

Science Online
Print out a worksheet of this page at booka.msscience.com

section 1

Ideas About Evolution

as you read

What You'll Learn

- **Describe** Lamarck's hypothesis of acquired characteristics and Darwin's theory of natural selection.
- **Identify** why variations in organisms are important.
- **Compare and contrast** gradualism and punctuated equilibrium.

Why It's Important

The theory of evolution suggests why there are so many different living things.

Review Vocabulary
hypothesis: an explanation that can be tested

New Vocabulary
- species
- evolution
- natural selection
- variation
- adaptation
- gradualism
- punctuated equilibrium

Early Models of Evolution

Millions of species of plants, animals, and other organisms live on Earth today. Do you suppose they are exactly the same as they were when they first appeared—or have any of them changed? A **species** is a group of organisms that share similar characteristics and can reproduce among themselves to produce fertile offspring. Many characteristics of a species are inherited when they pass from parent to offspring. Change in these inherited characteristics over time is **evolution. Figure 1** shows how the characteristics of the camel have changed over time.

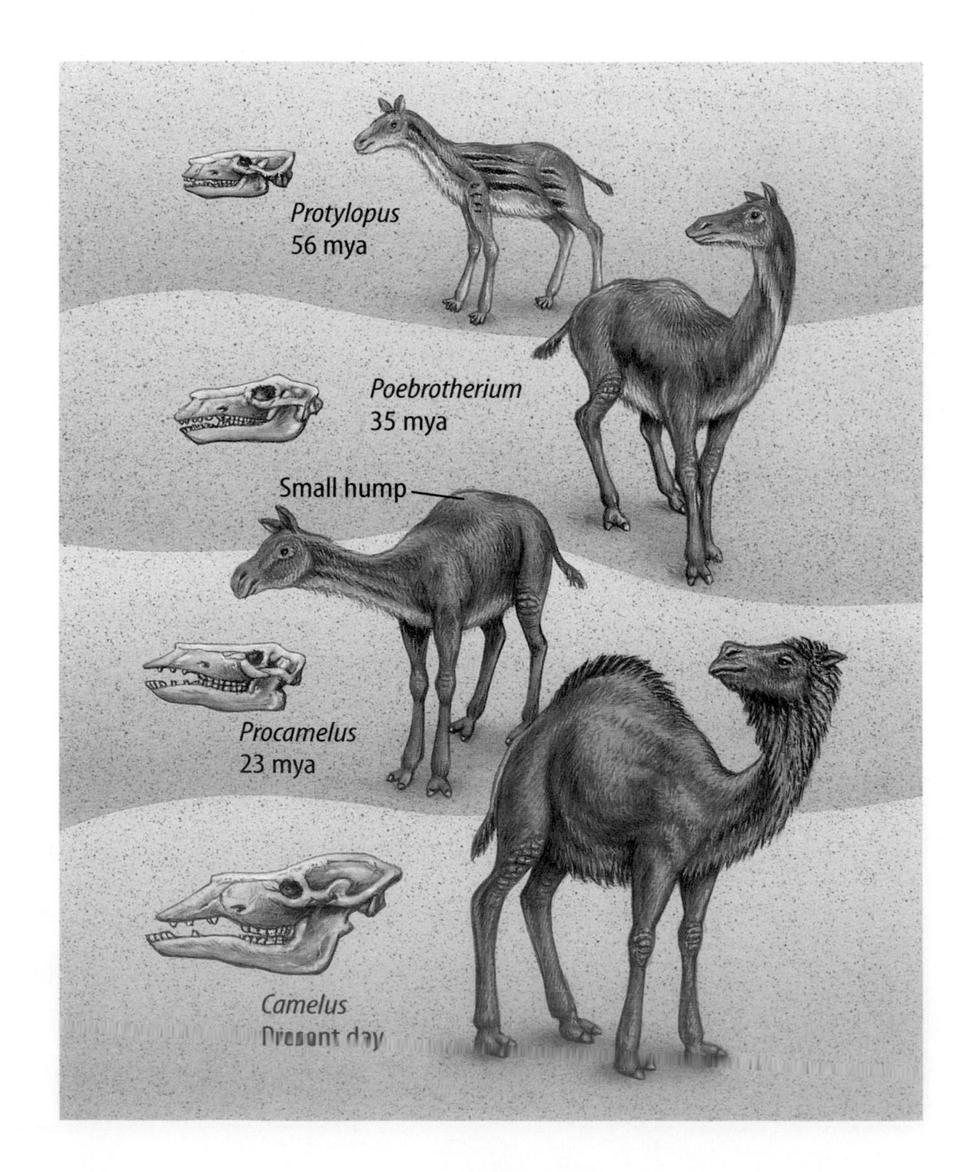

Figure 1 By studying fossils, scientists have traced the hypothesized evolution of the camel.
Discuss *the changes you observe in camels over time.*

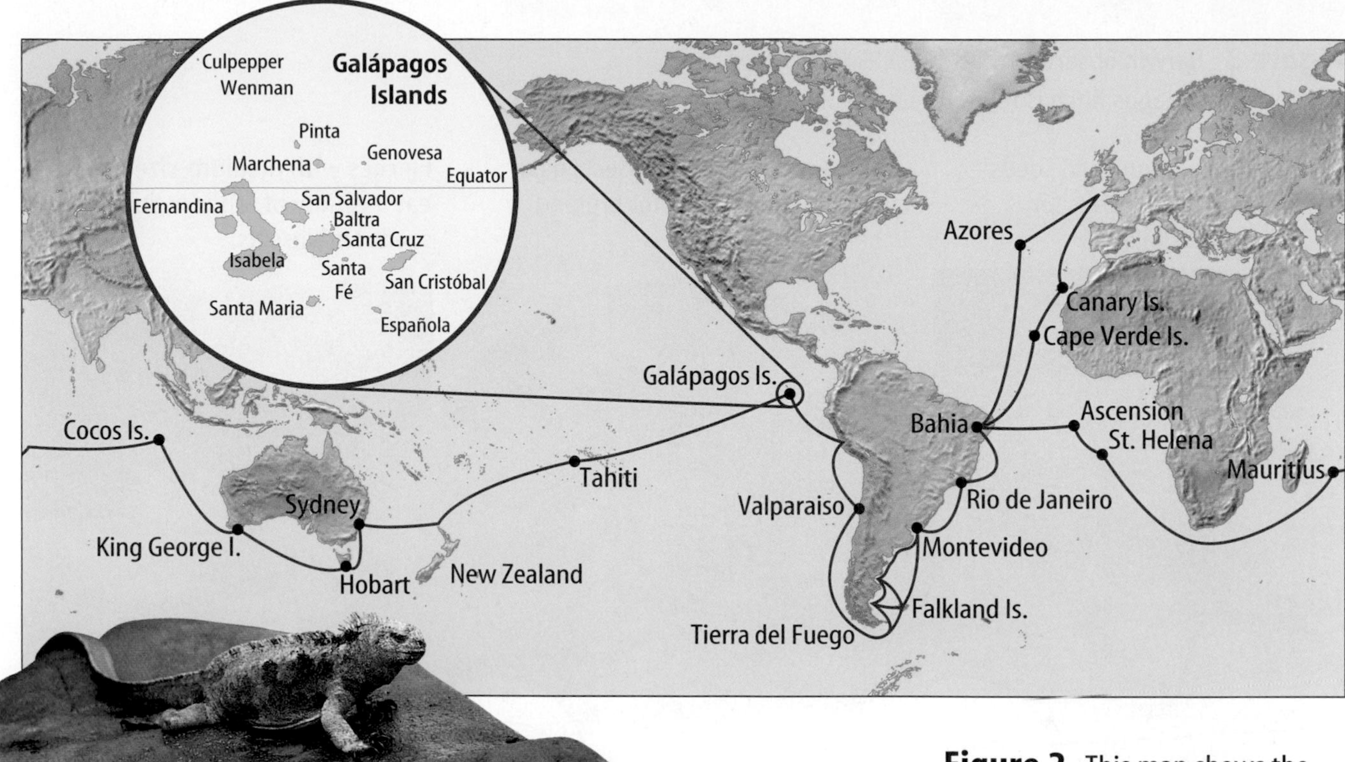

Figure 2 This map shows the route of Darwin's voyage on the HMS *Beagle*. Darwin noticed many species on the Galápagos Islands that he had not seen along the coast of South America, including the marine iguana. This species is the only lizard in the world known to enter the ocean and feed on seaweed.

Hypothesis of Acquired Characteristics In 1809, Jean Baptiste de Lamarck proposed a hypothesis to explain how species change over time. He suggested that characteristics, or traits, developed during a parent organism's lifetime are inherited by its offspring. His hypothesis is called the inheritance of acquired characteristics. Scientists collected data on traits that are passed from parents to offspring. The data showed that traits developed during a parent's lifetime, such as large muscles built by hard work or exercise, are not passed on to offspring. The evidence did not support Lamarck's hypothesis.

Reading Check *What was Lamarck's explanation of evolution?*

Darwin's Model of Evolution

In December 1831, the HMS *Beagle* sailed from England on a journey to explore the South American coast. On board was a young naturalist named Charles Darwin. During the journey, Darwin recorded observations about the plants and animals he saw. He was amazed by the variety of life on the Galápagos Islands, which are about 1,000 km from the coast of Ecuador. Darwin hypothesized that the plants and animals on the Galápagos Islands originally came from Central and South America. But the islands were home to many species he had not seen in South America, including giant cactus trees, huge land tortoises, and the iguana shown in **Figure 2.**

Figure 3 Darwin observed that the beak shape of each species of Galápagos finch is related to its eating habits.

Finches that eat nuts and seeds have short, strong beaks for breaking hard shells.

Finches that feed on insects have long, slender beaks for probing beneath tree bark.

Finches with medium-sized beaks eat a variety of foods including seeds and insects.

Topic: Darwin's Finches
Visit booka.msscience.com for Web links to information about the finches Darwin observed.

Activity In your Science Journal, describe the similarities and differences of any two species of Galápagos finches.

Darwin's Observations Darwin observed 13 species of finches on the Galápagos Islands. He noticed that all 13 species were similar, except for differences in body size, beak shape, and eating habits, as shown in **Figure 3.** He also noticed that all the Galápagos finch species were similar to one finch species he had seen on the South American coast.

Darwin reasoned that the Galápagos finches must have had to compete for food. Finches with beak shapes that allowed them to eat available food survived longer and produced more offspring than finches without those beak shapes. After many generations, these groups of finches became separate species.

How did Darwin explain the evolution of the different species of Galápagos finches?

Natural Selection

After the voyage, Charles Darwin returned to England and continued to think about his observations. He collected more evidence on inherited traits by breeding racing pigeons. He also studied breeds of dogs and varieties of flowers. In the mid 1800s, Darwin developed a theory of evolution that is accepted by most scientists today. He described his ideas in a book called *On the Origin of Species,* which was published in 1859.

Darwin's Theory Darwin's observations led many other scientists to conduct experiments on inherited characteristics. After many years, Darwin's ideas became known as the theory of evolution by natural selection. **Natural selection** means that organisms with traits best suited to their environment are more likely to survive and reproduce. Their traits are passed to more offspring. All living organisms produce more offspring than survive. Galápagos finches lay several eggs every few months. Darwin realized that in just a few years, several pairs of finches could produce a large population. A population is all of the individuals of a species living in the same area. Members of a large population compete for living space, food, and other resources. Those that are best able to survive are more likely to reproduce and pass on their traits to the next generation.

The principles that describe how natural selection works are listed in **Table 1.** Over time, as new data was gathered and reported, changes were made to Darwin's original ideas about evolution by natural selection. His theory remains one of the most important ideas in the study of life science.

Table 1 The Principles of Natural Selection
1. Organisms produce more offspring than can survive.
2. Differences, or variations, occur among individuals of a species.
3. Some variations are passed to offspring.
4. Some variations are helpful. Individuals with helpful variations survive and reproduce better than those without these variations.
5. Over time, the offspring of individuals with helpful variations make up more of a population and eventually may become a separate species.

Applying Science

Does natural selection take place in a fish tank?

Alejandro raises tropical fish as a hobby. Could the observations that he makes over several weeks illustrate the principles of natural selection?

Identifying the Problem

Alejandro keeps a detailed journal of his observations, some of which are given in the table to the right.

Fish Tank Observations

Date	Observation
June 6	6 fish are placed in aquarium tank.
July 22	16 new young appear.
July 24	3 young have short or missing tail fins. 13 young have normal tail fins.
July 28	Young with short or missing tail fins die.
August 1	2 normal fish die—from overcrowding?
August 12	30 new young appear.
August 15	5 young have short or missing tail fins. 25 young have normal tail fins.
August 18	Young with short or missing tail fins die.
August 20	Tank is overcrowded. Fish are divided equally into two tanks.

Solving the Problem

Refer to **Table 1** and match each of Alejandro's journal entries with the principle(s) it demonstrates. Here's a hint: *Some entries may not match any of the principles of natural selection. Some entries may match more than one principle.*

INTEGRATE Language Arts

Evolution of English
If someone from Shakespeare's time were to speak to you today, you probably would not understand her. Languages, like species, change over time. In your Science Journal, discuss some words or phrases that you use that your parents or teachers do not use correctly.

Variation and Adaptation

Darwin's theory of evolution by natural selection emphasizes the differences among individuals of a species. These differences are called variations. A **variation** is an inherited trait that makes an individual different from other members of its species. Variations result from permanent changes, or mutations, in an organism's genes. Some gene changes produce small variations, such as differences in the shape of human hairlines. Other gene changes produce large variations, such as an albino squirrel in a population of gray squirrels or fruit without seeds. Over time, more and more individuals of the species might inherit these variations. If individuals with these variations continue to survive and reproduce over many generations, a new species can evolve. It might take hundreds, thousands, or millions of generations for a new species to evolve.

Some variations are more helpful than others. An **adaptation** is any variation that makes an organism better suited to its environment. The variations that result in an adaptation can involve an organism's color, shape, behavior, or chemical makeup. Camouflage (KA muh flahj) is an adaptation. A camouflaged organism, like the one shown in **Figure 4,** blends into its environment and is more likely to survive and reproduce.

Figure 4 Variations that provide an advantage tend to increase in a population over time. Variations that result in a disadvantage tend to decrease in a population over time.

Albinism can prevent an organism from blending into its environment. **Infer** *what might happen to an albino lemur in its natural environment.*

Camouflage allows organisms to blend into their environments. **Infer** *how its coloration gives this scorpion fish a survival advantage.*

Figure 5 During the last ice age, Virginia white-tailed deer moved south ahead of an advancing ice sheet. When ice sheets melted worldwide about 4,000–10,000 years ago, ocean levels rose. Some deer were isolated on a chain of islands and evolved into a new subspecies, the Key deer. Key deer live only on approximately 30 islands in the subtropical lower keys of Florida.

A Virginia white-tailed deer can be 0.9 m to 1.1 m tall at the shoulder.

A Key deer can be 0.6 m to 0.7 m tall at the shoulder.
Infer *why Key deer are smaller than Virginia white-tailed deer.*

Changes in the Sources of Genes Over time, the genetic makeup of a species might change its appearance. For example, as the genetic makeup of a species of seed-eating Galápagos finch changed, so did the size and shape of its beak. Many kinds of environmental factors help bring about changes. When individuals of the same species move into or out of an area, they might bring in or remove genes and variations. Suppose a family from another country moves to your neighborhood. They might bring different foods, customs, and ways of speaking with them. In a similar way, when new individuals enter an existing population, they can bring in different genes and variations.

Geographic Isolation Sometimes mountains, lakes, or other geologic features isolate a small number of individuals from the rest of a population. Over several generations, variations that do not exist in the larger population might begin to be more common in the isolated population. Also, gene mutations can occur that add variations to populations. Over time, the two populations can become so different that they no longer can breed with each other. Key deer, like the one shown in **Figure 5,** evolved because of geographic isolation about 4,000–6,000 years ago.

Modeling Evolution

Procedure

1. On a piece of **paper,** print the word *train.*
2. Add, subtract, or change one letter to make a new word.
3. Repeat step 2 with the new word.
4. Repeat steps 2 and 3 two more times.
5. Make a "family tree" that shows how your first word changed over time.

Analysis

Compare your tree to those of other people. How is this process similar to evolution by natural selection?

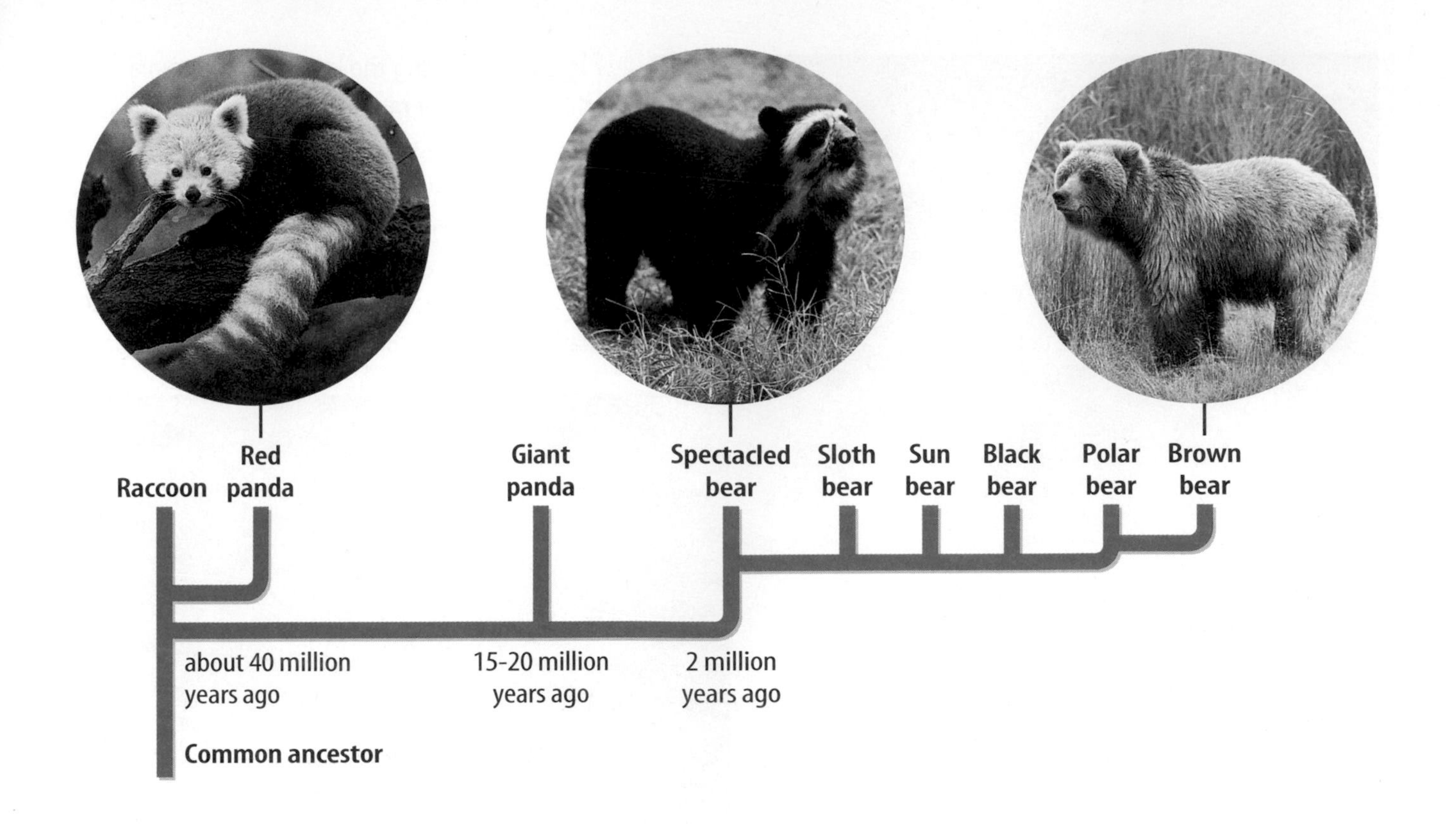

Figure 6 The hypothesized evolution of bears illustrates the punctuated equilibrium model of evolution.
Discuss *how the six species on the far right are explained better by punctuated equlibrium.*

The Speed of Evolution

Scientists do not agree on how quickly evolution occurs. Many scientists hypothesize that evolution occurs slowly, perhaps over tens or hundreds of millions of years. Other scientists hypothesize that evolution can occur quickly. Most scientists agree that evidence supports both of these models.

Gradualism Darwin hypothesized that evolution takes place slowly. The model that describes evolution as a slow, ongoing process by which one species changes to a new species is known as **gradualism.** According to the gradualism model, a continuing series of mutations and variations over time will result in a new species. Look back at **Figure 1,** which shows the evolution of the camel over tens of millions of years. Fossil evidence shows a series of intermediate forms that indicate a gradual change from the earliest camel species to today's species.

Punctuated Equilibrium Gradualism doesn't explain the evolution of all species. For some species, the fossil record shows few intermediate forms—one species suddenly changes to another. According to the **punctuated equilibrium** model, rapid evolution comes about when the mutation of a few genes results in the appearance of a new species over a relatively short period of time. The fossil record gives examples of this type of evolution, as you can see in **Figure 6.**

Punctuated Equilibrium Today Evolution by the punctuated equilibrium model can occur over a few thousand or million years, and sometimes even faster. For example, many bacteria have changed in a few decades. The antibiotic penicillin originally came from the fungus shown in **Figure 7.** But many bacteria species that were once easily killed by penicillin no longer are harmed by it. These bacteria have developed resistance to the drug. Penicillin has been in use since 1943. Just four years later, in 1947, a species of bacteria that causes pneumonia and other infections already had developed resistance to the drug. By the 1990s, several disease-producing bacteria had become resistant to penicillin and many other antibiotics.

How did penicillin-resistant bacteria evolve so quickly? As in any population, some organisms have variations that allow them to survive unfavorable living conditions when other organisms cannot. When penicillin was used to kill bacteria, those with the penicillin-resistant variation survived, reproduced, and passed this trait to their offspring. Over a period of time, this bacteria population became penicillin-resistant.

Figure 7 The fungus growing in this petri dish is *Penicillium,* the original source of penicillin. It produces an antibiotic substance that prevents the growth of certain bacteria.

section 1 review

Summary

Early Models of Evolution

- Evolution is change in the characteristics of a species over time.
- Lamarck proposed the hypothesis of inherited acquired characteristics.

Natural Selection

- Darwin proposed evolution by natural selection, a process by which organisms best suited to their environments are most likely to survive and reproduce.
- Organisms have more offspring than can survive, individuals of a species vary, and many of these variations are passed to offspring.

Variation and Adaptation

- Adaptations are variations that help an organism survive or reproduce in its environment.
- Mutations are the source of new variations.

The Speed of Evolution

- Evolution may be a slow or fast process depending on the species under study.

Self Check

1. **Compare** Lamarck's and Darwin's ideas about how evolution takes place.
2. **Explain** why variations are important to understanding change in a population over time.
3. **Discuss** how the gradualism model of evolution differs from the punctuated equilibrium model of evolution.
4. **Describe** how geographic isolation contributes to evolution.
5. **Think Critically** What adaptations would be helpful for an animal species that was moved to the Arctic?
6. **Concept Map** Use information given in **Figure 6** to make a map that shows how raccoons, red pandas, giant pandas, polar bears, and black bears are related to a common ancestor.

Applying Math

7. **Use Percentages** The evolution of the camel can be traced back at least 56 million years. Use **Figure 1** to estimate the percent of this time that the modern camel has existed.

Hidden Frogs

Through natural selection, animals become adapted for survival in their environment. Adaptations include shapes, colors, and even textures that help an animal blend into its surroundings. These adaptations are called camouflage. The red-eyed tree frog's mint green body blends in with tropical forest vegetation as shown in the photo on the right. Could you design camouflage for a desert frog? A temperate forest frog?

Real-World Question

What type of camouflage would best suit a frog living in a particular habitat?

Goals

- **Create** a frog model camouflaged to blend in with its surroundings.

Materials (for each group)

cardboard form of a frog
colored markers
crayons
colored pencils
glue
beads
sequins
modeling clay

Safety Precautions

Procedure

1. Choose one of the following habitats for your frog model: muddy shore of a pond, orchid flowers in a tropical rain forest, multicolored clay in a desert, or the leaves and branches of trees in a temperate forest.
2. **List** the features of your chosen habitat that will determine the camouflage your frog model will need.
3. **Brainstorm** with your group the body shape, coloring, and skin texture that would make the best camouflage for your model. Record your ideas in your Science Journal.
4. **Draw** in your Science Journal samples of colors, patterns, texture, and other features your frog model might have.
5. Show your design ideas to your teacher and ask for further input.
6. **Construct** your frog model.

Conclude and Apply

1. **Explain** how the characteristics of the habitat helped you decide on the specific frog features you chose.
2. **Infer** how the color patterns and other physical features of real frogs develop in nature.
3. **Explain** why it might be harmful to release a frog into a habitat for which it is not adapted.

Communicating Your Data

Create a poster or other visual display that represents the habitat you chose for this activity. Use your display to show classmates how your design helps camouflage your frog model. **For more help, refer to the Science Skill Handbook.**

section 2

Clues About Evolution

Clues from Fossils

Imagine going on a fossil hunt in Wyoming. Your companions are paleontologists—scientists who study the past by collecting and examining fossils. As you climb a low hill, you notice a curved piece of stone jutting out of the sandy soil. One of the paleontologists carefully brushes the soil away and congratulates you on your find. You've discovered part of the fossilized shell of a turtle like the one shown in **Figure 8.**

The Green River Formation covers parts of Wyoming, Utah, and Colorado. On your fossil hunt, you learn that about 50 million years ago, during the Eocene Epoch, this region was covered by lakes. The water was home to fish, crocodiles, lizards, and turtles. Palms, fig trees, willows, and cattails grew on the lakeshores. Insects and birds flew through the air. How do scientists know all this? After many of the plants and animals of that time died, they were covered with silt and mud. Over millions of years, they became the fossils that have made the Green River Formation one of the richest fossil deposits in the world.

The turtle *Cistemum undatum* is from the same fossil formation.

The most abundant fossils are of a freshwater herring, *Knightia oecaena*, which is Wyoming's state fossil.

Figure 8 The desert of the Green River Formation is home to pronghorn antelope, elks, coyotes, and eagles. Fossil evidence shows that about 50 million years ago the environment was much warmer and wetter than it is today.

as you read

What **You'll Learn**

- **Identify** the importance of fossils as evidence of evolution.
- **Explain** how relative and radiometric dating are used to estimate the age of fossils.
- **List** examples of five types of evidence for evolution.

Why **It's Important**

The scientific evidence for evolution helps you understand why this theory is so important to the study of biology.

Review Vocabulary

epoch: next-smaller division of geological time after a period; is characterized by differences in life-forms that may vary regionally

New Vocabulary

- sedimentary rock
- radioactive element
- embryology
- homologous
- vestigial structure

Figure 9 Examples of several different types of fossils are shown here.
Infer *which of these would most likely be found in a layer of sedimentary rock.*

Mineralized fossils Minerals can replace wood or bone to create a piece of petrified wood as shown to the left or a mineralized bone fossil.

Imprint fossils A leaf, feather, bones, or even the entire body of an organism can leave an imprint on sediment that later hardens to become rock.

Frozen fossils The remains of organisms like this mammoth can be trapped in ice that remains frozen for thousands of years.

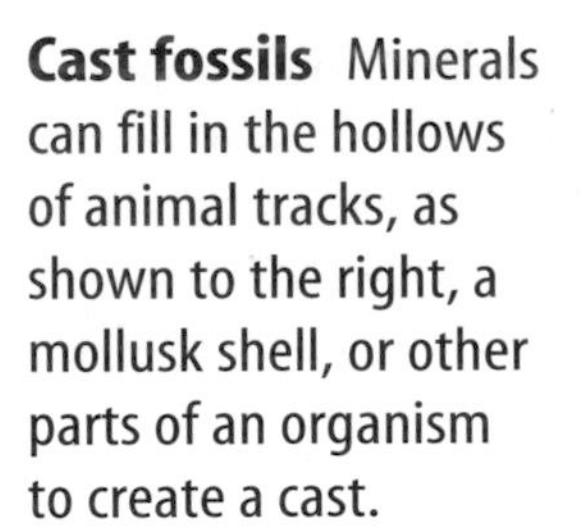

Cast fossils Minerals can fill in the hollows of animal tracks, as shown to the right, a mollusk shell, or other parts of an organism to create a cast.

Fossils in amber When the sticky resin of certain cone-bearing plants hardens over time, amber forms. It can contain the remains of trapped insects.

Types of Fossils

INTEGRATE Earth Science

Most of the evidence for evolution comes from fossils. A fossil is the remains, an imprint, or a trace of a prehistoric organism. Several types of fossils are shown in **Figure 9.** Most fossils are found in sedimentary rock. **Sedimentary rock** is formed when layers of sand, silt, clay, or mud are compacted and cemented together, or when minerals are deposited from a solution. Limestone, sandstone, and shale are all examples of sedimentary rock. Fossils are found more often in limestone than in any other kind of sedimentary rock. The fossil record provides evidence that living things have evolved.

Determining a Fossil's Age

Paleontologists use detective skills to determine the age of dinosaur fossils or the remains of other ancient organisms. They can use clues provided by unique rock layers and the fossils they contain. The clues provide information about the geology, weather, and life-forms that must have been present during each geologic time period. Two basic methods—relative dating and radiometric dating—can be used, alone or together, to estimate the ages of rocks and fossils.

Topic: Fossil Finds
Visit booka.msscience.com for Web links to information about recent fossil discoveries.

Activity Prepare a newspaper article describing how one of these discoveries was made, what it reveals about past life on Earth, and how it has impacted our understanding of what the past environments of Earth were like.

Relative Dating One way to find the approximate age of fossils found within a rock layer is relative dating. Relative dating is based on the idea that in undisturbed areas, younger rock layers are deposited on top of older rock layers, as shown in **Figure 10.** Relative dating provides only an estimate of a fossil's age. The estimate is made by comparing the ages of rock layers found above and below the fossil layer. For example, suppose a 50-million-year-old rock layer lies below a fossil, and a 35-million-year-old layer lies above it. According to relative dating, the fossil is between 35 million and 50 million years old.

Why can relative dating be used only to estimate the age of a fossil?

Radiometric Dating Scientists can obtain a more accurate estimate of the age of a rock layer by using radioactive elements. A **radioactive element** gives off a steady amount of radiation as it slowly changes to a nonradioactive element. Each radioactive element gives off radiation at a different rate. Scientists can estimate the age of the rock by comparing the amount of radioactive element with the amount of nonradioactive element in the rock. This method of dating does not always produce exact results, because the original amount of radioactive element in the rock can never be determined for certain.

Figure 10 In Bryce Canyon, erosion by water and wind has cut through the sedimentary rock, exposing the layers.
Infer *the relative age of rocks in the lowest layers compared to the top layer.*

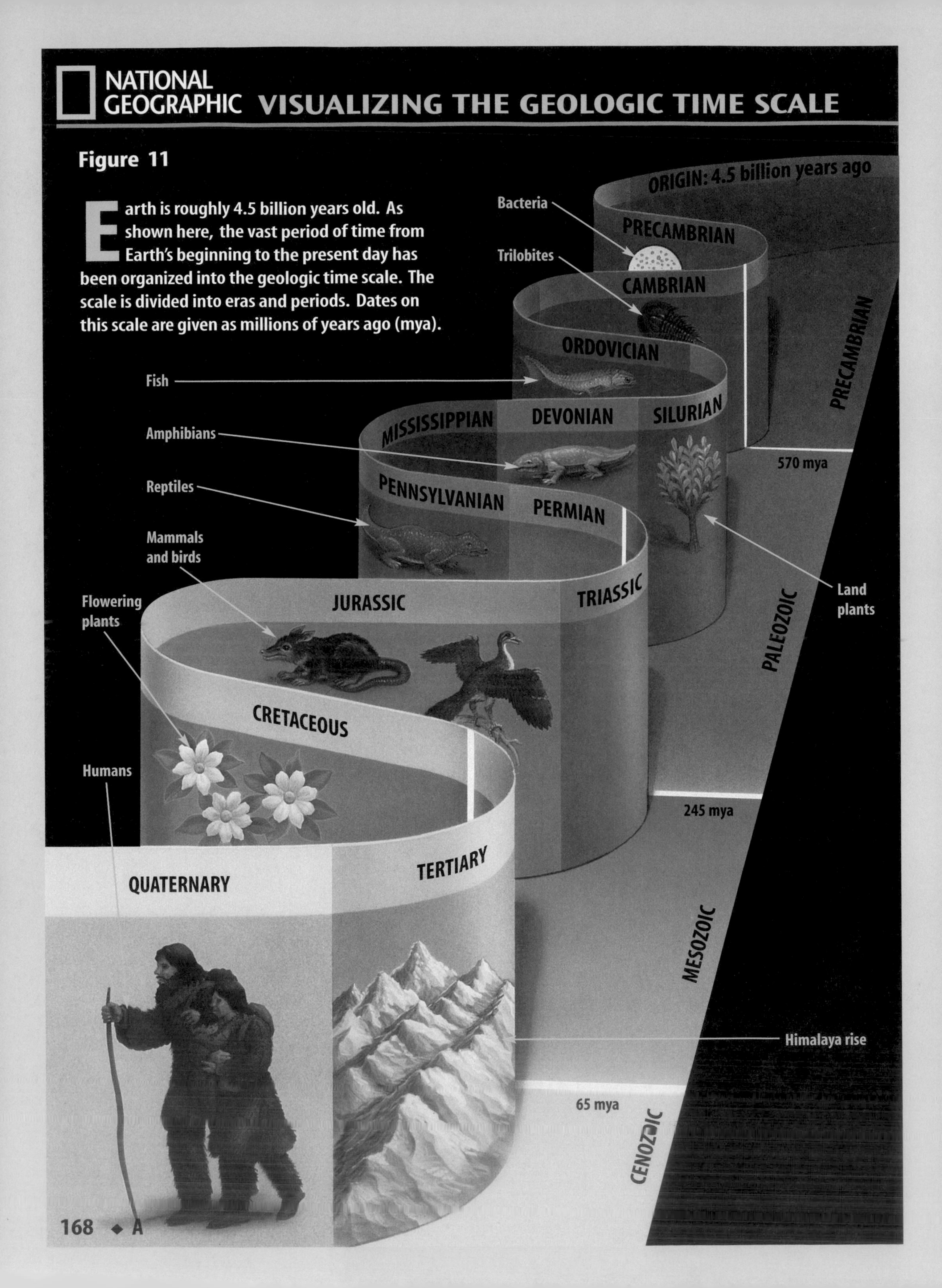
NATIONAL GEOGRAPHIC
VISUALIZING THE GEOLOGIC TIME SCALE
Figure 11
Earth is roughly 4.5 billion years old. As shown here, the vast period of time from Earth's beginning to the present day has been organized into the geologic time scale. The scale is divided into eras and periods. Dates on this scale are given as millions of years ago (mya).
ORIGIN: 4.5 billion years ago
Bacteria
PRECAMBRIAN
Trilobites
CAMBRIAN
ORDOVICIAN
Fish
MISSISSIPPIAN
DEVONIAN
SILURIAN
Amphibians
Reptiles
PENNSYLVANIAN
PERMIAN
Mammals and birds
Flowering plants
JURASSIC
TRIASSIC
CRETACEOUS
Humans
QUATERNARY
TERTIARY
PRECAMBRIAN
570 mya
PALEOZOIC
Land plants
245 mya
MESOZOIC
Himalaya rise
65 mya
CENOZOIC

Fossils and Evolution

Fossils provide a record of organisms that lived in the past. However, the fossil record is incomplete, or has gaps, much like a book with missing pages. The gaps exist because most organisms do not become fossils. By looking at fossils, scientists conclude that many simpler forms of life existed earlier in Earth's history, and more complex forms of life appeared later, as shown in **Figure 11.** Fossils provide indirect evidence that evolution has occurred on Earth.

Almost every week, fossil discoveries are made somewhere in the world. When fossils are found, they are used to help scientists understand the past. Scientists can use fossils to make models that show what the organisms might have looked like. From fossils, scientists can sometimes determine whether the organisms lived in family groups or alone, what types of food they ate, what kind of environment they lived in, and many other things about them. Most fossils represent extinct organisms. From a study of the fossil record, scientists have concluded that more than 99 percent of all organisms that have ever existed on Earth are now extinct.

Evolution in Fossils Many organisms have a history that has been preserved in sedimentary rock. Fossils show that the bones of animals such as horses and whales have become reduced in size or number over geologic time, as the species has evolved. In your Science Journal, explain what information can be gathered from changes in structures that occur over time.

More Clues About Evolution

Besides fossils, what other clues exist about evolution? Sometimes, evolution can be observed directly. Plant breeders observe evolution when they use cross-breeding to produce genetic changes in plants. The development of antibiotic resistance in bacteria is another direct observation of evolution. Entomologists have noted similar rapid evolution of pesticide-resistant insect species. These observations provide direct evidence that evolution occurs. Also, many examples of indirect evidence for evolution exist. They include similarities in embryo structures, the chemical makeup of organisms including DNA, and the way organisms develop into adults. Indirect evidence does not provide proof of evolution, but it does support the idea that evolution takes place over time.

Embryology The study of embryos and their development is called **embryology** (em bree AH luh jee). An embryo is the earliest growth stage of an organism. A tail and pharyngeal pouches are found at some point in the embryos of fish, reptiles, birds, and mammals, as **Figure 12** shows. Fish develop gills, but the other organisms develop other structures as their development continues. Fish, birds, and reptiles keep their tails, but many mammals lose theirs. These similarities suggest an evolutionary relationship among all vertebrate species.

Figure 12 Similarities in the embryos of fish, chickens, and rabbits show evidence of evolution. **Evaluate** *these embryos as evidence for evolution.*

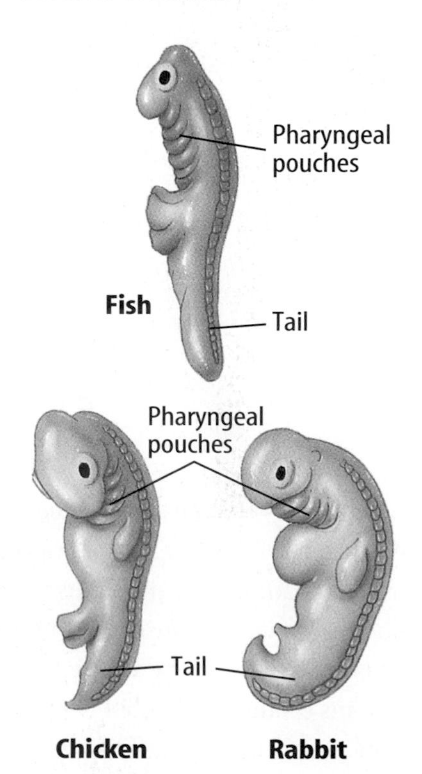

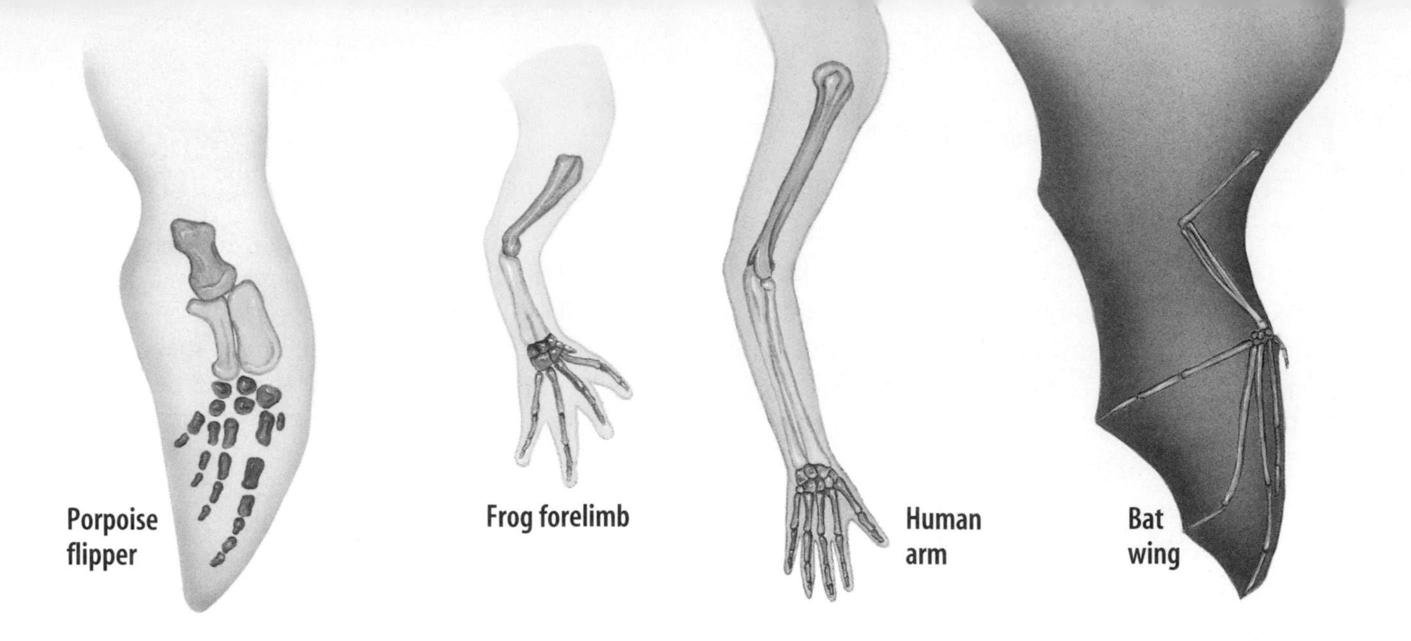

Figure 13 A porpoise flipper, frog forelimb, human arm, and bat wing are homologous. These structures show different arrangements and shapes of the bones of the forelimb. They have the same number of bones, muscles, and blood vessels, and they developed from similar tissues.

Homologous Structures What do the structures shown in **Figure 13** have in common? Although they have different functions, each of these structures is made up of the same kind of bones. Body parts that are similar in origin and structure are called **homologous** (hoh MAH luh gus). Homologous structures also can be similar in function. They often indicate that two or more species share common ancestors.

Reading Check *What do homologous structures indicate?*

Vestigial Structures The bodies of some organisms include **vestigial** (veh STIH jee ul) **structures**—structures that don't seem to have a function. Vestigial structures also provide evidence for evolution. For example, manatees, snakes, and whales no longer have back legs, but, like all animals with legs, they still have pelvic bones. The human appendix is a vestigial structure. The appendix appears to be a small version of the cecum, which is an important part of the digestive tract of many mammals. Scientists hypothesize that vestigial structures, like those shown in **Figure 14,** are body parts that once functioned in an ancestor.

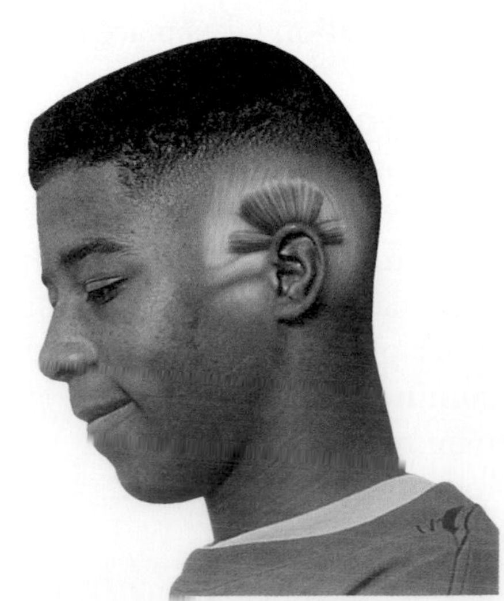

Figure 14 Humans have three small muscles around each ear that are vestigial. In some mammals, such as horses, these muscles are large. They allow a horse to turn its ears toward the source of a sound. Humans cannot rotate their ears, but some people can wiggle their ears.

DNA If you enjoy science fiction, you have read books or seen movies in which scientists re-create dinosaurs and other extinct organisms from DNA taken from fossils. DNA is the molecule that controls heredity and directs the development of every organism. In a cell with a nucleus, DNA is found in genes that make up the chromosomes. Scientists compare DNA from living organisms to identify similarities among species. Examinations of ancient DNA often provide additional evidence of how some species evolved from their extinct ancestors. By looking at DNA, scientists also can determine how closely related organisms are. For example, DNA studies indicate that dogs are the closest relatives of bears.

Figure 15 Gorillas have DNA and proteins that are similar to humans and other primates.

Similar DNA also can suggest common ancestry. Apes such as the gorillas shown in **Figure 15,** chimpanzees, and orangutans have 24 pairs of chromosomes. Humans have 23 pairs. When two of an ape's chromosomes are laid end to end, a match for human chromosome number 2 is formed. Also, similar proteins such as hemoglobin—the oxygen-carrying protein in red blood cells—are found in many primates. This can be further evidence that primates have a common ancestor.

section 2 review

Summary

Clues from Fossils

- Scientists learn about past life by studying fossils.

Determining a Fossil's Age

- The relative date of a fossil can be estimated from the ages of rocks in nearby layers.
- Radiometric dating using radioactive elements gives more accurate dates for fossils.

Fossils and Evolution

- The fossil record has gaps which may yet be filled with later discoveries.

More Clues About Evolution

- Homologous structures, similar embryos, or vestigial structures can show evolutionary relationships.
- Evolutionary relationships among organisms can be inferred from DNA comparisons.

Self Check

1. **Compare and contrast** relative dating and radiometric dating.
2. **Discuss** the importance of fossils as evidence of evolution and describe five different kinds of fossils.
3. **Explain** how DNA can provide some evidence of evolution.
4. **List** three examples of direct evidence for evolution.
5. **Interpret Scientific Illustrations** According to data in **Figure 11,** what was the longest geologic era? What was the shortest era? In what period did mammals appear?
6. **Think Critically** Compare and contrast the five types of evidence that support the theory of evolution.

Applying Math

7. **Use Percentages** The Cenozoic Era represents about 65 million years. Approximately what percent of Earth's 4.5-billion-year history does this era represent?

section 3

The Evolution of Primates

as you read

What You'll Learn

- **Describe** the differences among living primates.
- **Identify** the adaptations of primates.
- **Discuss** the evolutionary history of modern primates.

Why It's Important

Studying primate evolution will help you appreciate the differences among primates.

Review Vocabulary
opposable: can be placed against another digit of a hand or foot

New Vocabulary
- primate
- hominid
- *Homo sapiens*

Primates

Humans, monkeys, and apes belong to the group of mammals known as the **primates.** All primates have opposable thumbs, binocular vision, and flexible shoulders that allow the arms to rotate. These shared characteristics indicate that all primates may have evolved from a common ancestor.

Having an opposable thumb allows you to cross your thumb over your palm and touch your fingers. This means that you can grasp and hold things with your hands. An opposable thumb allows tree-dwelling primates to hold on to branches.

Binocular vision permits you to judge depth or distance with your eyes. In a similar way, it allows tree-dwelling primates to judge the distances as they move between branches. Flexible shoulders and rotating forelimbs also help tree-dwelling primates move from branch to branch. They also allow humans to do the backstroke, as shown in **Figure 16.**

Primates are divided into two major groups. The first group, the strepsirhines (STREP suh rines), includes lemurs and tarsiers like those shown in **Figure 17.** The second group, haplorhines (HAP luh rines), includes monkeys, apes, and humans.

Figure 16 The ability to rotate the shoulder in a complete circle allows humans to swim through water and tree-dwelling primates to travel through treetops.

Figure 17 Tarsiers and lemurs are active at night. Tarsiers are commonly found in the rain forests of Southeast Asia. Lemurs live on Madagascar and other nearby islands.
List *the traits that distinguish these animals as primates.*

Hominids About 4 million to 6 million years ago, humanlike primates appeared that were different from the other primates. These ancestors, called **hominids,** ate both meat and plants and walked upright on two legs. Hominids shared some characteristics with gorillas, orangutans, and chimpanzees, but a larger brain separated them from the apes.

African Origins In the early 1920s, a fossil skull was discovered in a quarry in South Africa. The skull had a small space for the brain, but it had a humanlike jaw and teeth. The fossil, named *Australopithecus,* was one of the oldest hominids discovered. An almost-complete skeleton of *Australopithecus* was found in northern Africa in 1974. This hominid fossil, shown in **Figure 18,** was called Lucy and had a small brain but is thought to have walked upright. This fossil indicates that modern hominids might have evolved from similar ancestors.

Figure 18 The fossil remains of Lucy are estimated to be 2.9 million to 3.4 million years old.

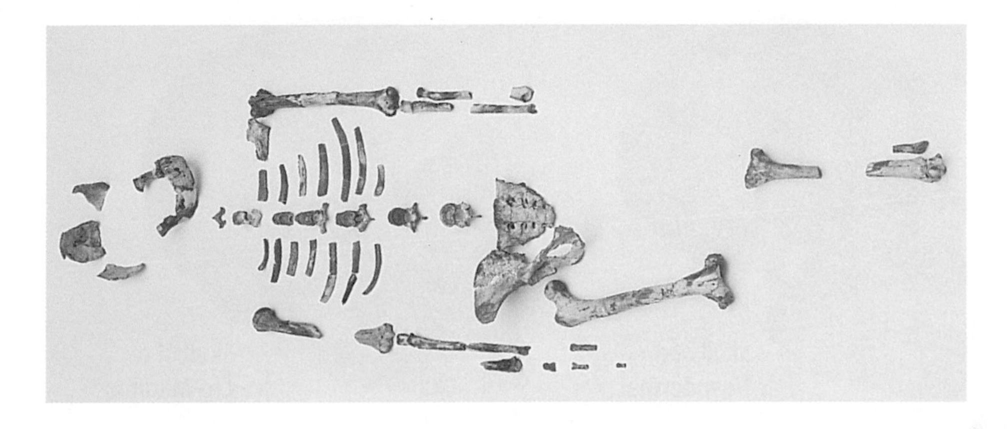

Living Without Thumbs

Procedure

1. Using **tape,** fasten down each of your thumbs next to the palm of each hand.
2. Leave your thumbs taped down for at least 1 h. During this time, do the following activities: eat a meal, change clothes, and brush your teeth. Be careful not to try anything that could be dangerous.
3. Untape your thumbs, then write about your experiences in your **Science Journal.**

Analysis

1. Did not having use of your thumbs significantly affect the way you did anything? Explain.
2. Infer how having opposable thumbs could have influenced primate evolution.

Try at Home

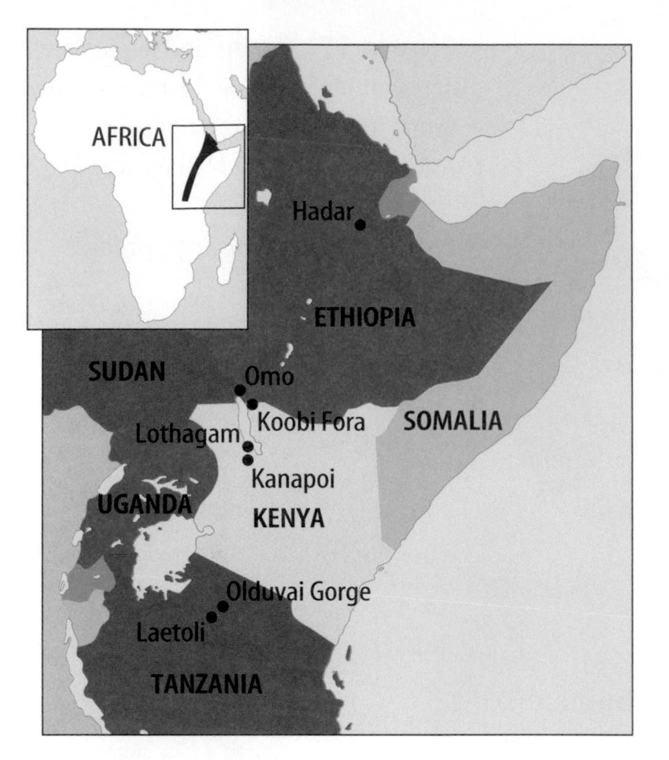

Figure 19 Many of the oldest humanlike skeletons have been found in this area of east Africa.

Early Humans In the 1960s in the region of Africa shown in **Figure 19,** a hominid fossil, which was more like present-day humans than *Australopithecus,* was discovered. The hominid was named *Homo habilis,* meaning "handy man," because simple stone tools were found near him. *Homo habilis* is estimated to be 1.5 million to 2 million years old. Based upon many fossil comparisons, scientists have suggested that *Homo habilis* gave rise to another species, *Homo erectus,* about 1.6 million years ago. This hominid had a larger brain than *Homo habilis. Homo erectus* traveled from Africa to Southeast Asia, China, and possibly Europe. *Homo habilis* and *Homo erectus* are thought to be ancestors of humans because they had larger brains and more humanlike features than *Australopithecus.*

Why was Homo habilis *given that name?*

Humans

The fossil record indicates that ***Homo sapiens*** evolved about 400,000 years ago. By about 125,000 years ago, two early human groups, Neanderthals (nee AN dur tawlz) and Cro-Magnon humans, as shown in **Figure 20,** probably lived at the same time in parts of Africa and Europe.

Neanderthals Short, heavy bodies with thick bones, small chins, and heavy browridges were physical characteristics of Neanderthals. Family groups lived in caves and used well-made stone tools to hunt large animals. Neanderthals disappeared from the fossil record about 30,000 years ago. They probably are not direct ancestors of modern humans, but represent a side branch of human evolution.

Figure 20 Compare the skull of a Neanderthal with the skull of a Cro-Magnon. **Describe** *what differences you can see between these two skulls.*

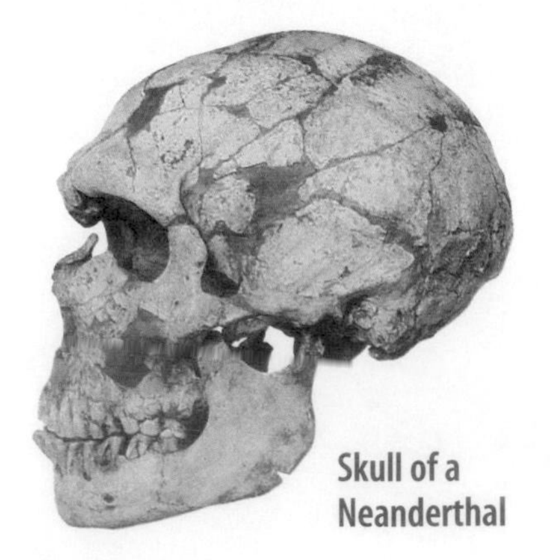

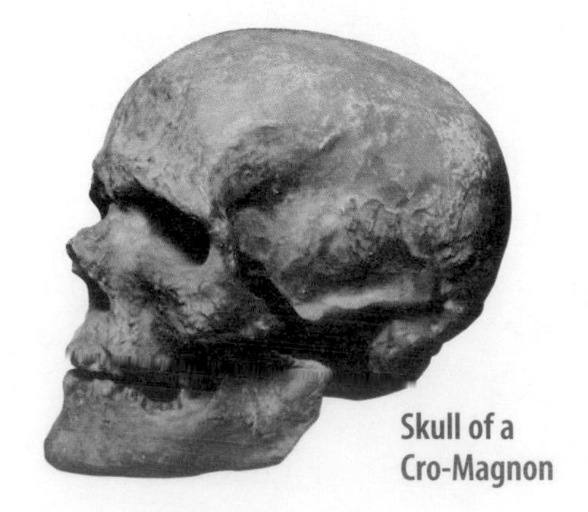

Figure 21 Paintings on cave walls have led scientists to hypothesize that Cro-Magnon humans had a well-developed culture.

Cro-Magnon Humans Cro-Magnon fossils have been found in Europe, Asia, and Australia and date from 10,000 to about 40,000 years in age. Standing about 1.6 m to 1.7 m tall, the physical appearance of Cro-Magnon people was almost the same as that of modern humans. They lived in caves, made stone carvings, and buried their dead. As shown in **Figure 21,** the oldest recorded art has been found on the walls of caves in France, where Cro-Magnon humans first painted bison, horses, and people carrying spears. Cro-Magnon humans are thought to be direct ancestors of early humans, *Homo sapiens,* which means "wise human." Evidence indicates that modern humans, *Homo sapiens sapiens,* evolved from *Homo sapiens.*

section 3 review

Summary

Primates

- Primates are an order of mammals characterized by opposable thumbs, binocular vision, and flexible shoulder joints.
- Primates are divided into strepsirrhines and haplorhines.
- Hominids are human ancestors that first appeared in Africa 4–6 million years ago.
- Hominids in the genus *Homo* first used tools and had larger brains than previous primates.

Humans

- *Homo sapiens* first appeared about 400,000 years ago.
- Cro-Magnon humans and Neanderthals coexisted in many places until Neanderthals disappeared about 30,000 years ago.
- *Homo sapiens* looked like modern humans and are believed to be our direct ancestors.

Self Check

1. **Describe** three kinds of evidence suggesting that all primates might have shared a common ancestor.
2. **Discuss** the importance of *Australopithecus.*
3. **Compare and contrast** Neanderthals, Cro-Magnon humans, and early humans.
4. **Identify** three groups most scientists consider to be direct ancestors of modern humans.
5. **Think Critically** Propose a hypothesis to explain why teeth are the most abundant fossil of hominids.

Applying Skills

6. **Concept Map** Make a concept map to show in what sequence hominids appeared. Use the following: *Homo sapiens sapiens,* Neanderthal, *Homo habilis, Australopithecus, Homo sapiens,* and Cro-Magnon human.
7. **Write** a story in your Science Journal about what life might have been like when both Neanderthals and Cro-Magnon humans were alive.

Recognizing Variation in a Population

Goals

- **Design** an experiment that will allow you to collect data about variation in a population.
- **Observe, measure, and analyze** variations in a population.

Possible Materials

fruit and seeds from one plant species
metric ruler
magnifying lens
graph paper

Safety Precautions

WARNING: *Do not put any fruit or seeds in your mouth.*

Real-World Question

When you first observe a flock of pigeons, you might think all the birds look alike. However, if you look closer, you will notice minor differences, or variations, among the individuals. Different pigeons might have different color markings, or some might be smaller or larger than others. Individuals of the same species—whether they're birds, plants, or worms—might look alike at first, but some variations undoubtedly exist. According to the principles of natural selection, evolution could not occur without variations. What kinds of variations have you noticed among species of plants or animals? How can you measure variation in a plant or animal population?

Form a Hypothesis

Make a hypothesis about the amount of variation in the fruit and seeds of one species of plant.

Test Your Hypothesis

Make a Plan

1. As a group, agree upon and write out the prediction.
2. **List** the steps you need to take to test your prediction. Be specific. Describe exactly what you will do at each step. List your materials.
3. **Decide** what characteristic of fruit and seeds you will study. For example, you could measure the length of fruit and seeds or count the number of seeds per fruit.
4. **Design** a data table in your Science Journal to collect data about two variations. Use the table to record the data your group collects.
5. **Identify** any constants, variables, and controls of the experiment.
6. How many fruit and seeds will you examine? Will your data be more accurate if you examine larger numbers?
7. **Summarize** your data in a graph or chart.

Follow Your Plan

1. Make sure your teacher approves your plan before you start.
2. Carry out the experiment as planned.
3. While the experiment is going on, write down any observations you make and complete the data table in your Science Journal.

Analyze Your Data

1. **Calculate** the mean and range of variation in your experiment. The range is the difference between the largest and the smallest measurements. The mean is the sum of all the data divided by the sample size.
2. **Graph** your group's results by making a line graph for the variations you measured. Place the range of variation on the *x*-axis and the number of organisms that had that measurement on the *y*-axis.

Conclude and Apply

1. **Explain** your results in terms of natural selection.
2. **Discuss** the factors you used to determine the amount of variation present.
3. **Infer** why one or more of the variations you observed in this activity might be helpful to the survival of the individual.

Create a poster or other exhibit that illustrates the variations you and your classmates observed.

TIME SCIENCE AND HISTORY

SCIENCE CAN CHANGE THE COURSE OF HISTORY!

Fighting HIV

The first cases of AIDS, or acquired immune deficiency syndrome, in humans were reported in the early 1980s. AIDS is caused by the human immunodeficiency virus, or HIV.

A major problem in AIDS research is the rapid evolution of HIV. When HIV multiplies inside a host cell, new versions of the virus are produced as well as identical copies of the virus that invaded the cell. New versions of the virus soon can outnumber the original version. A treatment that works against today's HIV might not work against tomorrow's version.

These rapid changes in HIV also mean that different strains of the virus exist in different places around the world. Treatments developed in the United States work only for people who contracted the virus in the United States. This leaves people in some parts of the world without effective treatments. So, researchers such as geneticist Flossie Wong-Staal at the University of California, San Diego, must look for new ways to fight the evolving virus.

Working Backwards

Flossie Wong-Staal is taking a new approach. First, her team identifies the parts of a human cell that HIV depends on and the parts of the human cell that HIV needs but the human cell doesn't need. Then the team looks for a way to remove—or inactivate—those unneeded parts. This technique limits the virus's ability to multiply.

Wong-Staal's research combines three important aspects of science—a deep understanding of how cells and genes operate, great skill in the techniques of genetics, and great ideas. Understanding, skill, and great ideas are the best weapons so far in the fight to conquer HIV.

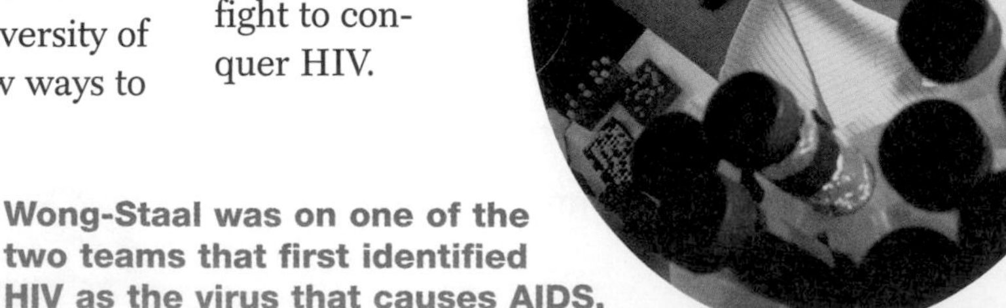

Wong-Staal was on one of the two teams that first identified HIV as the virus that causes AIDS.

Research Use the link to the right and other sources to determine which nations have the highest rates of HIV infection. Which nation has the highest rate? Where does the U.S. rank? Next, find data from ten years ago. Have the rankings changed?

For more information, visit booka.msscience.com/time

Reviewing Main Ideas

Section 1 Ideas About Evolution

1. Evolution is one of the central ideas of biology. It explains how living things have changed in the past and is a basis for predicting how they might change in the future.
2. Charles Darwin developed the theory of evolution by natural selection to explain how evolutionary changes account for the diversity of organisms on Earth.
3. Natural selection includes concepts of variation, overproduction, and competition.
4. According to natural selection, organisms with traits best suited to their environment are more likely to survive and reproduce.

Section 2 Clues About Evolution

1. Fossils provide evidence for evolution.
2. Relative dating and radiometric dating can be used to estimate the age of fossils.
3. The evolution of antibiotic-resistant bacteria, pesticide-resistant insects, and rapid genetic changes in plant species provides direct evidence that evolution occurs.
4. Homologous structures, vestigial structures, comparative embryology, and similarities in DNA provide indirect evidence of evolution.

Section 3 The Evolution of Primates

1. Primates include monkeys, apes, and humans. Hominids are humanlike primates.
2. The earliest known hominid fossil is *Australopithecus.*
3. *Homo sapiens* are thought to have evolved from Cro-Magnon humans about 400,000 years ago.

Visualizing Main Ideas

Copy and complete the following spider map on evolution.

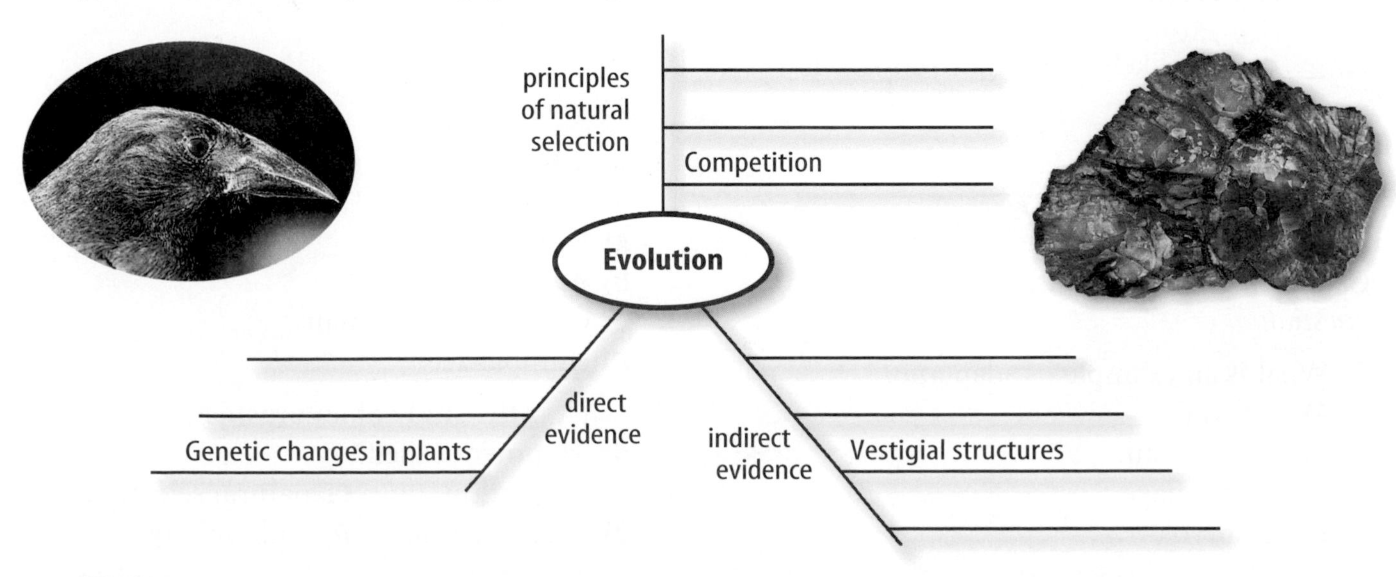

Using Vocabulary

adaptation p. 160
embryology p. 169
evolution p. 156
gradualism p. 162
hominid p. 173
Homo sapiens p. 174
homologous p. 170
natural selection p. 159
primate p. 172
punctuated equilibrium p. 162
radioactive element p. 167
sedimentary rock p. 166
species p. 156
variation p. 160
vestigial structure p. 170

Fill in the blanks with the correct vocabulary word or words.

1. _________ contains many different kinds of fossils.
2. The muscles that move the human ear appear to be _________.
3. Forelimbs of bats, humans, and seals are _________.
4. Opposable thumbs are a characteristic of _________.
5. The study of _________ can provide evidence of evolution.
6. The principles of _________ include variation and competition.
7. _________ likely evolved directly from Cro-Magnons.

Checking Concepts

Choose the word or phrase that best answers the question.

8. What is an example of adaptation?
   A) a fossil
   B) gradualism
   C) camouflage
   D) embryo
9. What method provides the most accurate estimate of a fossil's age?
   A) natural selection
   B) radiometric dating
   C) relative dating
   D) camouflage
10. What do homologous structures, vestigial structures, and fossils provide evidence of?
    A) gradualism
    B) food choice
    C) populations
    D) evolution
11. Which model of evolution shows change over a relatively short period of time?
    A) embryology
    B) adaptation
    C) gradualism
    D) punctuated equilibrium
12. What might a series of helpful variations in a species result in?
    A) adaptation
    B) fossils
    C) embryology
    D) climate change

Use the following chart to answer question 13.

Homo habilis
↓
Homo erectus
↓
_____?_____
↓
Homo sapiens

13. Which correctly fills the gap in the line of descent from *Homo habilis?*
    A) Neanderthal
    B) *Australopithecus*
    C) Cro-Magnon human
    D) chimpanzee
14. What is the study of an organism's early development called?
    A) adaptation
    B) relative dating
    C) natural selection
    D) embryology

Science Online bookA.msscience.com/vocabulary_puzzlemaker

Thinking Critically

15. **Predict** what type of bird the foot pictured at right would belong to. Explain your reasoning.

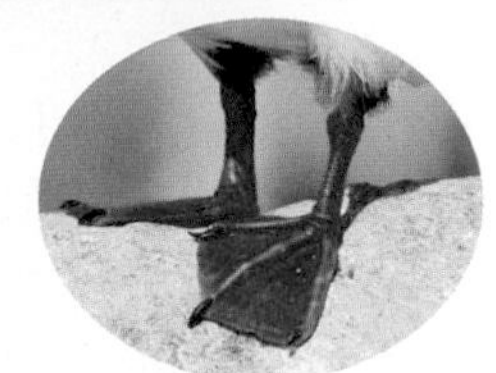

16. **Discuss** how Lamarck and Darwin would have explained the large eyes of an owl.
17. **Explain,** using an example, how a new species of organism could evolve.
18. **Identify** how the color-changing ability of chameleons is an adaptation.
19. **Form a hypothesis** as to why ponds are not overpopulated by frogs in summer. Use the concept of natural selection to help you.
20. **Sequence** Make an events-chain concept map of the events that led Charles Darwin to his theory of evolution by natural selection.

Use the table below to answer question 21.

Chemicals Present in Bacteria	
Species 1	A, G, T, C, L, E, S, H
Species 2	A, G, T, C, L, D, H
Species 3	A,G, T, C, L, D, P, U, S, R, I, V
Species 4	A, G, T, C, L, D, H

21. **Interpret Data** Each letter above represents a chemical found in a species of bacteria. Which species are most closely related?
22. **Discuss** the evidence you would use to determine whether the evolution of a group were best explained by gradualism. How would this differ from a group that followed a punctuated equilibrium model?
23. **Describe** the processes a scientist would use to figure out the age of a fossil.
24. **Evaluate** the possibility for each of the five types of fossils in **Figure 9** to yield a DNA sample. Remember that only biological tissue will contain DNA.

Performance Activities

25. **Collection** With permission, collect fossils from your area and identify them. Show your collection to your class.
26. **Brochure** Assume that you are head of an advertising company. Develop a brochure to explain Darwin's theory of evolution by natural selection.

Applying Math

27. **Relative Age** The rate of radioactive decay is measured in half-lives—the amount of time it takes for one half of a radioactive element to decay. Determine the relative age of a fossil given the following information:
    - Rock layers are undisturbed.
    - The layer below the fossil has potassium-40 with a half-life of 1 million years and only one half of the original potassium is left.
    - The layer above the fossil has carbon-14 with a half-life of 5,730 years and one-sixteenth of the carbon isotope remains.

Use the graph below to answer question 28.

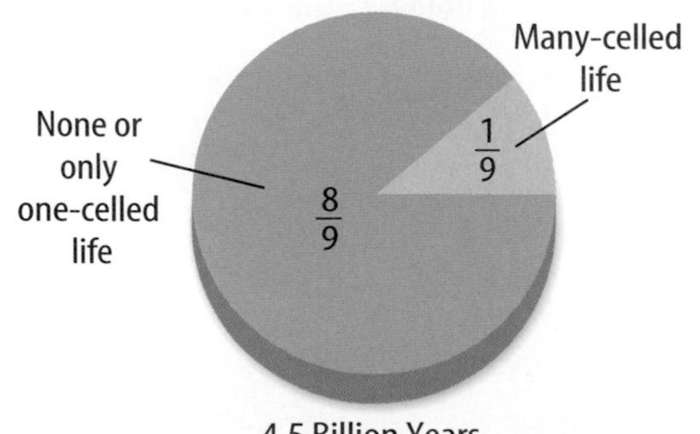

28. **First Appearances** If Earth is 4.5 billion years old, how long ago did the first many-celled life-forms appear?

chapter 6 Standardized Test Practice

Part 1 Multiple Choice

Record your answers on the answer sheet provided by your teacher or on a sheet of paper.

1. A species is a group of organisms
   - **A.** that lives together with similar characteristics.
   - **B.** that shares similar characteristics and can reproduce among themselves to produce fertile offspring.
   - **C.** across a wide area that cannot reproduce.
   - **D.** that chooses mates from among themselves.

2. Which of the following is considered an important factor in natural selection?
   - **A.** limited reproduction
   - **B.** competition for resources
   - **C.** no variations within a population
   - **D.** plentiful food and other resources

3. The marine iguana of the Galápagos Islands enters the ocean and feeds on seaweed. What is this an example of?
   - **A.** adaptation
   - **B.** gradualism
   - **C.** survival of the fittest
   - **D.** acquired characteristic

Use the illustration below to answer question 4.

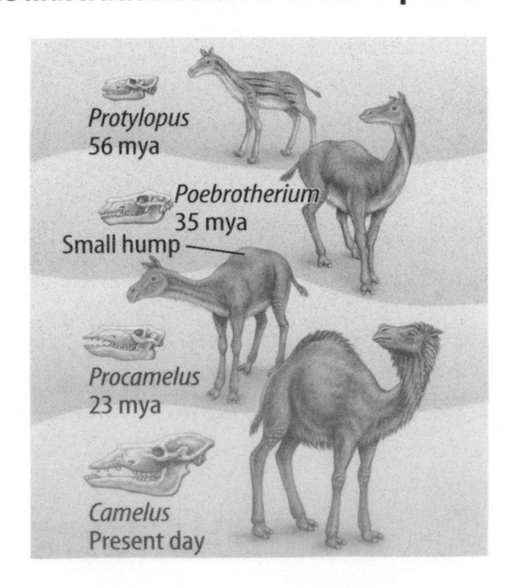

4. According to Lamarck's hypothesis of acquired characteristics, which statement best explains the changes in the camel over time?
   - **A.** All characteristics developed during an individual's lifetime are passed on to offspring.
   - **B.** Characteristics that do not help the animal survive are passed to offspring.
   - **C.** Variation of the species leads to adaptation.
   - **D.** Individuals moving from one area to another carry with them new characteristics.

Use the illustrations below to answer question 5.

5. What, besides competition for food, contributed to the evolution of the species of Darwin's finches?
   - **A.** predation
   - **B.** natural disaster
   - **C.** DNA
   - **D.** variation in beak shapes

6. Some harmless species imitate or mimic a poisonous species as a means for increased survival. What is this an example of?
   - **A.** acquired characteristics
   - **B.** adaptation
   - **C.** variation
   - **D.** geographic isolation

Part 2 Short Response/Grid In

Record your answers on the answer sheet provided by your teacher or on a sheet of paper.

7. How does camouflage benefit a species?

Use the photo below to answer question 8.

8. Describe an environment where the albino lemur would not be at a disadvantage.

9. Variation between members of a species plays an important role in Darwin's theory of evolution. What happens to variation in endangered species where the number of individuals is very low?

10. Describe what happens to an endangered species if a variation provides an advantage for the species. What would happen if the variation resulted in a disadvantage?

11. Using the theory of natural selection, hypothesize why the Cro-Magnon humans survived and the Neanderthals disappeared.

Test-Taking Tip

Never Leave Any Answer Blank Answer each question as best you can. You can receive partial credit for partially correct answers.

Question 16 If you cannot remember all primate characteristics, list as many as you can.

Part 3 Open Ended

Record your answers on a sheet of paper.

12. What are the two groups of early humans that lived about 125,000 years ago in Africa and Europe? Describe their general appearance and characteristics. Compare these characteristics to modern humans.

13. Explain how bacterial resistance to antibiotics is an example of punctuated equilibrium.

14. Why are radioactive elements useful in dating fossils? Does this method improve accuracy over relative dating?

Use the illustrations below to answer question 15.

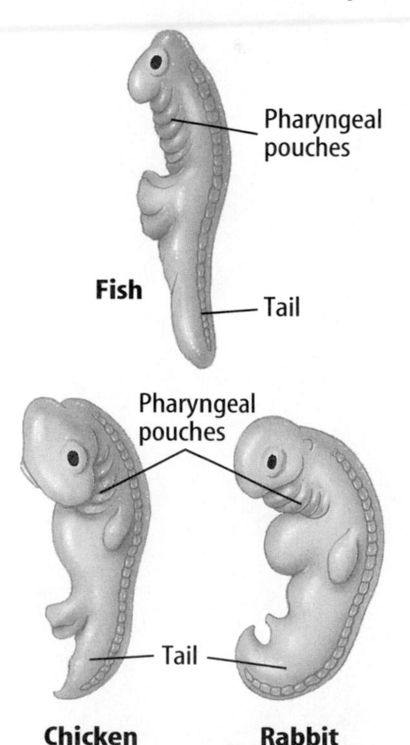

15. Why would scientists study embryos? What features of these three embryos support evolution?

16. How does DNA evidence provide support that primates have a common ancestor?

Student Resources

CONTENTS

Scientific Methods

Scientists use an orderly approach called the scientific method to solve problems. This includes organizing and recording data so others can understand them. Scientists use many variations in this method when they solve problems.

Identify a Question

The first step in a scientific investigation or experiment is to identify a question to be answered or a problem to be solved. For example, you might ask which gasoline is the most efficient.

Gather and Organize Information

After you have identified your question, begin gathering and organizing information. There are many ways to gather information, such as researching in a library, interviewing those knowledgeable about the subject, testing and working in the laboratory and field. Fieldwork is investigations and observations done outside of a laboratory.

Researching Information Before moving in a new direction, it is important to gather the information that already is known about the subject. Start by asking yourself questions to determine exactly what you need to know. Then you will look for the information in various reference sources, like the student is doing in **Figure 1.** Some sources may include textbooks, encyclopedias, government documents, professional journals, science magazines, and the Internet. Always list the sources of your information.

Figure 1 The Internet can be a valuable research tool.

Evaluate Sources of Information Not all sources of information are reliable. You should evaluate all of your sources of information, and use only those you know to be dependable. For example, if you are researching ways to make homes more energy efficient, a site written by the U.S. Department of Energy would be more reliable than a site written by a company that is trying to sell a new type of weatherproofing material. Also, remember that research always is changing. Consult the most current resources available to you. For example, a 1985 resource about saving energy would not reflect the most recent findings.

Sometimes scientists use data that they did not collect themselves, or conclusions drawn by other researchers. This data must be evaluated carefully. Ask questions about how the data were obtained, if the investigation was carried out properly, and if it has been duplicated exactly with the same results. Would you reach the same conclusion from the data? Only when you have confidence in the data can you believe it is true and feel comfortable using it.

Interpret Scientific Illustrations As you research a topic in science, you will see drawings, diagrams, and photographs to help you understand what you read. Some illustrations are included to help you understand an idea that you can't see easily by yourself, like the tiny particles in an atom in **Figure 2.** A drawing helps many people to remember details more easily and provides examples that clarify difficult concepts or give additional information about the topic you are studying. Most illustrations have labels or a caption to identify or to provide more information.

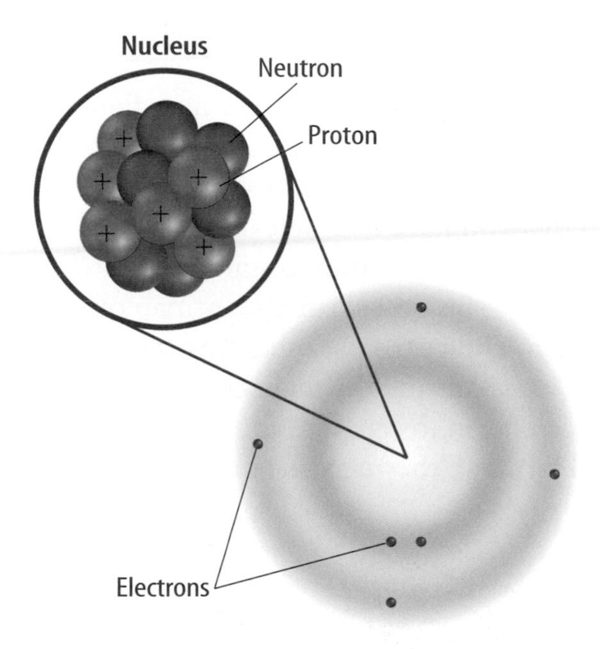

Figure 2 This drawing shows an atom of carbon with its six protons, six neutrons, and six electrons.

Concept Maps One way to organize data is to draw a diagram that shows relationships among ideas (or concepts). A concept map can help make the meanings of ideas and terms more clear, and help you understand and remember what you are studying. Concept maps are useful for breaking large concepts down into smaller parts, making learning easier.

Network Tree A type of concept map that not only shows a relationship, but how the concepts are related is a network tree, shown in **Figure 3.** In a network tree, the words are written in the ovals, while the description of the type of relationship is written across the connecting lines.

When constructing a network tree, write down the topic and all major topics on separate pieces of paper or notecards. Then arrange them in order from general to specific. Branch the related concepts from the major concept and describe the relationship on the connecting line. Continue to more specific concepts until finished.

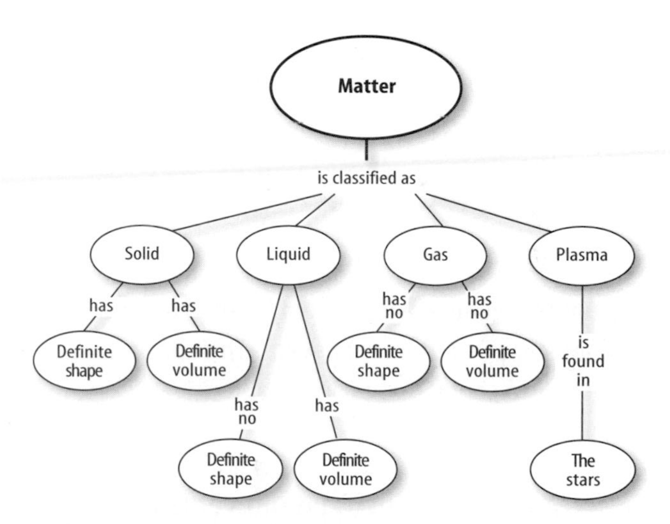

Figure 3 A network tree shows how concepts or objects are related.

Events Chain Another type of concept map is an events chain. Sometimes called a flow chart, it models the order or sequence of items. An events chain can be used to describe a sequence of events, the steps in a procedure, or the stages of a process.

When making an events chain, first find the one event that starts the chain. This event is called the initiating event. Then, find the next event and continue until the outcome is reached, as shown in **Figure 4.**

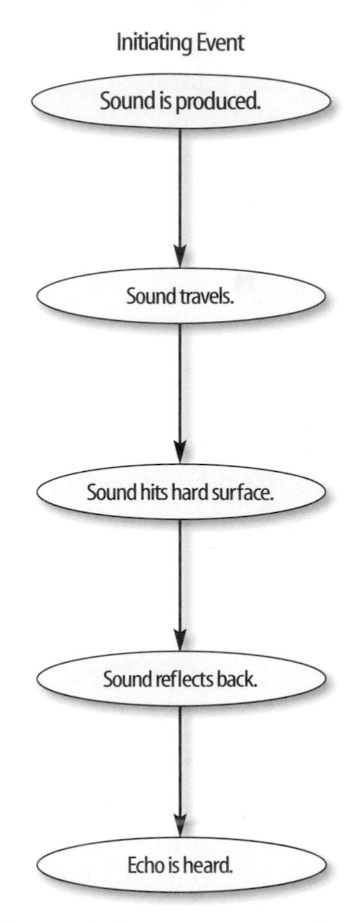

Figure 4 Events-chain concept maps show the order of steps in a process or event. This concept map shows how a sound makes an echo.

Cycle Map A specific type of events chain is a cycle map. It is used when the series of events do not produce a final outcome, but instead relate back to the beginning event, such as in **Figure 5.** Therefore, the cycle repeats itself.

To make a cycle map, first decide what event is the beginning event. This is also called the initiating event. Then list the next events in the order that they occur, with the last event relating back to the initiating event. Words can be written between the events that describe what happens from one event to the next. The number of events in a cycle map can vary, but usually contain three or more events.

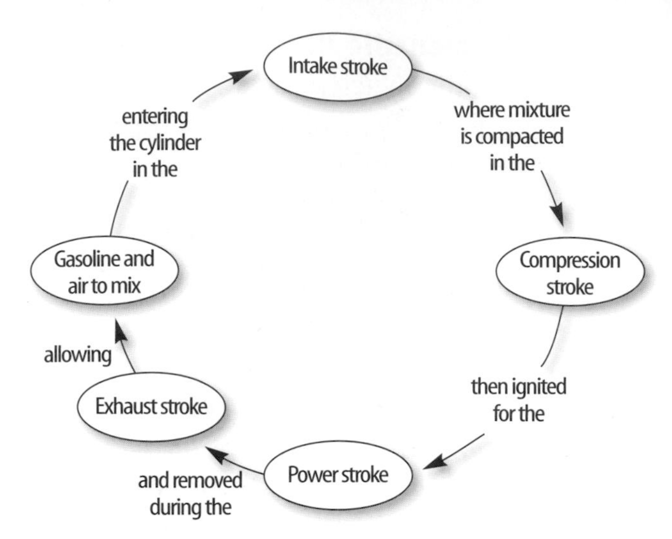

Figure 5 A cycle map shows events that occur in a cycle.

Spider Map A type of concept map that you can use for brainstorming is the spider map. When you have a central idea, you might find that you have a jumble of ideas that relate to it but are not necessarily clearly related to each other. The spider map on sound in **Figure 6** shows that if you write these ideas outside the main concept, then you can begin to separate and group unrelated terms so they become more useful.

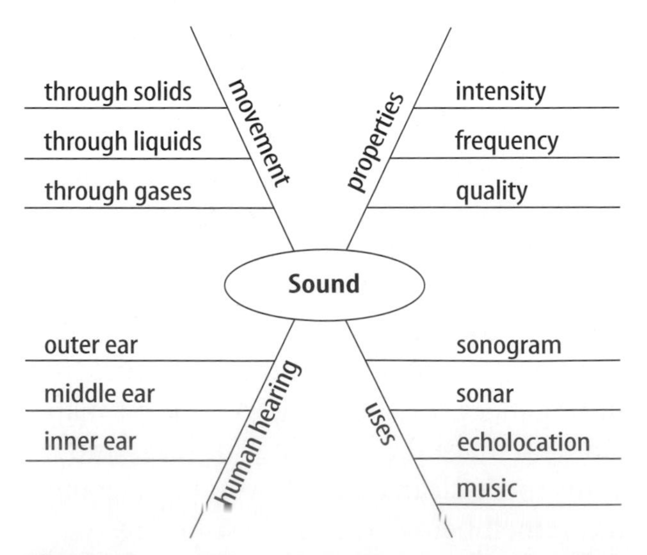

Figure 6 A spider map allows you to list ideas that relate to a central topic but not necessarily to one another.

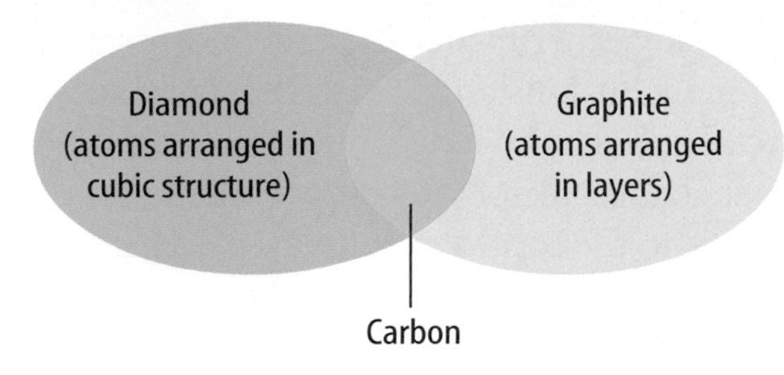

Figure 7 This Venn diagram compares and contrasts two substances made from carbon.

Venn Diagram To illustrate how two subjects compare and contrast you can use a Venn diagram. You can see the characteristics that the subjects have in common and those that they do not, shown in **Figure 7.**

To create a Venn diagram, draw two overlapping ovals that that are big enough to write in. List the characteristics unique to one subject in one oval, and the characteristics of the other subject in the other oval. The characteristics in common are listed in the overlapping section.

Make and Use Tables One way to organize information so it is easier to understand is to use a table. Tables can contain numbers, words, or both.

To make a table, list the items to be compared in the first column and the characteristics to be compared in the first row. The title should clearly indicate the content of the table, and the column or row heads should be clear. Notice that in **Table 1** the units are included.

Table 1 Recyclables Collected During Week

Day of Week	Paper (kg)	Aluminum (kg)	Glass (kg)
Monday	5.0	4.0	12.0
Wednesday	4.0	1.0	10.0
Friday	2.5	2.0	10.0

Make a Model One way to help you better understand the parts of a structure, the way a process works, or to show things too large or small for viewing is to make a model. For example, an atomic model made of a plastic-ball nucleus and pipe-cleaner electron shells can help you visualize how the parts of an atom relate to each other. Other types of models can by devised on a computer or represented by equations.

Form a Hypothesis

A possible explanation based on previous knowledge and observations is called a hypothesis. After researching gasoline types and recalling previous experiences in your family's car you form a hypothesis—our car runs more efficiently because we use premium gasoline. To be valid, a hypothesis has to be something you can test by using an investigation.

Predict When you apply a hypothesis to a specific situation, you predict something about that situation. A prediction makes a statement in advance, based on prior observation, experience, or scientific reasoning. People use predictions to make everyday decisions. Scientists test predictions by performing investigations. Based on previous observations and experiences, you might form a prediction that cars are more efficient with premium gasoline. The prediction can be tested in an investigation.

Design an Experiment A scientist needs to make many decisions before beginning an investigation. Some of these include: how to carry out the investigation, what steps to follow, how to record the data, and how the investigation will answer the question. It also is important to address any safety concerns.

Test the Hypothesis

Now that you have formed your hypothesis, you need to test it. Using an investigation, you will make observations and collect data, or information. This data might either support or not support your hypothesis. Scientists collect and organize data as numbers and descriptions.

Follow a Procedure In order to know what materials to use, as well as how and in what order to use them, you must follow a procedure. **Figure 8** shows a procedure you might follow to test your hypothesis.

Procedure

1. Use regular gasoline for two weeks.
2. Record the number of kilometers between fill-ups and the amount of gasoline used.
3. Switch to premium gasoline for two weeks.
4. Record the number of kilometers between fill-ups and the amount of gasoline used.

Figure 8 A procedure tells you what to do step by step.

Identify and Manipulate Variables and Controls In any experiment, it is important to keep everything the same except for the item you are testing. The one factor you change is called the independent variable. The change that results is the dependent variable. Make sure you have only one independent variable, to assure yourself of the cause of the changes you observe in the dependent variable. For example, in your gasoline experiment the type of fuel is the independent variable. The dependent variable is the efficiency.

Many experiments also have a control—an individual instance or experimental subject for which the independent variable is not changed. You can then compare the test results to the control results. To design a control you can have two cars of the same type. The control car uses regular gasoline for four weeks. After you are done with the test, you can compare the experimental results to the control results.

Collect Data

Whether you are carrying out an investigation or a short observational experiment, you will collect data, as shown in **Figure 9.** Scientists collect data as numbers and descriptions and organize it in specific ways.

Observe Scientists observe items and events, then record what they see. When they use only words to describe an observation, it is called qualitative data. Scientists' observations also can describe how much there is of something. These observations use numbers, as well as words, in the description and are called quantitative data. For example, if a sample of the element gold is described as being "shiny and very dense" the data are qualitative. Quantitative data on this sample of gold might include "a mass of 30 g and a density of 19.3 g/cm^3."

Figure 9 Collecting data is one way to gather information directly.

Figure 10 Record data neatly and clearly so it is easy to understand.

When you make observations you should examine the entire object or situation first, and then look carefully for details. It is important to record observations accurately and completely. Always record your notes immediately as you make them, so you do not miss details or make a mistake when recording results from memory. Never put unidentified observations on scraps of paper. Instead they should be recorded in a notebook, like the one in **Figure 10.** Write your data neatly so you can easily read it later. At each point in the experiment, record your observations and label them. That way, you will not have to determine what the figures mean when you look at your notes later. Set up any tables that you will need to use ahead of time, so you can record any observations right away. Remember to avoid bias when collecting data by not including personal thoughts when you record observations. Record only what you observe.

Estimate Scientific work also involves estimating. To estimate is to make a judgment about the size or the number of something without measuring or counting. This is important when the number or size of an object or population is too large or too difficult to accurately count or measure.

Sample Scientists may use a sample or a portion of the total number as a type of estimation. To sample is to take a small, representative portion of the objects or organisms of a population for research. By making careful observations or manipulating variables within that portion of the group, information is discovered and conclusions are drawn that might apply to the whole population. A poorly chosen sample can be unrepresentative of the whole. If you were trying to determine the rainfall in an area, it would not be best to take a rainfall sample from under a tree.

Measure You use measurements everyday. Scientists also take measurements when collecting data. When taking measurements, it is important to know how to use measuring tools properly. Accuracy also is important.

Length To measure length, the distance between two points, scientists use meters. Smaller measurements might be measured in centimeters or millimeters.

Length is measured using a metric ruler or meter stick. When using a metric ruler, line up the 0-cm mark with the end of the object being measured and read the number of the unit where the object ends. Look at the metric ruler shown in **Figure 11.** The centimeter lines are the long, numbered lines, and the shorter lines are millimeter lines. In this instance, the length would be 4.50 cm.

Figure 11 This metric ruler has centimeter and millimeter divisions.

Mass The SI unit for mass is the kilogram (kg). Scientists can measure mass using units formed by adding metric prefixes to the unit gram (g), such as milligram (mg). To measure mass, you might use a triple-beam balance similar to the one shown in **Figure 12.** The balance has a pan on one side and a set of beams on the other side. Each beam has a rider that slides on the beam.

When using a triple-beam balance, place an object on the pan. Slide the largest rider along its beam until the pointer drops below zero. Then move it back one notch. Repeat the process for each rider proceeding from the larger to smaller until the pointer swings an equal distance above and below the zero point. Sum the masses on each beam to find the mass of the object. Move all riders back to zero when finished.

Instead of putting materials directly on the balance, scientists often take a tare of a container. A tare is the mass of a container into which objects or substances are placed for measuring their masses. To mass objects or substances, find the mass of a clean container. Remove the container from the pan, and place the object or substances in the container. Find the mass of the container with the materials in it. Subtract the mass of the empty container from the mass of the filled container to find the mass of the materials you are using.

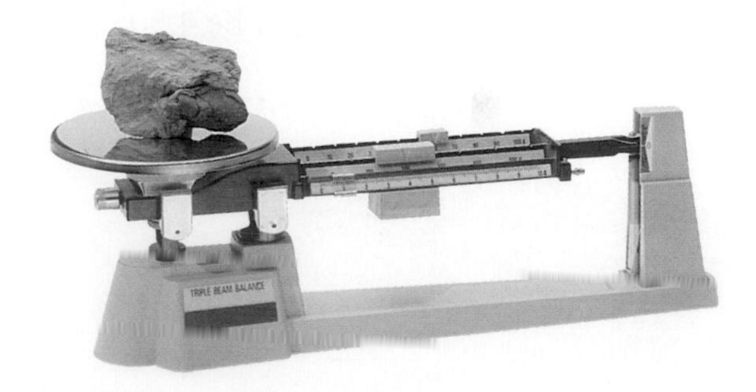

Figure 12 A triple-beam balance is used to determine the mass of an object.

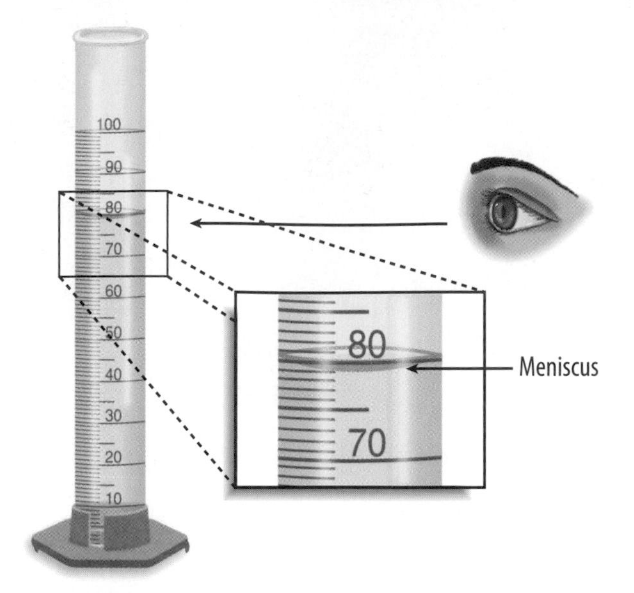

Figure 13 Graduated cylinders measure liquid volume.

Liquid Volume To measure liquids, the unit used is the liter. When a smaller unit is needed, scientists might use a milliliter. Because a milliliter takes up the volume of a cube measuring 1 cm on each side it also can be called a cubic centimeter ($cm^3 = cm \times cm \times cm$).

You can use beakers and graduated cylinders to measure liquid volume. A graduated cylinder, shown in **Figure 13,** is marked from bottom to top in milliliters. In lab, you might use a 10-mL graduated cylinder or a 100-mL graduated cylinder. When measuring liquids, notice that the liquid has a curved surface. Look at the surface at eye level, and measure the bottom of the curve. This is called the meniscus. The graduated cylinder in **Figure 13** contains 79.0 mL, or 79.0 cm^3, of a liquid.

Temperature Scientists often measure temperature using the Celsius scale. Pure water has a freezing point of 0°C and boiling point of 100°C. The unit of measurement is degree Celsius. Two other scales often used are the Fahrenheit and Kelvin scales.

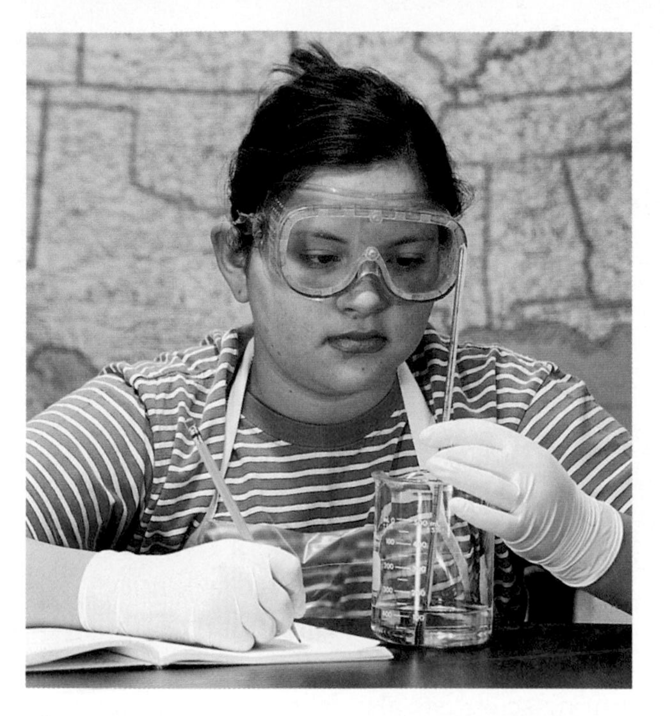

Figure 14 A thermometer measures the temperature of an object.

Scientists use a thermometer to measure temperature. Most thermometers in a laboratory are glass tubes with a bulb at the bottom end containing a liquid such as colored alcohol. The liquid rises or falls with a change in temperature. To read a glass thermometer like the thermometer in **Figure 14,** rotate it slowly until a red line appears. Read the temperature where the red line ends.

Form Operational Definitions An operational definition defines an object by how it functions, works, or behaves. For example, when you are playing hide and seek and a tree is home base, you have created an operational definition for a tree.

Objects can have more than one operational definition. For example, a ruler can be defined as a tool that measures the length of an object (how it is used). It can also be a tool with a series of marks used as a standard when measuring (how it works).

Analyze the Data

To determine the meaning of your observations and investigation results, you will need to look for patterns in the data. Then you must think critically to determine what the data mean. Scientists use several approaches when they analyze the data they have collected and recorded. Each approach is useful for identifying specific patterns.

Interpret Data The word *interpret* means "to explain the meaning of something." When analyzing data from an experiment, try to find out what the data show. Identify the control group and the test group to see whether or not changes in the independent variable have had an effect. Look for differences in the dependent variable between the control and test groups.

Classify Sorting objects or events into groups based on common features is called classifying. When classifying, first observe the objects or events to be classified. Then select one feature that is shared by some members in the group, but not by all. Place those members that share that feature in a subgroup. You can classify members into smaller and smaller subgroups based on characteristics. Remember that when you classify, you are grouping objects or events for a purpose. Keep your purpose in mind as you select the features to form groups and subgroups.

Compare and Contrast Observations can be analyzed by noting the similarities and differences between two more objects or events that you observe. When you look at objects or events to see how they are similar, you are comparing them. Contrasting is looking for differences in objects or events.

Recognize Cause and Effect A cause is a reason for an action or condition. The effect is that action or condition. When two events happen together, it is not necessarily true that one event caused the other. Scientists must design a controlled investigation to recognize the exact cause and effect.

Draw Conclusions

When scientists have analyzed the data they collected, they proceed to draw conclusions about the data. These conclusions are sometimes stated in words similar to the hypothesis that you formed earlier. They may confirm a hypothesis, or lead you to a new hypothesis.

Infer Scientists often make inferences based on their observations. An inference is an attempt to explain observations or to indicate a cause. An inference is not a fact, but a logical conclusion that needs further investigation. For example, you may infer that a fire has caused smoke. Until you investigate, however, you do not know for sure.

Apply When you draw a conclusion, you must apply those conclusions to determine whether the data supports the hypothesis. If your data do not support your hypothesis, it does not mean that the hypothesis is wrong. It means only that the result of the investigation did not support the hypothesis. Maybe the experiment needs to be redesigned, or some of the initial observations on which the hypothesis was based were incomplete or biased. Perhaps more observation or research is needed to refine your hypothesis. A successful investigation does not always come out the way you originally predicted.

Avoid Bias Sometimes a scientific investigation involves making judgments. When you make a judgment, you form an opinion. It is important to be honest and not to allow any expectations of results to bias your judgments. This is important throughout the entire investigation, from researching to collecting data to drawing conclusions.

Communicate

The communication of ideas is an important part of the work of scientists. A discovery that is not reported will not advance the scientific community's understanding or knowledge. Communication among scientists also is important as a way of improving their investigations.

Scientists communicate in many ways, from writing articles in journals and magazines that explain their investigations and experiments, to announcing important discoveries on television and radio. Scientists also share ideas with colleagues on the Internet or present them as lectures, like the student is doing in **Figure 15.**

Figure 15 A student communicates to his peers about his investigation.

SAFETY SYMBOLS	HAZARD	EXAMPLES	PRECAUTION	REMEDY
DISPOSAL	Special disposal procedures need to be followed.	certain chemicals, living organisms	Do not dispose of these materials in the sink or trash can.	Dispose of wastes as directed by your teacher.
BIOLOGICAL	Organisms or other biological materials that might be harmful to humans	bacteria, fungi, blood, unpreserved tissues, plant materials	Avoid skin contact with these materials. Wear mask or gloves.	Notify your teacher if you suspect contact with material. Wash hands thoroughly.
EXTREME TEMPERATURE	Objects that can burn skin by being too cold or too hot	boiling liquids, hot plates, dry ice, liquid nitrogen	Use proper protection when handling.	Go to your teacher for first aid.
SHARP OBJECT	Use of tools or glassware that can easily puncture or slice skin	razor blades, pins, scalpels, pointed tools, dissecting probes, broken glass	Practice common-sense behavior and follow guidelines for use of the tool.	Go to your teacher for first aid.
FUME	Possible danger to respiratory tract from fumes	ammonia, acetone, nail polish remover, heated sulfur, moth balls	Make sure there is good ventilation. Never smell fumes directly. Wear a mask.	Leave foul area and notify your teacher immediately.
ELECTRICAL	Possible danger from electrical shock or burn	improper grounding, liquid spills, short circuits, exposed wires	Double-check setup with teacher. Check condition of wires and apparatus.	Do not attempt to fix electrical problems. Notify your teacher immediately.
IRRITANT	Substances that can irritate the skin or mucous membranes of the respiratory tract	pollen, moth balls, steel wool, fiberglass, potassium permanganate	Wear dust mask and gloves. Practice extra care when handling these materials.	Go to your teacher for first aid.
CHEMICAL	Chemicals can react with and destroy tissue and other materials	bleaches such as hydrogen peroxide; acids such as sulfuric acid, hydrochloric acid; bases such as ammonia, sodium hydroxide	Wear goggles, gloves, and an apron.	Immediately flush the affected area with water and notify your teacher.
TOXIC	Substance may be poisonous if touched, inhaled, or swallowed.	mercury, many metal compounds, iodine, poinsettia plant parts	Follow your teacher's instructions.	Always wash hands thoroughly after use. Go to your teacher for first aid.
FLAMMABLE	Flammable chemicals may be ignited by open flame, spark, or exposed heat.	alcohol, kerosene, potassium permanganate	Avoid open flames and heat when using flammable chemicals.	Notify your teacher immediately. Use fire safety equipment if applicable.
OPEN FLAME	Open flame in use, may cause fire.	hair, clothing, paper, synthetic materials	Tie back hair and loose clothing. Follow teacher's instruction on lighting and extinguishing flames.	Notify your teacher immediately. Use fire safety equipment if applicable.

Eye Safety
Proper eye protection should be worn at all times by anyone performing or observing science activities.

Clothing Protection
This symbol appears when substances could stain or burn clothing.

Animal Safety
This symbol appears when safety of animals and students must be ensured.

Handwashing
After the lab, wash hands with soap and water before removing goggles.

Safety in the Science Laboratory

The science laboratory is a safe place to work if you follow standard safety procedures. Being responsible for your own safety helps to make the entire laboratory a safer place for everyone. When performing any lab, read and apply the caution statements and safety symbol listed at the beginning of the lab.

General Safety Rules

1. Obtain your teacher's permission to begin all investigations and use laboratory equipment.
2. Study the procedure. Ask your teacher any questions. Be sure you understand safety symbols shown on the page.
3. Notify your teacher about allergies or other health conditions which can affect your participation in a lab.
4. Learn and follow use and safety procedures for your equipment. If unsure, ask your teacher.
5. Never eat, drink, chew gum, apply cosmetics, or do any personal grooming in the lab. Never use lab glassware as food or drink containers. Keep your hands away from your face and mouth.
6. Know the location and proper use of the safety shower, eye wash, fire blanket, and fire alarm.

Prevent Accidents

1. Use the safety equipment provided to you. Goggles and a safety apron should be worn during investigations.
2. Do NOT use hair spray, mousse, or other flammable hair products. Tie back long hair and tie down loose clothing.
3. Do NOT wear sandals or other open-toed shoes in the lab.
4. Remove jewelry on hands and wrists. Loose jewelry, such as chains and long necklaces, should be removed to prevent them from getting caught in equipment.
5. Do not taste any substances or draw any material into a tube with your mouth.
6. Proper behavior is expected in the lab. Practical jokes and fooling around can lead to accidents and injury.
7. Keep your work area uncluttered.

Laboratory Work

1. Collect and carry all equipment and materials to your work area before beginning a lab.
2. Remain in your own work area unless given permission by your teacher to leave it.

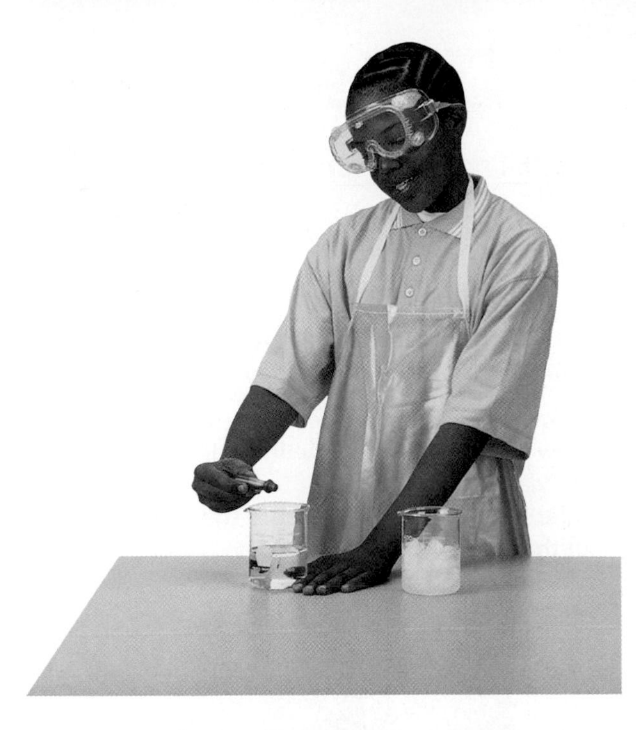

3. Always slant test tubes away from yourself and others when heating them, adding substances to them, or rinsing them.
4. If instructed to smell a substance in a container, hold the container a short distance away and fan vapors towards your nose.
5. Do NOT substitute other chemicals/substances for those in the materials list unless instructed to do so by your teacher.
6. Do NOT take any materials or chemicals outside of the laboratory.
7. Stay out of storage areas unless instructed to be there and supervised by your teacher.

Laboratory Cleanup

1. Turn off all burners, water, and gas, and disconnect all electrical devices.
2. Clean all pieces of equipment and return all materials to their proper places.
3. Dispose of chemicals and other materials as directed by your teacher. Place broken glass and solid substances in the proper containers. Never discard materials in the sink.
4. Clean your work area.
5. Wash your hands with soap and water thoroughly BEFORE removing your goggles.

Emergencies

1. Report any fire, electrical shock, glassware breakage, spill, or injury, no matter how small, to your teacher immediately. Follow his or her instructions.
2. If your clothing should catch fire, STOP, DROP, and ROLL. If possible, smother it with the fire blanket or get under a safety shower. NEVER RUN.
3. If a fire should occur, turn off all gas and leave the room according to established procedures.
4. In most instances, your teacher will clean up spills. Do NOT attempt to clean up spills unless you are given permission and instructions to do so.
5. If chemicals come into contact with your eyes or skin, notify your teacher immediately. Use the eyewash or flush your skin or eyes with large quantities of water.
6. The fire extinguisher and first-aid kit should only be used by your teacher unless it is an extreme emergency and you have been given permission.
7. If someone is injured or becomes ill, only a professional medical provider or someone certified in first aid should perform first-aid procedures.

From Your Kitchen, Junk Drawer, or Yard

1 Your Daily Drink

Real-World Question
How much liquid do you consume in a day?

Possible Materials
- 500-mL measuring cup
- calculator

Procedure
1. When you drink a bottle or can of juice, soda, water, or other beverage, look on the label of the container to find the volume in milliliters.
2. Record the volumes of all the canned and bottled drinks you consume in one day in your Science Journal.
3. Use a measuring cup to measure the liquids that you pour from larger containers. Record these volumes in your Science Journal.
4. Add up the volumes of all the drinks you consumed during the day.

Conclude and Apply
1. How much liquid did you drink during the day?
2. Infer how you would measure the mass of the foods you ate in one day.

2 Cell Sizes

Real-World Question
How do different cells compare in size?

Possible Materials
- meterstick
- metric ruler
- white paper
- pencil
- pen
- masking tape

Procedure
1. Make a dot on a white sheet of paper with a pencil.
2. Use a metric ruler to make a second dot 1 mm away from the first dot. This distance represents the average length of a bacteria cell.
3. Measure a distance 8 mm away from the first dot and make a third dot. This distance represents the average length of a red blood cell.
4. Mark a spot on the floor with a piece of tape and use the meterstick to measure a distance of 7 m. Mark this distance with a second piece of tape. This distance represents the average length of an amoeba cell.

Conclude and Apply
1. The distance between the first and second dot is 1,000 times longer than the actual size of a bacterium cell. Calculate the length of an actual bacterium cell.
2. A large chicken egg is just one cell, and it is 100 times longer than an amoeba cell. Using your measurement from step 4, calculate the distance you would have to measure to represent the average length of a hen's egg.

Adult supervision required for all labs.

3 Expanding Eggs

Real-World Question

How can you observe liquids passing through a cell membrane?

Possible Materials

- glass jar with lid
- white vinegar
- medium chicken egg
- tape measure or string and ruler
- tongs
- measuring cup

Procedure

1. Obtain a glass jar with a lid and a medium egg.
2. Make certain your egg easily fits into your jar.
3. Measure the circumference of your egg.
4. Pour 250 mL of white vinegar into the jar.
5. Carefully place your egg in the jar so that it is submerged in the vinegar. Be careful not to crack or break the egg.
6. Observe your egg each day for three days. Measure the circumference of the egg after three days.

Conclude and Apply

1. Describe the changes that happened to your egg.
2. Infer why the egg's circumference changed. *HINT: A hen's egg is a single cell.*

4 Putting Down Roots

Real-World Question

Can cells from a plant's stem produce root cells for a new plant?

Possible Materials

- houseplant
- scissors
- metric ruler
- glasses or jars (3)
- water
- magnifying lens

Procedure

1. Examine the stems of a houseplant, such as *Pothos,* and locate a node on three different stems. A node looks like a small bump.
2. Cut 3 stems off the plant at a 45° angle about 3–4 mm below the node.
3. Place the end of each stem into a separate glass of water and observe them for a week.

Conclude and Apply

1. Describe what happened to the ends of the stems.
2. Infer how plant stem cells can produce root cells.

Adult supervision required for all labs.

5 Do your ears hang low?

Real-World Question

Is ear lobe attachment a dominant or recessive trait?

Possible Materials

- Science Journal
- pencil
- pen
- calculator

Procedure

1. Ask your friends, family members, and other people you know if their ear lobes are attached or free.
2. Try to interview as many people as possible to collect a large sample of data.
3. Record the number of people who have attached ear lobes and the number with free ear lobes in your Science Journal.

Conclude and Apply

1. Calculate the percentage of people who have attached ear lobes and the percentage of people who have free ear lobes.
2. Infer whether or not attached ear lobes is a dominant or recessive trait. Do research to confirm your results.
3. Infer how many children in a family would have free ear lobes if their parents had attached ear lobes.

6 Frozen Fossils

Real-World Question

How can we model the formation of an amber fossil?

Possible Materials

- small glass jar with lid
- honey
- ruler
- dead insect or spider or small rubber insect or spider
- freezer

Procedure

1. Thoroughly wash and dry a small glass jar and its lid.
2. Pour 3 cm of honey into the jar. Do not pour honey down the sides of the jar.
3. Search for a dead insect or spider around your home or school and drop it into the center of the honey's surface.
4. Pour another 3 cm of honey into the jar to cover the organism.
5. Place the jar in the freezer overnight.

Conclude and Apply

1. Explain how you modeled the formation of an amber fossil.
2. Infer how amber fossils help scientists observe adaptations of organisms over time.

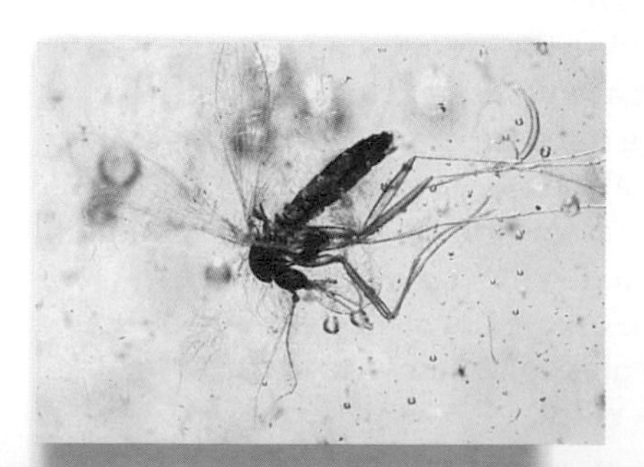

Adult supervision required for all labs.

Computer Skills

People who study science rely on computers, like the one in **Figure 16,** to record and store data and to analyze results from investigations. Whether you work in a laboratory or just need to write a lab report with tables, good computer skills are a necessity.

Using the computer comes with responsibility. Issues of ownership, security, and privacy can arise. Remember, if you did not author the information you are using, you must provide a source for your information. Also, anything on a computer can be accessed by others. Do not put anything on the computer that you would not want everyone to know. To add more security to your work, use a password.

Use a Word Processing Program

A computer program that allows you to type your information, change it as many times as you need to, and then print it out is called a word processing program. Word processing programs also can be used to make tables.

Figure 16 A computer will make reports neater and more professional looking.

Learn the Skill To start your word processing program, a blank document, sometimes called "Document 1," appears on the screen. To begin, start typing. To create a new document, click the *New* button on the standard tool bar. These tips will help you format the document.

- The program will automatically move to the next line; press *Enter* if you wish to start a new paragraph.
- Symbols, called non-printing characters, can be hidden by clicking the *Show/Hide* button on your toolbar.
- To insert text, move the cursor to the point where you want the insertion to go, click on the mouse once, and type the text.
- To move several lines of text, select the text and click the *Cut* button on your toolbar. Then position your cursor in the location that you want to move the cut text and click *Paste.* If you move to the wrong place, click *Undo.*
- The spell check feature does not catch words that are misspelled to look like other words, like "cold" instead of "gold." Always reread your document to catch all spelling mistakes.
- To learn about other word processing methods, read the user's manual or click on the *Help* button.
- You can integrate databases, graphics, and spreadsheets into documents by copying from another program and pasting it into your document, or by using desktop publishing (DTP). DTP software allows you to put text and graphics together to finish your document with a professional look. This software varies in how it is used and its capabilities.

Use a Database

A collection of facts stored in a computer and sorted into different fields is called a database. A database can be reorganized in any way that suits your needs.

Learn the Skill A computer program that allows you to create your own database is a database management system (DBMS). It allows you to add, delete, or change information. Take time to get to know the features of your database software.

- Determine what facts you would like to include and research to collect your information.
- Determine how you want to organize the information.
- Follow the instructions for your particular DBMS to set up fields. Then enter each item of data in the appropriate field.
- Follow the instructions to sort the information in order of importance.
- Evaluate the information in your database, and add, delete, or change as necessary.

Use the Internet

The Internet is a global network of computers where information is stored and shared. To use the Internet, like the students in **Figure 17,** you need a modem to connect your computer to a phone line and an Internet Service Provider account.

Learn the Skill To access internet sites and information, use a "Web browser," which lets you view and explore pages on the World Wide Web. Each page is its own site, and each site has its own address, called a URL. Once you have found a Web browser, follow these steps for a search (this also is how you search a database).

Figure 17 The Internet allows you to search a global network for a variety of information.

- Be as specific as possible. If you know you want to research "gold," don't type in "elements." Keep narrowing your search until you find what you want.
- Web sites that end in *.com* are commercial Web sites; *.org, .edu,* and *.gov* are non-profit, educational, or government Web sites.
- Electronic encyclopedias, almanacs, indexes, and catalogs will help locate and select relevant information.
- Develop a "home page" with relative ease. When developing a Web site, NEVER post pictures or disclose personal information such as location, names, or phone numbers. Your school or community usually can host your Web site. A basic understanding of HTML (hypertext mark-up language), the language of Web sites, is necessary. Software that creates HTML code is called authoring software, and can be downloaded free from many Web sites. This software allows text and pictures to be arranged as the software is writing the HTML code.

Use a Spreadsheet

A spreadsheet, shown in **Figure 18,** can perform mathematical functions with any data arranged in columns and rows. By entering a simple equation into a cell, the program can perform operations in specific cells, rows, or columns.

Learn the Skill Each column (vertical) is assigned a letter, and each row (horizontal) is assigned a number. Each point where a row and column intersect is called a cell, and is labeled according to where it is located—Column A, Row 1 (A1).

- Decide how to organize the data, and enter it in the correct row or column.
- Spreadsheets can use standard formulas or formulas can be customized to calculate cells.
- To make a change, click on a cell to make it activate, and enter the edited data or formula.
- Spreadsheets also can display your results in graphs. Choose the style of graph that best represents the data.

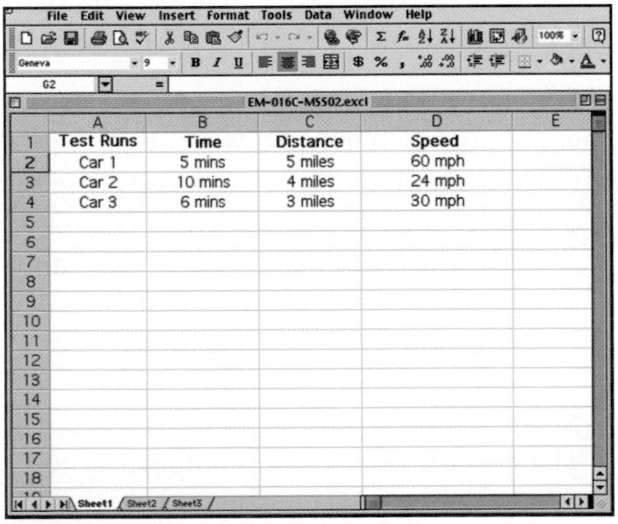

	A	B	C	D	E
1	Test Runs	Time	Distance	Speed	
2	Car 1	5 mins	5 miles	60 mph	
3	Car 2	10 mins	4 miles	24 mph	
4	Car 3	6 mins	3 miles	30 mph	

Figure 18 A spreadsheet allows you to perform mathematical operations on your data.

Use Graphics Software

Adding pictures, called graphics, to your documents is one way to make your documents more meaningful and exciting. This software adds, edits, and even constructs graphics. There is a variety of graphics software programs. The tools used for drawing can be a mouse, keyboard, or other specialized devices. Some graphics programs are simple. Others are complicated, called computer-aided design (CAD) software.

Learn the Skill It is important to have an understanding of the graphics software being used before starting. The better the software is understood, the better the results. The graphics can be placed in a word-processing document.

- Clip art can be found on a variety of internet sites, and on CDs. These images can be copied and pasted into your document.
- When beginning, try editing existing drawings, then work up to creating drawings.
- The images are made of tiny rectangles of color called pixels. Each pixel can be altered.
- Digital photography is another way to add images. The photographs in the memory of a digital camera can be downloaded into a computer, then edited and added to the document.
- Graphics software also can allow animation. The software allows drawings to have the appearance of movement by connecting basic drawings automatically. This is called in-betweening, or tweening.
- Remember to save often.

Presentation Skills

Develop Multimedia Presentations

Most presentations are more dynamic if they include diagrams, photographs, videos, or sound recordings, like the one shown in **Figure 19.** A multimedia presentation involves using stereos, overhead projectors, televisions, computers, and more.

Learn the Skill Decide the main points of your presentation, and what types of media would best illustrate those points.

- Make sure you know how to use the equipment you are working with.
- Practice the presentation using the equipment several times.
- Enlist the help of a classmate to push play or turn lights out for you. Be sure to practice your presentation with him or her.
- If possible, set up all of the equipment ahead of time, and make sure everything is working properly.

Figure 19 These students are engaging the audience using a variety of tools.

Computer Presentations

There are many different interactive computer programs that you can use to enhance your presentation. Most computers have a compact disc (CD) drive that can play both CDs and digital video discs (DVDs). Also, there is hardware to connect a regular CD, DVD, or VCR. These tools will enhance your presentation.

Another method of using the computer to aid in your presentation is to develop a slide show using a computer program. This can allow movement of visuals at the presenter's pace, and can allow for visuals to build on one another.

Learn the Skill In order to create multimedia presentations on a computer, you need to have certain tools. These may include traditional graphic tools and drawing programs, animation programs, and authoring systems that tie everything together. Your computer will tell you which tools it supports. The most important step is to learn about the tools that you will be using.

- Often, color and strong images will convey a point better than words alone. Use the best methods available to convey your point.
- As with other presentations, practice many times.
- Practice your presentation with the tools you and any assistants will be using.
- Maintain eye contact with the audience. The purpose of using the computer is not to prompt the presenter, but to help the audience understand the points of the presentation.

Math Review

Use Fractions

A fraction compares a part to a whole. In the fraction $\frac{2}{3}$, the 2 represents the part and is the numerator. The 3 represents the whole and is the denominator.

Reduce Fractions To reduce a fraction, you must find the largest factor that is common to both the numerator and the denominator, the greatest common factor (GCF). Divide both numbers by the GCF. The fraction has then been reduced, or it is in its simplest form.

Example Twelve of the 20 chemicals in the science lab are in powder form. What fraction of the chemicals used in the lab are in powder form?

Step 1 Write the fraction.

$$\frac{\text{part}}{\text{whole}} = \frac{12}{20}$$

Step 2 To find the GCF of the numerator and denominator, list all of the factors of each number.

Factors of 12: 1, 2, 3, 4, 6, 12 (the numbers that divide evenly into 12)

Factors of 20: 1, 2, 4, 5, 10, 20 (the numbers that divide evenly into 20)

Step 3 List the common factors.

1, 2, 4.

Step 4 Choose the greatest factor in the list.

The GCF of 12 and 20 is 4.

Step 5 Divide the numerator and denominator by the GCF.

$$\frac{12 \div 4}{20 \div 4} = \frac{3}{5}$$

In the lab, $\frac{3}{5}$ of the chemicals are in powder form.

Practice Problem At an amusement park, 66 of 90 rides have a height restriction. What fraction of the rides, in its simplest form, has a height restriction?

Add and Subtract Fractions To add or subtract fractions with the same denominator, add or subtract the numerators and write the sum or difference over the denominator. After finding the sum or difference, find the simplest form for your fraction.

Example 1 In the forest outside your house, $\frac{1}{8}$ of the animals are rabbits, $\frac{3}{8}$ are squirrels, and the remainder are birds and insects. How many are mammals?

Step 1 Add the numerators.

$$\frac{1}{8} + \frac{3}{8} = \frac{(1 + 3)}{8} = \frac{4}{8}$$

Step 2 Find the GCF.

$$\frac{4}{8} \text{ (GCF, 4)}$$

Step 3 Divide the numerator and denominator by the GCF.

$$\frac{4}{4} = 1, \ \frac{8}{4} = 2$$

$\frac{1}{2}$ of the animals are mammals.

Example 2 If $\frac{7}{16}$ of the Earth is covered by freshwater, and $\frac{1}{16}$ of that is in glaciers, how much freshwater is not frozen?

Step 1 Subtract the numerators.

$$\frac{7}{16} - \frac{1}{16} = \frac{(7 - 1)}{16} = \frac{6}{16}$$

Step 2 Find the GCF.

$$\frac{6}{16} \text{ (GCF, 2)}$$

Step 3 Divide the numerator and denominator by the GCF.

$$\frac{6}{2} = 3, \ \frac{16}{2} = 8$$

$\frac{3}{8}$ of the freshwater is not frozen.

Practice Problem A bicycle rider is going 15 km/h for $\frac{4}{9}$ of his ride, 10 km/h for $\frac{2}{9}$ of his ride, and 8 km/h for the remainder of the ride. How much of his ride is he going over 8 km/h?

Unlike Denominators To add or subtract fractions with unlike denominators, first find the least common denominator (LCD). This is the smallest number that is a common multiple of both denominators. Rename each fraction with the LCD, and then add or subtract. Find the simplest form if necessary.

Example 1 A chemist makes a paste that is $\frac{1}{2}$ table salt (NaCl), $\frac{1}{3}$ sugar ($C_6H_{12}O_6$), and the rest water (H_2O). How much of the paste is a solid?

Step 1 Find the LCD of the fractions.

$\frac{1}{2} + \frac{1}{3}$ (LCD, 6)

Step 2 Rename each numerator and each denominator with the LCD.

$1 \times 3 = 3, \; 2 \times 3 = 6$

$1 \times 2 = 2, \; 3 \times 2 = 6$

Step 3 Add the numerators.

$\frac{3}{6} + \frac{2}{6} = \frac{(3 + 2)}{6} = \frac{5}{6}$

$\frac{5}{6}$ of the paste is a solid.

Example 2 The average precipitation in Grand Junction, CO, is $\frac{7}{10}$ inch in November, and $\frac{3}{5}$ inch in December. What is the total average precipitation?

Step 1 Find the LCD of the fractions.

$\frac{7}{10} + \frac{3}{5}$ (LCD, 10)

Step 2 Rename each numerator and each denominator with the LCD.

$7 \times 1 = 7, \; 10 \times 1 = 10$

$3 \times 2 = 6, \; 5 \times 2 = 10$

Step 3 Add the numerators.

$\frac{7}{10} + \frac{6}{10} = \frac{(7 + 6)}{10} = \frac{13}{10}$

$\frac{13}{10}$ inches total precipitation, or $1\frac{3}{10}$ inches.

Practice Problem On an electric bill, about $\frac{1}{8}$ of the energy is from solar energy and about $\frac{1}{10}$ is from wind power. How much of the total bill is from solar energy and wind power combined?

Example 3 In your body, $\frac{7}{10}$ of your muscle contractions are involuntary (cardiac and smooth muscle tissue). Smooth muscle makes $\frac{3}{15}$ of your muscle contractions. How many of your muscle contractions are made by cardiac muscle?

Step 1 Find the LCD of the fractions.

$\frac{7}{10} - \frac{3}{15}$ (LCD, 30)

Step 2 Rename each numerator and each denominator with the LCD.

$7 \times 3 = 21, \; 10 \times 3 = 30$

$3 \times 2 = 6, \; 15 \times 2 = 30$

Step 3 Subtract the numerators.

$\frac{21}{30} - \frac{6}{30} = \frac{(21 - 6)}{30} = \frac{15}{30}$

Step 4 Find the GCF.

$\frac{15}{30}$ (GCF, 15)

$\frac{1}{2}$

$\frac{1}{2}$ of all muscle contractions are cardiac muscle.

Example 4 Tony wants to make cookies that call for $\frac{3}{4}$ of a cup of flour, but he only has $\frac{1}{3}$ of a cup. How much more flour does he need?

Step 1 Find the LCD of the fractions.

$\frac{3}{4} - \frac{1}{3}$ (LCD, 12)

Step 2 Rename each numerator and each denominator with the LCD.

$3 \times 3 = 9, \; 4 \times 3 = 12$

$1 \times 4 = 4, \; 3 \times 4 = 12$

Step 3 Subtract the numerators.

$\frac{9}{12} - \frac{4}{12} = \frac{(9 - 4)}{12} = \frac{5}{12}$

$\frac{5}{12}$ of a cup of flour.

Practice Problem Using the information provided to you in Example 3 above, determine how many muscle contractions are voluntary (skeletal muscle).

Multiply Fractions To multiply with fractions, multiply the numerators and multiply the denominators. Find the simplest form if necessary.

Example Multiply $\frac{3}{5}$ by $\frac{1}{3}$.

Step 1 Multiply the numerators and denominators.

$$\frac{3}{5} \times \frac{1}{3} = \frac{(3 \times 1)}{(5 \times 3)} = \frac{3}{15}$$

Step 2 Find the GCF.

$\frac{3}{15}$ (GCF, 3)

Step 3 Divide the numerator and denominator by the GCF.

$\frac{3}{3} = 1, \ \frac{15}{3} = 5$

$\frac{1}{5}$

$\frac{3}{5}$ multiplied by $\frac{1}{3}$ is $\frac{1}{5}$.

Practice Problem Multiply $\frac{3}{14}$ by $\frac{5}{16}$.

Find a Reciprocal Two numbers whose product is 1 are called multiplicative inverses, or reciprocals.

Example Find the reciprocal of $\frac{3}{8}$.

Step 1 Inverse the fraction by putting the denominator on top and the numerator on the bottom.

$\frac{8}{3}$

The reciprocal of $\frac{3}{8}$ is $\frac{8}{3}$.

Practice Problem Find the reciprocal of $\frac{4}{9}$.

Divide Fractions To divide one fraction by another fraction, multiply the dividend by the reciprocal of the divisor. Find the simplest form if necessary.

Example 1 Divide $\frac{1}{9}$ by $\frac{1}{3}$.

Step 1 Find the reciprocal of the divisor.

The reciprocal of $\frac{1}{3}$ is $\frac{3}{1}$.

Step 2 Multiply the dividend by the reciprocal of the divisor.

$$\frac{\frac{1}{9}}{\frac{1}{3}} = \frac{1}{9} \times \frac{3}{1} = \frac{(1 \times 3)}{(9 \times 1)} = \frac{3}{9}$$

Step 3 Find the GCF.

$\frac{3}{9}$ (GCF, 3)

Step 4 Divide the numerator and denominator by the GCF.

$\frac{3}{3} = 1, \ \frac{9}{3} = 3$

$\frac{1}{3}$

$\frac{1}{9}$ divided by $\frac{1}{3}$ is $\frac{1}{3}$.

Example 2 Divide $\frac{3}{5}$ by $\frac{1}{4}$.

Step 1 Find the reciprocal of the divisor.

The reciprocal of $\frac{1}{4}$ is $\frac{4}{1}$.

Step 2 Multiply the dividend by the reciprocal of the divisor.

$$\frac{\frac{3}{5}}{\frac{1}{4}} = \frac{3}{5} \times \frac{4}{1} = \frac{(3 \times 4)}{(5 \times 1)} = \frac{12}{5}$$

$\frac{3}{5}$ divided by $\frac{1}{4}$ is $\frac{12}{5}$ or $2\frac{2}{5}$.

Practice Problem Divide $\frac{3}{11}$ by $\frac{7}{10}$.

Use Ratios

When you compare two numbers by division, you are using a ratio. Ratios can be written 3 to 5, 3:5, or $\frac{3}{5}$. Ratios, like fractions, also can be written in simplest form.

Ratios can represent probabilities, also called odds. This is a ratio that compares the number of ways a certain outcome occurs to the number of outcomes. For example, if you flip a coin 100 times, what are the odds that it will come up heads? There are two possible outcomes, heads or tails, so the odds of coming up heads are 50:100. Another way to say this is that 50 out of 100 times the coin will come up heads. In its simplest form, the ratio is 1:2.

Example 1 A chemical solution contains 40 g of salt and 64 g of baking soda. What is the ratio of salt to baking soda as a fraction in simplest form?

Step 1 Write the ratio as a fraction.

$$\frac{\text{salt}}{\text{baking soda}} = \frac{40}{64}$$

Step 2 Express the fraction in simplest form.

The GCF of 40 and 64 is 8.

$$\frac{40}{64} = \frac{40 \div 8}{64 \div 8} = \frac{5}{8}$$

The ratio of salt to baking soda in the sample is 5:8.

Example 2 Sean rolls a 6-sided die 6 times. What are the odds that the side with a 3 will show?

Step 1 Write the ratio as a fraction.

$$\frac{\text{number of sides with a 3}}{\text{number of sides}} = \frac{1}{6}$$

Step 2 Multiply by the number of attempts.

$$\frac{1}{6} \times 6 \text{ attempts} = \frac{6}{6} \text{ attempts} = 1 \text{ attempt}$$

1 attempt out of 6 will show a 3.

Practice Problem Two metal rods measure 100 cm and 144 cm in length. What is the ratio of their lengths in simplest form?

Use Decimals

A fraction with a denominator that is a power of ten can be written as a decimal. For example, 0.27 means $\frac{27}{100}$. The decimal point separates the ones place from the tenths place.

Any fraction can be written as a decimal using division. For example, the fraction $\frac{5}{8}$ can be written as a decimal by dividing 5 by 8. Written as a decimal, it is 0.625.

Add or Subtract Decimals When adding and subtracting decimals, line up the decimal points before carrying out the operation.

Example 1 Find the sum of 47.68 and 7.80.

Step 1 Line up the decimal places when you write the numbers.

$$\begin{array}{r} 47.68 \\ +\ 7.80 \\ \hline \end{array}$$

Step 2 Add the decimals.

$$\begin{array}{r} 47.68 \\ +\ 7.80 \\ \hline 55.48 \end{array}$$

The sum of 47.68 and 7.80 is 55.48.

Example 2 Find the difference of 42.17 and 15.85.

Step 1 Line up the decimal places when you write the number.

$$\begin{array}{r} 42.17 \\ -15.85 \\ \hline \end{array}$$

Step 2 Subtract the decimals.

$$\begin{array}{r} 42.17 \\ -15.85 \\ \hline 26.32 \end{array}$$

The difference of 42.17 and 15.85 is 26.32.

Practice Problem Find the sum of 1.245 and 3.842.

Math Skill Handbook

Multiply Decimals To multiply decimals, multiply the numbers like any other number, ignoring the decimal point. Count the decimal places in each factor. The product will have the same number of decimal places as the sum of the decimal places in the factors.

Example Multiply 2.4 by 5.9.

Step 1 Multiply the factors like two whole numbers.
$24 \times 59 = 1416$

Step 2 Find the sum of the number of decimal places in the factors. Each factor has one decimal place, for a sum of two decimal places.

Step 3 The product will have two decimal places.
14.16

The product of 2.4 and 5.9 is 14.16.

Practice Problem Multiply 4.6 by 2.2.

Divide Decimals When dividing decimals, change the divisor to a whole number. To do this, multiply both the divisor and the dividend by the same power of ten. Then place the decimal point in the quotient directly above the decimal point in the dividend. Then divide as you do with whole numbers.

Example Divide 8.84 by 3.4.

Step 1 Multiply both factors by 10.
$3.4 \times 10 = 34$, $8.84 \times 10 = 88.4$

Step 2 Divide 88.4 by 34.

$$\begin{array}{r} 2.6 \\ 34\overline{)88.4} \\ -68 \\ \hline 204 \\ -204 \\ \hline 0 \end{array}$$

8.84 divided by 3.4 is 2.6.

Practice Problem Divide 75.6 by 3.6.

Use Proportions

An equation that shows that two ratios are equivalent is a proportion. The ratios $\frac{2}{4}$ and $\frac{5}{10}$ are equivalent, so they can be written as $\frac{2}{4} = \frac{5}{10}$. This equation is a proportion.

When two ratios form a proportion, the cross products are equal. To find the cross products in the proportion $\frac{2}{4} = \frac{5}{10}$, multiply the 2 and the 10, and the 4 and the 5. Therefore $2 \times 10 = 4 \times 5$, or $20 = 20$.

Because you know that both proportions are equal, you can use cross products to find a missing term in a proportion. This is known as solving the proportion.

Example The heights of a tree and a pole are proportional to the lengths of their shadows. The tree casts a shadow of 24 m when a 6-m pole casts a shadow of 4 m. What is the height of the tree?

Step 1 Write a proportion.
$$\frac{\text{height of tree}}{\text{height of pole}} = \frac{\text{length of tree's shadow}}{\text{length of pole's shadow}}$$

Step 2 Substitute the known values into the proportion. Let h represent the unknown value, the height of the tree.
$$\frac{h}{6} = \frac{24}{4}$$

Step 3 Find the cross products.
$h \times 4 = 6 \times 24$

Step 4 Simplify the equation.
$4h = 144$

Step 5 Divide each side by 4.
$$\frac{4h}{4} = \frac{144}{4}$$
$h = 36$

The height of the tree is 36 m.

Practice Problem The ratios of the weights of two objects on the Moon and on Earth are in proportion. A rock weighing 3 N on the Moon weighs 18 N on Earth. How much would a rock that weighs 5 N on the Moon weigh on Earth?

Use Percentages

The word *percent* means "out of one hundred." It is a ratio that compares a number to 100. Suppose you read that 77 percent of the Earth's surface is covered by water. That is the same as reading that the fraction of the Earth's surface covered by water is $\frac{77}{100}$. To express a fraction as a percent, first find the equivalent decimal for the fraction. Then, multiply the decimal by 100 and add the percent symbol.

Example Express $\frac{13}{20}$ as a percent.

Step 1 Find the equivalent decimal for the fraction.

$$\begin{array}{r} 0.65 \\ 20\overline{)13.00} \\ \underline{12\ 0} \\ 1\ 00 \\ \underline{1\ 00} \\ 0 \end{array}$$

Step 2 Rewrite the fraction $\frac{13}{20}$ as 0.65.

Step 3 Multiply 0.65 by 100 and add the % sign.

$0.65 \times 100 = 65 = 65\%$

So, $\frac{13}{20} = 65\%$.

This also can be solved as a proportion.

Example Express $\frac{13}{20}$ as a percent.

Step 1 Write a proportion.

$\frac{13}{20} = \frac{x}{100}$

Step 2 Find the cross products.

$1300 = 20x$

Step 3 Divide each side by 20.

$\frac{1300}{20} = \frac{20x}{20}$

$65\% = x$

Practice Problem In one year, 73 of 365 days were rainy in one city. What percent of the days in that city were rainy?

Solve One-Step Equations

A statement that two things are equal is an equation. For example, $A = B$ is an equation that states that A is equal to B.

An equation is solved when a variable is replaced with a value that makes both sides of the equation equal. To make both sides equal the inverse operation is used. Addition and subtraction are inverses, and multiplication and division are inverses.

Example 1 Solve the equation $x - 10 = 35$.

Step 1 Find the solution by adding 10 to each side of the equation.

$x - 10 = 35$

$x - 10 + 10 = 35 + 10$

$x = 45$

Step 2 Check the solution.

$x - 10 = 35$

$45 - 10 = 35$

$35 = 35$

Both sides of the equation are equal, so $x = 45$.

Example 2 In the formula $a = bc$, find the value of c if $a = 20$ and $b = 2$.

Step 1 Rearrange the formula so the unknown value is by itself on one side of the equation by dividing both sides by b.

$a = bc$

$\frac{a}{b} = \frac{bc}{b}$

$\frac{a}{b} = c$

Step 2 Replace the variables a and b with the values that are given.

$\frac{a}{b} = c$

$\frac{20}{2} = c$

$10 = c$

Step 3 Check the solution.

$a = bc$

$20 = 2 \times 10$

$20 = 20$

Both sides of the equation are equal, so $c = 10$ is the solution when $a = 20$ and $b = 2$.

Practice Problem In the formula $h = gd$, find the value of d if $g = 12.3$ and $h = 17.4$.

Use Statistics

The branch of mathematics that deals with collecting, analyzing, and presenting data is statistics. In statistics, there are three common ways to summarize data with a single number—the mean, the median, and the mode.

The **mean** of a set of data is the arithmetic average. It is found by adding the numbers in the data set and dividing by the number of items in the set.

The **median** is the middle number in a set of data when the data are arranged in numerical order. If there were an even number of data points, the median would be the mean of the two middle numbers.

The **mode** of a set of data is the number or item that appears most often.

Another number that often is used to describe a set of data is the range. The **range** is the difference between the largest number and the smallest number in a set of data.

A **frequency table** shows how many times each piece of data occurs, usually in a survey. **Table 2** below shows the results of a student survey on favorite color.

Table 2 Student Color Choice

Color	Tally	Frequency
red	||||	4
blue	~~||||~~	5
black	||	2
green	|||	3
purple	~~||||~~ ||	7
yellow	~~||||~~ |	6

Based on the frequency table data, which color is the favorite?

Example The speeds (in m/s) for a race car during five different time trials are 39, 37, 44, 36, and 44.

To find the mean:

Step 1 Find the sum of the numbers.

$39 + 37 + 44 + 36 + 44 = 200$

Step 2 Divide the sum by the number of items, which is 5.

$200 \div 5 = 40$

The mean is 40 m/s.

To find the median:

Step 1 Arrange the measures from least to greatest.

36, 37, 39, 44, 44

Step 2 Determine the middle measure.

36, 37, <u>39</u>, 44, 44

The median is 39 m/s.

To find the mode:

Step 1 Group the numbers that are the same together.

44, 44, 36, 37, 39

Step 2 Determine the number that occurs most in the set.

<u>44, 44</u>, 36, 37, 39

The mode is 44 m/s.

To find the range:

Step 1 Arrange the measures from largest to smallest.

44, 44, 39, 37, 36

Step 2 Determine the largest and smallest measures in the set.

<u>44</u>, 44, 39, 37, <u>36</u>

Step 3 Find the difference between the largest and smallest measures.

$44 - 36 = 8$

The range is 8 m/s.

Practice Problem Find the mean, median, mode, and range for the data set 8, 4, 12, 8, 11, 14, 16.

Use Geometry

The branch of mathematics that deals with the measurement, properties, and relationships of points, lines, angles, surfaces, and solids is called geometry.

Perimeter The **perimeter** (P) is the distance around a geometric figure. To find the perimeter of a rectangle, add the length and width and multiply that sum by two, or $2(l + w)$. To find perimeters of irregular figures, add the length of the sides.

Example 1 Find the perimeter of a rectangle that is 3 m long and 5 m wide.

Step 1 You know that the perimeter is 2 times the sum of the width and length.
$P = 2(3 \text{ m} + 5 \text{ m})$

Step 2 Find the sum of the width and length.
$P = 2(8 \text{ m})$

Step 3 Multiply by 2.
$P = 16 \text{ m}$

The perimeter is 16 m.

Example 2 Find the perimeter of a shape with sides measuring 2 cm, 5 cm, 6 cm, 3 cm.

Step 1 You know that the perimeter is the sum of all the sides.
$P = 2 + 5 + 6 + 3$

Step 2 Find the sum of the sides.
$P = 2 + 5 + 6 + 3$
$P = 16$

The perimeter is 16 cm.

Practice Problem Find the perimeter of a rectangle with a length of 18 m and a width of 7 m.

Practice Problem Find the perimeter of a triangle measuring 1.6 cm by 2.4 cm by 2.4 cm.

Area of a Rectangle The **area** (A) is the number of square units needed to cover a surface. To find the area of a rectangle, multiply the length times the width, or $l \times w$. When finding area, the units also are multiplied. Area is given in square units.

Example Find the area of a rectangle with a length of 1 cm and a width of 10 cm.

Step 1 You know that the area is the length multiplied by the width.
$A = (1 \text{ cm} \times 10 \text{ cm})$

Step 2 Multiply the length by the width. Also multiply the units.
$A = 10 \text{ cm}^2$

The area is 10 cm^2.

Practice Problem Find the area of a square whose sides measure 4 m.

Area of a Triangle To find the area of a triangle, use the formula:

$$A = \frac{1}{2}(\text{base} \times \text{height})$$

The base of a triangle can be any of its sides. The height is the perpendicular distance from a base to the opposite endpoint, or vertex.

Example Find the area of a triangle with a base of 18 m and a height of 7 m.

Step 1 You know that the area is $\frac{1}{2}$ the base times the height.
$A = \frac{1}{2}(18 \text{ m} \times 7 \text{ m})$

Step 2 Multiply $\frac{1}{2}$ by the product of 18×7. Multiply the units.
$A = \frac{1}{2}(126 \text{ m}^2)$
$A = 63 \text{ m}^2$

The area is 63 m^2.

Practice Problem Find the area of a triangle with a base of 27 cm and a height of 17 cm.

Circumference of a Circle The **diameter** (d) of a circle is the distance across the circle through its center, and the **radius** (r) is the distance from the center to any point on the circle. The radius is half of the diameter. The distance around the circle is called the **circumference** (C). The formula for finding the circumference is:

$$C = 2\pi r \text{ or } C = \pi d$$

The circumference divided by the diameter is always equal to 3.1415926... This nonterminating and nonrepeating number is represented by the Greek letter π (pi). An approximation often used for π is 3.14.

Example 1 Find the circumference of a circle with a radius of 3 m.

Step 1 You know the formula for the circumference is 2 times the radius times π.
$C = 2\pi(3)$

Step 2 Multiply 2 times the radius.
$C = 6\pi$

Step 3 Multiply by π.
$C = 19$ m

The circumference is 19 m.

Example 2 Find the circumference of a circle with a diameter of 24.0 cm.

Step 1 You know the formula for the circumference is the diameter times π.
$C = \pi(24.0)$

Step 2 Multiply the diameter by π.
$C = 75.4$ cm

The circumference is 75.4 cm.

Practice Problem Find the circumference of a circle with a radius of 19 cm.

Area of a Circle The formula for the area of a circle is:
$A = \pi r^2$

Example 1 Find the area of a circle with a radius of 4.0 cm.

Step 1 $A = \pi(4.0)^2$

Step 2 Find the square of the radius.
$A = 16\pi$

Step 3 Multiply the square of the radius by π.
$A = 50 \text{ cm}^2$

The area of the circle is 50 cm^2.

Example 2 Find the area of a circle with a radius of 225 m.

Step 1 $A = \pi(225)^2$

Step 2 Find the square of the radius.
$A = 50625\pi$

Step 3 Multiply the square of the radius by π.
$A = 158962.5$

The area of the circle is 158,962 m^2.

Example 3 Find the area of a circle whose diameter is 20.0 mm.

Step 1 You know the formula for the area of a circle is the square of the radius times π, and that the radius is half of the diameter.
$A = \pi\left(\frac{20.0}{2}\right)^2$

Step 2 Find the radius.
$A = \pi(10.0)^2$

Step 3 Find the square of the radius.
$A = 100\pi$

Step 4 Multiply the square of the radius by π.
$A = 314 \text{ mm}^2$

The area is 314 mm^2.

Practice Problem Find the area of a circle with a radius of 16 m.

Volume The measure of space occupied by a solid is the **volume** (V). To find the volume of a rectangular solid multiply the length times width times height, or $V = l \times w \times h$. It is measured in cubic units, such as cubic centimeters (cm^3).

Example Find the volume of a rectangular solid with a length of 2.0 m, a width of 4.0 m, and a height of 3.0 m.

Step 1 You know the formula for volume is the length times the width times the height.
$V = 2.0\text{ m} \times 4.0\text{ m} \times 3.0\text{ m}$

Step 2 Multiply the length times the width times the height.
$V = 24\text{ m}^3$

The volume is 24 m^3.

Practice Problem Find the volume of a rectangular solid that is 8 m long, 4 m wide, and 4 m high.

To find the volume of other solids, multiply the area of the base times the height.

Example 1 Find the volume of a solid that has a triangular base with a length of 8.0 m and a height of 7.0 m. The height of the entire solid is 15.0 m.

Step 1 You know that the base is a triangle, and the area of a triangle is $\frac{1}{2}$ the base times the height, and the volume is the area of the base times the height.
$V = \left[\frac{1}{2}(b \times h)\right] \times 15$

Step 2 Find the area of the base.
$V = \left[\frac{1}{2}(8 \times 7)\right] \times 15$
$V = \left(\frac{1}{2} \times 56\right) \times 15$

Step 3 Multiply the area of the base by the height of the solid.
$V = 28 \times 15$
$V = 420\text{ m}^3$

The volume is 420 m^3.

Example 2 Find the volume of a cylinder that has a base with a radius of 12.0 cm, and a height of 21.0 cm.

Step 1 You know that the base is a circle, and the area of a circle is the square of the radius times π, and the volume is the area of the base times the height.
$V = (\pi r^2) \times 21$
$V = (\pi 12^2) \times 21$

Step 2 Find the area of the base.
$V = 144\pi \times 21$
$V = 452 \times 21$

Step 3 Multiply the area of the base by the height of the solid.
$V = 9490\text{ cm}^3$

The volume is 9490 cm^3.

Example 3 Find the volume of a cylinder that has a diameter of 15 mm and a height of 4.8 mm.

Step 1 You know that the base is a circle with an area equal to the square of the radius times π. The radius is one-half the diameter. The volume is the area of the base times the height.
$V = (\pi r^2) \times 4.8$
$V = \left[\pi\left(\frac{1}{2} \times 15\right)^2\right] \times 4.8$
$V = (\pi 7.5^2) \times 4.8$

Step 2 Find the area of the base.
$V = 56.25\pi \times 4.8$
$V = 176.63 \times 4.8$

Step 3 Multiply the area of the base by the height of the solid.
$V = 847.8$

The volume is 847.8 mm^3.

Practice Problem Find the volume of a cylinder with a diameter of 7 cm in the base and a height of 16 cm.

Science Applications

Measure in SI

The metric system of measurement was developed in 1795. A modern form of the metric system, called the International System (SI), was adopted in 1960 and provides the standard measurements that all scientists around the world can understand.

The SI system is convenient because unit sizes vary by powers of 10. Prefixes are used to name units. Look at **Table 3** for some common SI prefixes and their meanings.

Table 3 Common SI Prefixes

Prefix	Symbol	Meaning	
kilo-	k	1,000	thousand
hecto-	h	100	hundred
deka-	da	10	ten
deci-	d	0.1	tenth
centi-	c	0.01	hundredth
milli-	m	0.001	thousandth

Example How many grams equal one kilogram?

Step 1 Find the prefix *kilo* in **Table 3.**

Step 2 Using **Table 3,** determine the meaning of *kilo.* According to the table, it means 1,000. When the prefix *kilo* is added to a unit, it means that there are 1,000 of the units in a "*kilo*unit."

Step 3 Apply the prefix to the units in the question. The units in the question are grams. There are 1,000 grams in a kilogram.

Practice Problem Is a milligram larger or smaller than a gram? How many of the smaller units equal one larger unit? What fraction of the larger unit does one smaller unit represent?

Dimensional Analysis

Convert SI Units In science, quantities such as length, mass, and time sometimes are measured using different units. A process called dimensional analysis can be used to change one unit of measure to another. This process involves multiplying your starting quantity and units by one or more conversion factors. A conversion factor is a ratio equal to one and can be made from any two equal quantities with different units. If 1,000 mL equal 1 L then two ratios can be made.

$$\frac{1{,}000 \text{ mL}}{1 \text{ L}} = \frac{1 \text{ L}}{1{,}000 \text{ mL}} = 1$$

One can covert between units in the SI system by using the equivalents in **Table 3** to make conversion factors.

Example 1 How many cm are in 4 m?

Step 1 Write conversion factors for the units given. From **Table 3,** you know that 100 cm = 1 m. The conversion factors are

$$\frac{100 \text{ cm}}{1 \text{ m}} \text{ and } \frac{1 \text{ m}}{100 \text{ cm}}$$

Step 2 Decide which conversion factor to use. Select the factor that has the units you are converting from (m) in the denominator and the units you are converting to (cm) in the numerator.

$$\frac{100 \text{ cm}}{1 \text{ m}}$$

Step 3 Multiply the starting quantity and units by the conversion factor. Cancel the starting units with the units in the denominator. There are 400 cm in 4 m.

$$4 \cancel{\text{m}} \times \frac{100 \text{ cm}}{1 \cancel{\text{m}}} = 400 \text{ cm}$$

Practice Problem How many milligrams are in one kilogram? (Hint: You will need to use two conversion factors from **Table 3.**)

Table 4 Unit System Equivalents

Type of Measurement	Equivalent
Length	1 in = 2.54 cm 1 yd = 0.91 m 1 mi = 1.61 km
Mass and Weight*	1 oz = 28.35 g 1 lb = 0.45 kg 1 ton (short) = 0.91 tonnes (metric tons) 1 lb = 4.45 N
Volume	1 in^3 = 16.39 cm^3 1 qt = 0.95 L 1 gal = 3.78 L
Area	1 in^2 = 6.45 cm^2 1 yd^2 = 0.83 m^2 1 mi^2 = 2.59 km^2 1 acre = 0.40 hectares
Temperature	$°C = \frac{(°F - 32)}{1.8}$ K = °C + 273

*Weight is measured in standard Earth gravity.

Convert Between Unit Systems **Table 4** gives a list of equivalents that can be used to convert between English and SI units.

Example If a meterstick has a length of 100 cm, how long is the meterstick in inches?

Step 1 Write the conversion factors for the units given. From **Table 4,** 1 in = 2.54 cm.

$$\frac{1 \text{ in}}{2.54 \text{ cm}} \text{ and } \frac{2.54 \text{ cm}}{1 \text{ in}}$$

Step 2 Determine which conversion factor to use. You are converting from cm to in. Use the conversion factor with cm on the bottom.

$$\frac{1 \text{ in}}{2.54 \text{ cm}}$$

Step 3 Multiply the starting quantity and units by the conversion factor. Cancel the starting units with the units in the denominator. Round your answer based on the number of significant figures in the conversion factor.

$$100 \cancel{\text{cm}} \times \frac{1 \text{ in}}{2.54 \cancel{\text{cm}}} = 39.37 \text{ in}$$

The meterstick is 39.4 in long.

Practice Problem A book has a mass of 5 lbs. What is the mass of the book in kg?

Practice Problem Use the equivalent for in and cm (1 in = 2.54 cm) to show how 1 in^3 = 16.39 cm^3.

Precision and Significant Digits

When you make a measurement, the value you record depends on the precision of the measuring instrument. This precision is represented by the number of significant digits recorded in the measurement. When counting the number of significant digits, all digits are counted except zeros at the end of a number with no decimal point such as 2,050, and zeros at the beginning of a decimal such as 0.03020. When adding or subtracting numbers with different precision, round the answer to the smallest number of decimal places of any number in the sum or difference. When multiplying or dividing, the answer is rounded to the smallest number of significant digits of any number being multiplied or divided.

Example The lengths 5.28 and 5.2 are measured in meters. Find the sum of these lengths and record your answer using the correct number of significant digits.

Step 1 Find the sum.

5.28 m	2 digits after the decimal
+ 5.2 m	1 digit after the decimal
10.48 m	

Step 2 Round to one digit after the decimal because the least number of digits after the decimal of the numbers being added is 1.

The sum is 10.5 m.

Practice Problem How many significant digits are in the measurement 7,071,301 m? How many significant digits are in the measurement 0.003010 g?

Practice Problem Multiply 5.28 and 5.2 using the rule for multiplying and dividing. Record the answer using the correct number of significant digits.

Scientific Notation

Many times numbers used in science are very small or very large. Because these numbers are difficult to work with scientists use scientific notation. To write numbers in scientific notation, move the decimal point until only one non-zero digit remains on the left. Then count the number of places you moved the decimal point and use that number as a power of ten. For example, the average distance from the Sun to Mars is 227,800,000,000 m. In scientific notation, this distance is 2.278×10^{11} m. Because you moved the decimal point to the left, the number is a positive power of ten.

The mass of an electron is about 0.000 000 000 000 000 000 000 000 000 000 911 kg. Expressed in scientific notation, this mass is 9.11×10^{-31} kg. Because the decimal point was moved to the right, the number is a negative power of ten.

Example Earth is 149,600,000 km from the Sun. Express this in scientific notation.

Step 1 Move the decimal point until one non-zero digit remains on the left.

1.496 000 00

Step 2 Count the number of decimal places you have moved. In this case, eight.

Step 3 Show that number as a power of ten, 10^8.

The Earth is 1.496×10^8 km from the Sun.

Practice Problem How many significant digits are in 149,600,000 km? How many significant digits are in 1.496×10^8 km?

Practice Problem Parts used in a high performance car must be measured to 7×10^{-6} m. Express this number as a decimal.

Practice Problem A CD is spinning at 539 revolutions per minute. Express this number in scientific notation.

Make and Use Graphs

Data in tables can be displayed in a graph—a visual representation of data. Common graph types include line graphs, bar graphs, and circle graphs.

Line Graph A line graph shows a relationship between two variables that change continuously. The independent variable is changed and is plotted on the x-axis. The dependent variable is observed, and is plotted on the y-axis.

Example Draw a line graph of the data below from a cyclist in a long-distance race.

Table 5 Bicycle Race Data

Time (h)	Distance (km)
0	0
1	8
2	16
3	24
4	32
5	40

Step 1 Determine the x-axis and y-axis variables. Time varies independently of distance and is plotted on the x-axis. Distance is dependent on time and is plotted on the y-axis.

Step 2 Determine the scale of each axis. The x-axis data ranges from 0 to 5. The y-axis data ranges from 0 to 40.

Step 3 Using graph paper, draw and label the axes. Include units in the labels.

Step 4 Draw a point at the intersection of the time value on the x-axis and corresponding distance value on the y-axis. Connect the points and label the graph with a title, as shown in **Figure 20.**

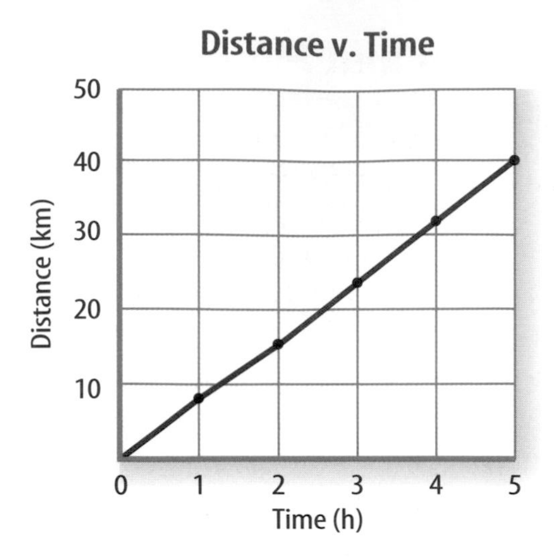

Figure 20 This line graph shows the relationship between distance and time during a bicycle ride.

Practice Problem A puppy's shoulder height is measured during the first year of her life. The following measurements were collected: (3 mo, 52 cm), (6 mo, 72 cm), (9 mo, 83 cm), (12 mo, 86 cm). Graph this data.

Find a Slope The slope of a straight line is the ratio of the vertical change, rise, to the horizontal change, run.

$$\text{Slope} = \frac{\text{vertical change (rise)}}{\text{horizontal change (run)}} = \frac{\text{change in } y}{\text{change in } x}$$

Example Find the slope of the graph in **Figure 20.**

Step 1 You know that the slope is the change in y divided by the change in x.

$$\text{Slope} = \frac{\text{change in } y}{\text{change in } x}$$

Step 2 Determine the data points you will be using. For a straight line, choose the two sets of points that are the farthest apart.

$$\text{Slope} = \frac{(40\text{–}0)\text{ km}}{(5\text{–}0)\text{ hr}}$$

Step 3 Find the change in y and x.

$$\text{Slope} = \frac{40\text{ km}}{5\text{h}}$$

Step 4 Divide the change in y by the change in x.

$$\text{Slope} = \frac{8\text{ km}}{\text{h}}$$

The slope of the graph is 8 km/h.

Bar Graph To compare data that does not change continuously you might choose a bar graph. A bar graph uses bars to show the relationships between variables. The x-axis variable is divided into parts. The parts can be numbers such as years, or a category such as a type of animal. The y-axis is a number and increases continuously along the axis.

Example A recycling center collects 4.0 kg of aluminum on Monday, 1.0 kg on Wednesday, and 2.0 kg on Friday. Create a bar graph of this data.

Step 1 Select the x-axis and y-axis variables. The measured numbers (the masses of aluminum) should be placed on the y-axis. The variable divided into parts (collection days) is placed on the x-axis.

Step 2 Create a graph grid like you would for a line graph. Include labels and units.

Step 3 For each measured number, draw a vertical bar above the x-axis value up to the y-axis value. For the first data point, draw a vertical bar above Monday up to 4.0 kg.

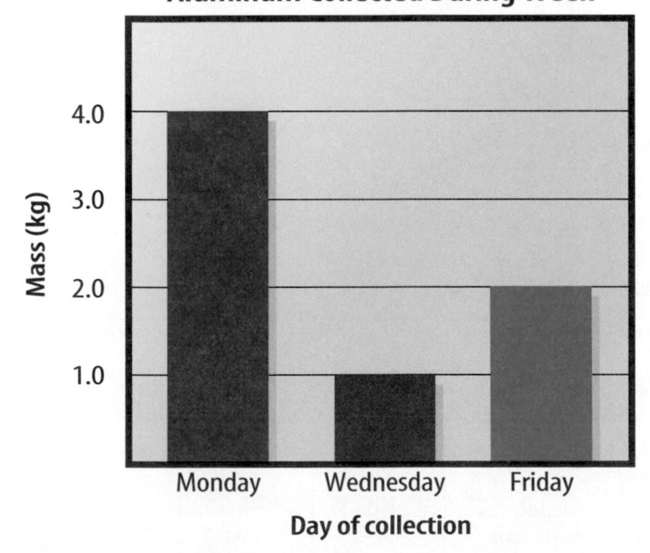

Practice Problem Draw a bar graph of the gases in air: 78% nitrogen, 21% oxygen, 1% other gases.

Circle Graph To display data as parts of a whole, you might use a circle graph. A circle graph is a circle divided into sections that represent the relative size of each piece of data. The entire circle represents 100%, half represents 50%, and so on.

Example Air is made up of 78% nitrogen, 21% oxygen, and 1% other gases. Display the composition of air in a circle graph.

Step 1 Multiply each percent by 360° and divide by 100 to find the angle of each section in the circle.

$$78\% \times \frac{360°}{100} = 280.8°$$

$$21\% \times \frac{360°}{100} = 75.6°$$

$$1\% \times \frac{360°}{100} = 3.6°$$

Step 2 Use a compass to draw a circle and to mark the center of the circle. Draw a straight line from the center to the edge of the circle.

Step 3 Use a protractor and the angles you calculated to divide the circle into parts. Place the center of the protractor over the center of the circle and line the base of the protractor over the straight line.

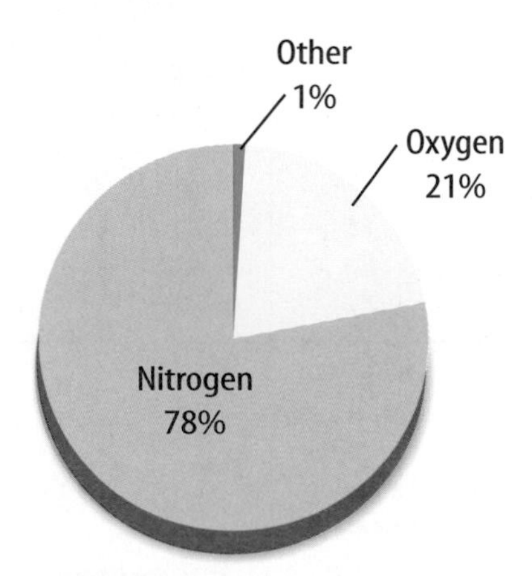

Practice Problem Draw a circle graph to represent the amount of aluminum collected during the week shown in the bar graph to the left.

PERIODIC TABLE OF THE ELEMENTS

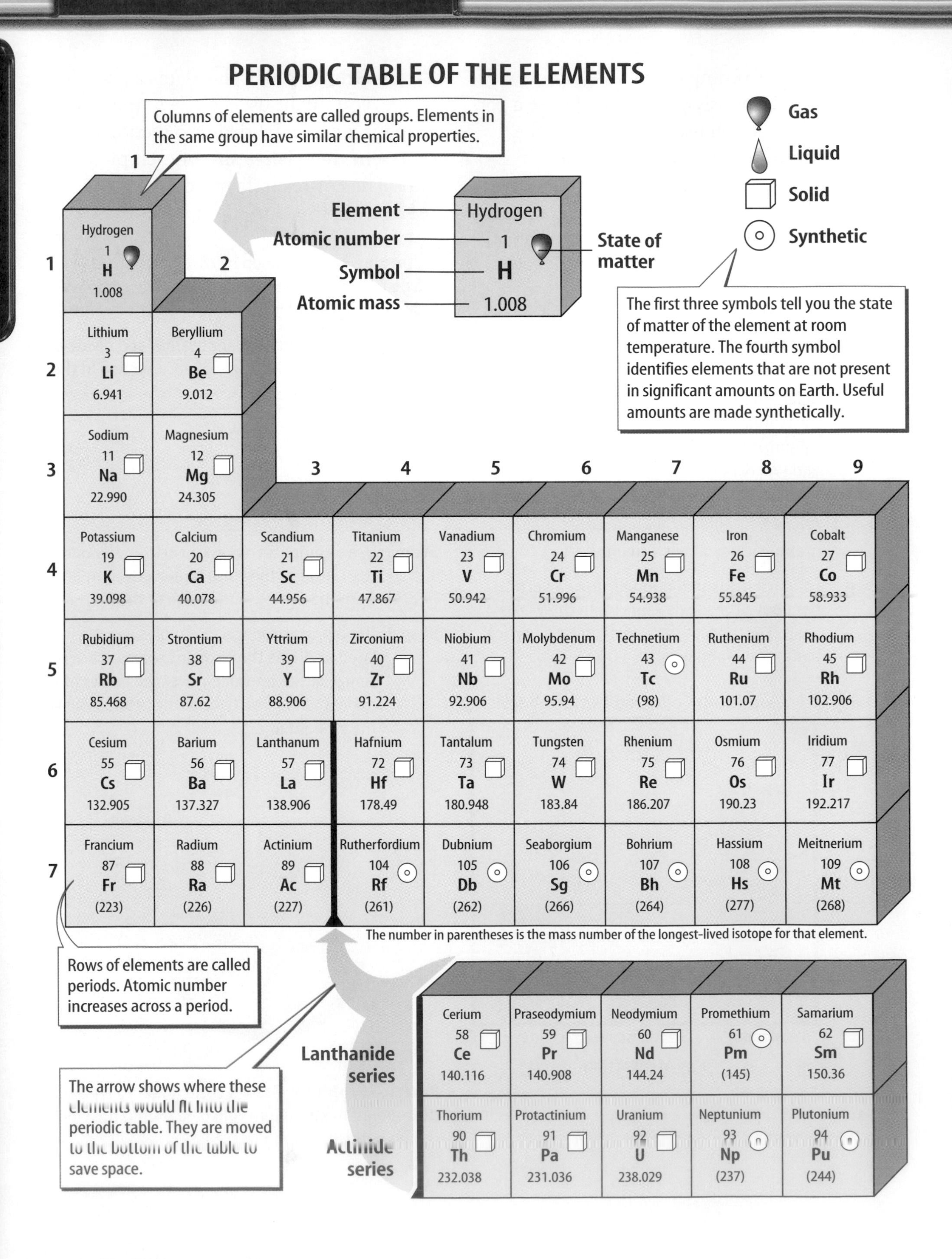

Period	1	2	3	4	5	6	7	8	9
1	Hydrogen 1 H 1.008								
2	Lithium 3 Li 6.941	Beryllium 4 Be 9.012							
3	Sodium 11 Na 22.990	Magnesium 12 Mg 24.305							
4	Potassium 19 K 39.098	Calcium 20 Ca 40.078	Scandium 21 Sc 44.956	Titanium 22 Ti 47.867	Vanadium 23 V 50.942	Chromium 24 Cr 51.996	Manganese 25 Mn 54.938	Iron 26 Fe 55.845	Cobalt 27 Co 58.933
5	Rubidium 37 Rb 85.468	Strontium 38 Sr 87.62	Yttrium 39 Y 88.906	Zirconium 40 Zr 91.224	Niobium 41 Nb 92.906	Molybdenum 42 Mo 95.94	Technetium 43 Tc (98)	Ruthenium 44 Ru 101.07	Rhodium 45 Rh 102.906
6	Cesium 55 Cs 132.905	Barium 56 Ba 137.327	Lanthanum 57 La 138.906	Hafnium 72 Hf 178.49	Tantalum 73 Ta 180.948	Tungsten 74 W 183.84	Rhenium 75 Re 186.207	Osmium 76 Os 190.23	Iridium 77 Ir 192.217
7	Francium 87 Fr (223)	Radium 88 Ra (226)	Actinium 89 Ac (227)	Rutherfordium 104 Rf (261)	Dubnium 105 Db (262)	Seaborgium 106 Sg (266)	Bohrium 107 Bh (264)	Hassium 108 Hs (277)	Meitnerium 109 Mt (268)

Series					
Lanthanide series	Cerium 58 Ce 140.116	Praseodymium 59 Pr 140.908	Neodymium 60 Nd 144.24	Promethium 61 Pm (145)	Samarium 62 Sm 150.36
Actinide series	Thorium 90 Th 232.038	Protactinium 91 Pa 231.036	Uranium 92 U 238.029	Neptunium 93 Np (237)	Plutonium 94 Pu (244)

Metal

Metalloid

Nonmetal

The color of an element's block tells you if the element is a metal, nonmetal, or metalloid.

Science Online
Visit booka.msscience.com for the updates to the periodic table.

10	11	12	13	14	15	16	17	18
								Helium 2 He 4.003
			Boron 5 B 10.811	Carbon 6 C 12.011	Nitrogen 7 N 14.007	Oxygen 8 O 15.999	Fluorine 9 F 18.998	Neon 10 Ne 20.180
			Aluminum 13 Al 26.982	Silicon 14 Si 28.086	Phosphorus 15 P 30.974	Sulfur 16 S 32.065	Chlorine 17 Cl 35.453	Argon 18 Ar 39.948
Nickel 28 Ni 58.693	Copper 29 Cu 63.546	Zinc 30 Zn 65.409	Gallium 31 Ga 69.723	Germanium 32 Ge 72.64	Arsenic 33 As 74.922	Selenium 34 Se 78.96	Bromine 35 Br 79.904	Krypton 36 Kr 83.798
Palladium 46 Pd 106.42	Silver 47 Ag 107.868	Cadmium 48 Cd 112.411	Indium 49 In 114.818	Tin 50 Sn 118.710	Antimony 51 Sb 121.760	Tellurium 52 Te 127.60	Iodine 53 I 126.904	Xenon 54 Xe 131.293
Platinum 78 Pt 195.078	Gold 79 Au 196.967	Mercury 80 Hg 200.59	Thallium 81 Tl 204.383	Lead 82 Pb 207.2	Bismuth 83 Bi 208.980	Polonium 84 Po (209)	Astatine 85 At (210)	Radon 86 Rn (222)
Darmstadtium 110 Ds (281)	Unununium * 111 Uuu (272)	Ununbium * 112 Uub (285)		Ununquadium * 114 Uuq (289)		** 116		** 118

* The names and symbols for elements 111–114 are temporary. Final names will be selected when the elements' discoveries are verified.

** Elements 116 and 118 were thought to have been created. The claim was retracted because the experimental results could not be repeated.

Europium 63 Eu 151.964	Gadolinium 64 Gd 157.25	Terbium 65 Tb 158.925	Dysprosium 66 Dy 162.500	Holmium 67 Ho 164.930	Erbium 68 Er 167.259	Thulium 69 Tm 168.934	Ytterbium 70 Yb 173.04	Lutetium 71 Lu 174.967
Americium 95 Am (243)	Curium 96 Cm (247)	Berkelium 97 Bk (247)	Californium 98 Cf (251)	Einsteinium 99 Es (252)	Fermium 100 Fm (257)	Mendelevium 101 Md (258)	Nobelium 102 No (259)	Lawrencium 103 Lr (262)

Use and Care of a Microscope

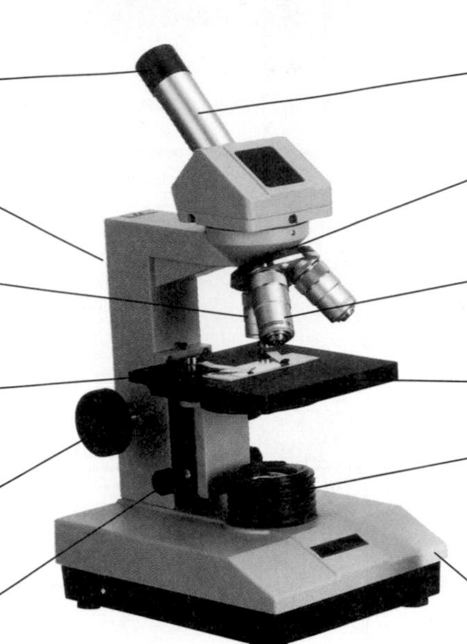

Caring for a Microscope

1. Always carry the microscope holding the arm with one hand and supporting the base with the other hand.
2. Don't touch the lenses with your fingers.
3. The coarse adjustment knob is used only when looking through the lowest-power objective lens. The fine adjustment knob is used when the high-power objective is in place.
4. Cover the microscope when you store it.

Using a Microscope

1. Place the microscope on a flat surface that is clear of objects. The arm should be toward you.
2. Look through the eyepiece. Adjust the diaphragm so light comes through the opening in the stage.
3. Place a slide on the stage so the specimen is in the field of view. Hold it firmly in place by using the stage clips.
4. Always focus with the coarse adjustment and the low-power objective lens first. After the object is in focus on low power, turn the nosepiece until the high-power objective is in place. Use ONLY the fine adjustment to focus with the high-power objective lens.

Making a Wet-Mount Slide

1. Carefully place the item you want to look at in the center of a clean, glass slide. Make sure the sample is thin enough for light to pass through.
2. Use a dropper to place one or two drops of water on the sample.
3. Hold a clean coverslip by the edges and place it at one edge of the water. Slowly lower the coverslip onto the water until it lies flat.
4. If you have too much water or a lot of air bubbles, touch the edge of a paper towel to the edge of the coverslip to draw off extra water and draw out unwanted air.

Diversity of Life: Classification of Living Organisms

A six-kingdom system of classification of organisms is used today. Two kingdoms—Kingdom Archaebacteria and Kingdom Eubacteria—contain organisms that do not have a nucleus and that lack membrane-bound structures in the cytoplasm of their cells. The members of the other four kingdoms have a cell or cells that contain a nucleus and structures in the cytoplasm, some of which are surrounded by membranes. These kingdoms are Kingdom Protista, Kingdom Fungi, Kingdom Plantae, and Kingdom Animalia.

Kingdom Archaebacteria

one-celled; some absorb food from their surroundings; some are photosynthetic; some are chemosynthetic; many are found in extremely harsh environments including salt ponds, hot springs, swamps, and deep-sea hydrothermal vents

Kingdom Eubacteria

one-celled; most absorb food from their surroundings; some are photosynthetic; some are chemosynthetic; many are parasites; many are round, spiral, or rod-shaped; some form colonies

Kingdom Protista

Phylum Euglenophyta one-celled; photosynthetic or take in food; most have one flagellum; euglenoids

Phylum Bacillariophyta one-celled; photosynthetic; have unique double shells made of silica; diatoms

Phylum Dinoflagellata one-celled; photosynthetic; contain red pigments; have two flagella; dinoflagellates

Phylum Chlorophyta one-celled, many-celled, or colonies; photosynthetic; contain chlorophyll; live on land, in freshwater, or salt water; green algae

Phylum Rhodophyta most are many-celled; photosynthetic; contain red pigments; most live in deep, saltwater environments; red algae

Phylum Phaeophyta most are many-celled; photosynthetic; contain brown pigments; most live in saltwater environments; brown algae

Phylum Rhizopoda one-celled; take in food; are free-living or parasitic; move by means of pseudopods; amoebas

Kingdom Eubacteria
Bacillus anthracis

Phylum Chlorophyta
Desmids

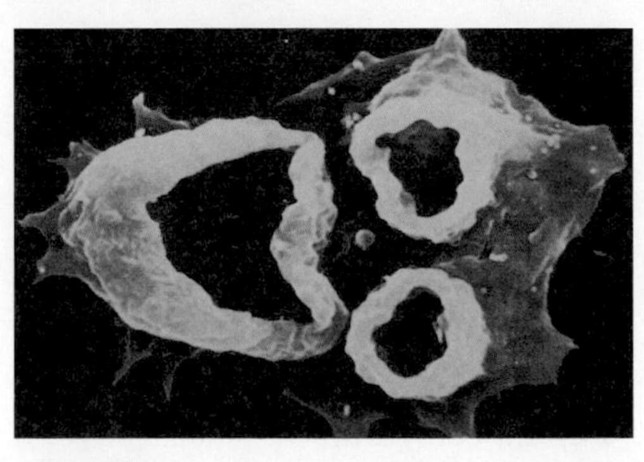

Amoeba

Phylum Zoomastigina one-celled; take in food; free-living or parasitic; have one or more flagella; zoomastigotes

Phylum Ciliophora one-celled; take in food; have large numbers of cilia; ciliates

Phylum Sporozoa one-celled; take in food; have no means of movement; are parasites in animals; sporozoans

Phylum Myxomycota
Slime mold

Phylum Oomycota
Phytophthora infestans

Phyla Myxomycota and Acrasiomycota one- or many-celled; absorb food; change form during life cycle; cellular and plasmodial slime molds

Phylum Oomycota many-celled; are either parasites or decomposers; live in freshwater or salt water; water molds, rusts and downy mildews

Kingdom Fungi

Phylum Zygomycota many-celled; absorb food; spores are produced in sporangia; zygote fungi; bread mold

Phylum Ascomycota one- and many-celled; absorb food; spores produced in asci; sac fungi; yeast

Phylum Basidiomycota many-celled; absorb food; spores produced in basidia; club fungi; mushrooms

Phylum Deuteromycota members with unknown reproductive structures; imperfect fungi; *Penicillium*

Phylum Mycophycota organisms formed by symbiotic relationship between an ascomycote or a basidiomycote and green alga or cyanobacterium; lichens

Lichens

Kingdom Plantae

Divisions Bryophyta (mosses), **Anthocerophyta** (hornworts), **Hepaticophyta** (liverworts), **Psilophyta** (whisk ferns) many-celled nonvascular plants; reproduce by spores produced in capsules; green; grow in moist, land environments

Division Lycophyta many-celled vascular plants; spores are produced in conelike structures; live on land; are photosynthetic; club mosses

Division Arthrophyta vascular plants; ribbed and jointed stems; scalelike leaves; spores produced in conelike structures; horsetails

Division Pterophyta vascular plants; leaves called fronds; spores produced in clusters of sporangia called sori; live on land or in water; ferns

Division Ginkgophyta deciduous trees; only one living species; have fan-shaped leaves with branching veins and fleshy cones with seeds; ginkgoes

Division Cycadophyta palmlike plants; have large, featherlike leaves; produces seeds in cones; cycads

Division Coniferophyta deciduous or evergreen; trees or shrubs; have needlelike or scale-like leaves; seeds produced in cones; conifers

Division Anthophyta
Tomato plant

Division Gnetophyta shrubs or woody vines; seeds are produced in cones; division contains only three genera; gnetum

Division Anthophyta dominant group of plants; flowering plants; have fruits with seeds

Kingdom Animalia

Phylum Porifera aquatic organisms that lack true tissues and organs; are asymmetrical and sessile; sponges

Phylum Cnidaria radially symmetrical organisms; have a digestive cavity with one opening; most have tentacles armed with stinging cells; live in aquatic environments singly or in colonies; includes jellyfish, corals, hydra, and sea anemones

Phylum Platyhelminthes bilaterally symmetrical worms; have flattened bodies; digestive system has one opening; parasitic and free-living species; flatworms

Division Bryophyta
Liverwort

Phylum Platyhelminthes
Flatworm

Phylum Chordata

Phylum Nematoda round, bilaterally symmetrical body; have digestive system with two openings; free-living forms and parasitic forms; roundworms

Phylum Mollusca soft-bodied animals, many with a hard shell and soft foot or footlike appendage; a mantle covers the soft body; aquatic and terrestrial species; includes clams, snails, squid, and octopuses

Phylum Annelida bilaterally symmetrical worms; have round, segmented bodies; terrestrial and aquatic species; includes earthworms, leeches, and marine polychaetes

Phylum Arthropoda largest animal group; have hard exoskeletons, segmented bodies, and pairs of jointed appendages; land and aquatic species; includes insects, crustaceans, and spiders

Phylum Echinodermata marine organisms; have spiny or leathery skin and a water-vascular system with tube feet; are radially symmetrical; includes sea stars, sand dollars, and sea urchins

Phylum Chordata organisms with internal skeletons and specialized body systems; most have paired appendages; all at some time have a notochord, nerve cord, gill slits, and a postanal tail; include fish, amphibians, reptiles, birds, and mammals

Cómo usar el glosario en español:
1. Busca el término en inglés que desees encontrar.
2. El término en español, junto con la definición, se encuentran en la columna de la derecha.

Pronunciation Key

Use the following key to help you sound out words in the glossary.

a	**b**ack (BAK)	**ew**	f**oo**d (FEWD)
ay	d**ay** (DAY)	**yoo**	p**ure** (PYOOR)
ah	f**a**ther (FAH thur)	**yew**	f**ew** (FYEW)
ow	fl**ow**er (FLOW ur)	**uh**	comm**a** (CAH muh)
ar	c**ar** (CAR)	**u** (+ con)	r**u**b (RUB)
e	l**e**ss (LES)	**sh**	**sh**elf (SHELF)
ee	l**ea**f (LEEF)	**ch**	na**t**ure (NAY chur)
ih	tr**i**p (TRIHP)	**g**	**g**ift (GIHFT)
i (i + con + e)	**i**dea (i DEE uh)	**j**	**g**em (JEM)
oh	g**o** (GOH)	**ing**	s**ing** (SING)
aw	s**o**ft (SAWFT)	**zh**	vi**s**ion (VIH zhun)
or	**or**bit (OR buht)	**k**	**c**ake (KAYK)
oy	c**oi**n (COYN)	**s**	**s**eed, **c**ent (SEED, SENT)
oo	f**oo**t (FOOT)	**z**	**z**one, rai**s**e (ZOHN, RAYZ)

English — A — Español

active transport: energy-requiring process in which transport proteins bind with particles and move them through a cell membrane. (p. 79)

adaptation: any variation that makes an organism better suited to its environment. (p. 160)

allele (uh LEEL): an alternate form that a gene may have for a single trait; can be dominant or recessive. (p. 128)

asexual reproduction: a type of reproduction—fission, budding, and regeneration—in which a new organism is produced from one organism and has DNA identical to the parent organism. (p. 103)

transporte activo: proceso que requiere energía y en el cual las proteínas de transporte se unen con partículas y las trasladan a través de la membrana celular. (p. 79)

adaptación: cualquier variación que haga que un organismo se adapte mejor a su medio ambiente. (p. 160)

alelo: forma alternativa que un gen puede tener para un rasgo único; puede ser dominante o recesivo. (p. 128)

reproducción asexual: tipo de reproducción –fisión, gemación y regeneración– en el que un organismo da origen a uno nuevo de ADN idéntico al organismo progenitor. (p. 103)

B

binomial nomenclature (bi NOH mee ul • NOH mun klay chur): two-word naming system that gives all organisms their scientific name. (p. 26)

biogenesis (bi oh JEH nuh sus): theory that living things come only from other living things. (p. 21)

nomenclatura binomial: sistema de denominación de dos palabras que da a todos los organismos su nombre científico. (p. 26)

biogénesis: teoría que sostiene que los seres vivos sólo provienen de otros seres vivos. (p. 21)

C

cell: smallest unit of an organism that can carry on life functions. (p. 16)

cell membrane: protective outer covering of all cells that regulates the interaction between the cell and the environment. (p. 40)

cell theory: states that all organisms are made up of one or more cells, the cell is the basic unit of life, and all cells come from other cells. (p. 53)

cell wall: rigid structure that encloses, supports, and protects the cells of plants, algae, fungi, and most bacteria. (p. 41)

chloroplast: green, chlorophyll-containing, plant-cell organelle that uses light energy to produce sugar from carbon dioxide and water. (p. 44)

chromosome: structure in a cell's nucleus that contains hereditary material. (p. 100)

control: standard to which the outcome of a test is compared. (p. 11)

cytoplasm: constantly moving gel-like mixture inside the cell membrane that contains heredity material and is the location of most of a cell's life processes. (p. 40)

célula: la unidad más pequeña de un organismo que puede continuar con sus funciones vitales. (p. 16)

membrana celular: capa externa protectora de todas las células y reguladora de la interacción entre la célula y su entorno. (p. 40)

teoría celular: establece que todos los organismos están formados por una o más células, que la célula es la unidad básica de la vida y que las células provienen de otras células. (p. 53)

pared celular: estructura rígida que envuelve, sostiene y protege a las células de las plantas, algas, hongos y de la mayoría de las bacterias. (p. 41)

cloroplasto: organelo de las células vegetales, de color verde, que contiene clorofila y que usa la luz solar para convertir el dióxido de carbono y el agua en azúcar. (p. 44)

cromosoma: estructura en el núcleo celular que contiene el material hereditario. (p. 100)

control: estándar contra el que se compara el resultado de una prueba. (p. 11)

citoplasma: mezcla parecida al gel y que está en constante movimiento dentro de la membrana celular, contiene material hereditario y es en donde tiene lugar la mayor parte de los procesos vitales de la célula. (p. 40)

D

diffusion: a type of passive transport in cells in which molecules move from areas where there are more of them to areas where there are fewer of them. (p. 77)

diploid (DIHP loyd): cell whose similar chromosomes occur in pairs. (p. 106)

DNA: deoxyribonucleic acid; the genetic material of all organisms; made up of two twisted strands of sugar-phosphate molecules and nitrogen bases. (p. 112)

dominant (DAH muh nunt): describes a trait that covers over, or dominates, another form of that trait. (p. 130)

difusión: tipo de transporte pasivo en las células en el que las moléculas se mueven de áreas de mayor concentración de éstas hacia áreas de menor concentración. (p. 77)

diploide: célula cuyos cromosomas similares están en pares. (p. 106)

ADN: ácido desoxirribonucleico; material genético de todos los organismos constituido por dos cadenas trenzadas de moléculas de azúcar-fosfato y bases de nitrógeno (p. 112)

dominante: describe un rasgo que encubre o domina a otra forma de ese rasgo. (p. 130)

E

egg: haploid sex cell formed in the female reproductive organs. (p. 106)

embryology (em bree AH luh jee): study of embryos and their development. (p. 169)

óvulo: célula sexual haploide que se forma en los órganos reproductivos femeninos. (p. 106)

embriología: el estudio de los embriones y su desarrollo. (p. 169)

endocytosis (en duh si TOH sus): process by which a cell takes in a substance by surrounding it with the cell membrane. (p. 80)

endoplasmic reticulum (ER): cytoplasmic organelle that moves materials around in a cell and is made up of a complex series of folded membranes; can be rough (with attached ribosomes) or smooth (without attached ribosomes). (p. 45)

enzyme: a type of protein that regulates nearly all chemical reactions in cells. (p. 73)

equilibrium: occurs when molecules of one substance are spread evenly throughout another substance. (p. 77)

evolution: change in inherited characteristics over time. (p. 156)

exocytosis (ek soh si TOH sus): process by which vesicles release their contents outside the cell. (p. 80)

endocitosis: proceso mediante el cual una célula capta una sustancia rodeándola con su membrana celular. (p. 80)

retículo endoplásmático (RE): organelo citoplasmático que transporta materiales dentro de una célula y está formado por una serie compleja de membranas plegadas; puede ser rugoso (con ribosomas adosados) o liso (sin ribosomas adosados). (p. 45)

enzima: tipo de proteína que regula casi todas las clases de reacciones químicas en las células. (p. 73)

equilibrio: ocurre cuando las moléculas de una sustancia están diseminadas completa y uniformemente a lo largo de otra sustancia. (p. 77)

evolución: cambio en las características heredadas a través del tiempo. (p. 156)

exocitosis: proceso mediante el cual las vesículas liberan su contenido fuera de la célula. (p. 80)

F

fermentation: process by which oxygen-lacking cells and some one-celled organisms release small amounts of energy from glucose molecules and produce wastes such as alcohol, carbon dioxide, and lactic acid. (p. 86)

fertilization: in sexual reproduction, the joining of a sperm and egg. (p. 106)

fermentación: proceso mediante el cual las células carentes de oxígeno y algunos organismos unicelulares liberan pequeñas cantidades de energía a partir de moléculas de glucosa y producen desechos como alcohol, dióxido de carbono y ácido láctico. (p. 86)

fertilización: en la reproducción sexual, la unión de un óvulo y un espermatozoide. (p. 106)

G

gene: section of DNA on a chromosome that contains instructions for making specific proteins. (p. 114)

genetic engineering: biological and chemical methods to change the arrangement of a gene's DNA to improve crop production, produce large volumes of medicine, and change how cells perform their normal functions. (p. 143)

genetics (juh NEH tihks): the study of how traits are inherited through the actions of alleles. (p. 128)

genotype (JEE nuh tipe): the genetic makeup of an organism. (p. 132)

genus: first word of the two-word scientific name used to identify a group of similar species. (p. 26)

Golgi bodies: organelles that package cellular materials and transport them within the cell or out of the cell. (p. 45)

gen: sección de ADN en un cromosoma, el cual contiene instrucciones para la formación de proteínas específicas. (p. 114)

ingeniería genética: métodos biológicos y químicos para cambiar la disposición del ADN de un gen y así mejorar la producción de cosechas, producir grandes volúmenes de un medicamento, o cambiar la forma en que las células realizan sus funciones normales. (p. 143)

genética: estudio de la forma como se heredan los rasgos a través de las acciones de los alelos. (p. 128)

genotipo: composición genética de un organismo. (p. 132)

género: primera palabra, de las dos palabras del nombre científico, que se usa para identificar a un grupo de especies similares. (p. 26)

aparato de Golgi: organelo que concentra los materiales celulares y los transporta hacia adentro o afuera de la célula. (p. 45)

Glossary/Glosario

gradualism: model describing evolution as a slow process by which one species changes into a new species through a continuing series of mutations and variations over time. (p. 162)

gradualismo: modelo que describe la evolución como un proceso lento mediante el cual una especie existente se convierte en una especie nueva a través de series continuas de mutaciones y variaciones a través del tiempo. (p. 162)

H

haploid (HAP loyd): cell that has half the number of chromosomes as body cells. (p. 107)

heredity (huh REH duh tee): the passing of traits from parent to offspring. (p. 128)

heterozygous (he tuh roh ZI gus): describes an organism with two different alleles for a trait. (p. 132)

homeostasis: regulation of an organism's internal, life-maintaining conditions. (p. 17)

hominid: humanlike primate that appeared about 4 million to 6 million years ago, ate both plants and meat, and walked upright on two legs. (p. 173)

Homo sapiens: early humans that likely evolved from Cro-Magnons. (p. 175)

homologous (huh MAH luh gus): body parts that are similar in structure and origin and can be similar in function. (p. 170)

homozygous (hoh muh ZI gus): describes an organism with two alleles that are the same for a trait. (p. 132)

host cell: living cell in which a virus can actively multiply or in which a virus can hide until activated by environmental stimuli. (p. 54)

hybrid (HI brud): an offspring that was given different genetic information for a trait from each parent. (p. 130)

hypothesis: prediction that can be tested. (p. 10)

haploide: célula que posee la mitad del número de cromosomas que tienen las células somáticas. (p. 107)

herencia: transferencia de rasgos de un progenitor a su descendencia. (p. 128)

heterocigoto: describe a un organismo con dos alelos diferentes para un rasgo. (p. 132)

homeostasis: control de las condiciones internas que mantienen la vida de un organismo. (p. 17)

homínido: primate con forma de humano que apareció entre 4 y 6 millones de años atrás, se alimentaba de plantas y carne, y caminaba erguido sobre sus dos pies. (p. 173)

Homo sapiens: humanos primitivos que probablemente evolucionaron a partir de los CroMagnon. (p. 175)

homólogos: partes del cuerpo que son similares en estructura y origen y que pueden tener funciones similares. (p. 170)

homocigoto: describe a un organismo con dos alelos iguales para un rasgo. (p. 132)

célula huésped: célula viva en la que un virus puede reproducirse activamente o en la que un virus puede ocultarse hasta que es activado por estímulos del medio ambiente. (p. 54)

híbrido: un descendiente que recibe de cada progenitor información genética diferente para un rasgo. (p. 130)

hipótesis: predicción que puede probarse. (p. 10)

I

incomplete dominance: production of a phenotype that is intermediate between the two homozygous parents. (p. 136)

inorganic compound: compound, such as H_2O, that is made from elements other than carbon and whose atoms usually can be arranged in only one structure. (p. 73)

dominancia incompleta: producción de un fenotipo intermedio entre dos progenitores homocigotos. (p. 136)

compuesto inorgánico: compuesto, como H_2O, formado por elementos distintos al carbono y cuyos átomos generalmente pueden estar organizados en sólo una estructura. (p. 73)

K

kingdom: first and largest category used to classify organisms. (p. 25)

reino: la primera y más grande categoría utilizada para clasificar a los organismos. (p. 25)

L

law: statement about how things work in nature that seems to be true consistently. (p. 12)

ley: enunciado acerca de cómo funciona todo en la naturaleza y que constantemente parece ser verdadero. (p. 12)

M

meiosis (mi OH sus): reproductive process that produces four haploid sex cells from one diploid cell and ensures offspring will have the same number of chromosomes as the parent organisms. (p. 107)

metabolism: the total of all chemical reactions in an organism. (p. 83)

mitochondrion: cell organelle that breaks down food and releases energy. (p. 44)

mitosis (mi TOH sus): cell process in which the nucleus divides to form two nuclei identical to each other, and identical to the original nucleus, in a series of steps (prophase, metaphase, anaphase, and telophase). (p. 100)

mixture: a combination of substances in which the individual substances do not change or combine chemically but instead retain their own individual properties; can be gases, solids, liquids, or any combination of them. (p. 71)

mutation: any permanent change in a gene or chromosome of a cell; may be beneficial, harmful, or have little effect on an organism. (p. 116)

meiosis: proceso reproductivo que produce cuatro células sexuales haploides a partir de una célula diploide y asegura que la descendencia tendrá el mismo número de cromosomas que los organismos progenitores. (p. 107)

metabolismo: el conjunto de todas las reacciones químicas en un organismo. (p. 83)

mitocondria: organelo celular que degrada nutrientes y libera energía. (p. 44)

mitosis: proceso celular en el que el núcleo se divide para formar dos núcleos idénticos entre sí e idénticos al núcleo original, a través de varias etapas (profase, metafase, anafase y telofase). (p. 100)

mezcla: una combinación de sustancias en la que las sustancias individuales no cambian ni se combinan químicamente pero mantienen sus propiedades individuales; pueden ser gases, sólidos, líquidos o una combinación de ellos. (p. 71)

mutación: cualquier cambio permanente en un gen o cromosoma de una célula; puede ser benéfica, perjudicial o tener un pequeño efecto sobre un organismo. (p. 116)

N

natural selection: a process by which organisms with traits best suited to their environment are more likely to survive and reproduce; includes concepts of variation, overproduction, and competition. (p. 158)

nucleus: organelle that controls all the activities of a cell and contains hereditary material made of proteins and DNA. (p. 42)

selección natural: proceso mediante el cual los organismos con rasgos mejor adaptados a su ambiente tienen mayor probabilidad de sobrevivir y reproducirse; incluye los conceptos de variación, sobreproducción y competencia. (p. 158)

núcleo: organelo que controla todas las actividades de una célula y que contiene el material hereditario formado por proteínas y ADN. (p. 42)

organ: structure, such as the heart, made up of different types of tissues that all work together. (p. 47)

órgano: estructura, como el corazón, que consiste en diferentes tipos de tejidos que trabajan conjuntamente. (p. 47)

organelle: structure in the cytoplasm of a eukaryotic cell that can act as a storage site, process energy, move materials, or manufacture substances. (p. 42)

organelo: estructura del citoplasma de una célula eucariota que puede actuar como sitio de almacenamiento, procesamiento de energía, movimiento de materiales o elaboración de sustancias. (p. 42)

organic compounds: compounds that always contain hydrogen and carbon; carbohydrates, lipids, proteins, and nucleic acids are organic compounds found in living things. (p. 72)

compuestos orgánicos: compuestos que siempre contienen hidrógeno y carbono; los carbohidratos, lípidos, proteínas y ácidos nucleicos son compuestos orgánicos que se encuentran en los seres vivos. (p. 72)

organism: any living thing. (p. 16)

organismo: cualquier ser vivo. (p. 16)

osmosis: a type of passive transport that occurs when water diffuses through a cell membrane. (p. 78)

ósmosis: tipo de transporte pasivo que ocurre cuando el agua se difunde a través de una membrana celular. (p. 78)

P

passive transport: movement of substances through a cell membrane without the use of cellular energy; includes diffusion, osmosis, and facilitated diffusion. (p. 76)

transporte pasivo: movimiento de sustancias a través de la membrana celular sin usar energía celular; incluye difusión, ósmosis y difusión facilitada. (p. 76)

phenotype (FEE nuh tipe): outward physical appearance and behavior of an organism as a result of its genotype. (p. 132)

fenotipo: apariencia física externa y comportamiento de un organismo como resultado de su genotipo. (p. 132)

photosynthesis: process by which plants and many other producers use light energy to produce a simple sugar from carbon dioxide and water and give off oxygen. (p. 84)

fotosíntesis: proceso mediante el cual las plantas y muchos otros organismos productores usan la energía solar para producir azúcares simples a partir de dióxido de carbono y agua y desprender oxígeno. (p. 84)

phylogeny (fi LAH juh nee): evolutionary history of an organism; used today to group organisms into six kingdoms. (p. 25)

filogenia: historia evolutiva de un organismo; usada hoy para agrupar a los organismos en seis reinos. (p. 25)

polygenic (pah lih JEH nihk) inheritance: occurs when a group of gene pairs acts together and produces a specific trait, such as human eye color, skin color, or height. (p. 138)

herencia poligénica: ocurre cuando un grupo de pares de genes actúa conjuntamente y produce un rasgo específico, tal como el color de los ojos , el color de la piel, o la estatura en los humanos. (p. 138)

primates: group of mammals including humans, monkeys, and apes that share characteristics such as opposable thumbs, binocular vision, and flexible shoulders. (p. 172)

primates: grupo de mamíferos que incluye a los humanos, monos y simios, los cuales comparten características como pulgares opuestos, visión binocular y hombros flexibles. (p. 172)

punctuated equilibrium: model describing the rapid evolution that occurs when mutation of a few genes results in a species suddenly changing into a new species. (p. 162)

equilibrio punteado: modelo que describe la evolución rápida que ocurre cuando la mutación de unos pocos genes resulta en que una especie cambie rápidamente para convertirse en otra especie. (p. 162)

Punnett (PUH nut) square: a tool to predict the probability of certain traits in offspring that shows the different ways alleles can combine. (p. 132)

Cuadrado de Punnett: herramienta para predecir la probabilidad de ciertos rasgos en la descendencia mostrando las diferentes formas en que los alelos pueden combinarse. (p. 132)

R

radioactive element: element that gives off a steady amount of radiation as it slowly changes to a nonradioactive element. (p. 167)

elemento radiactivo: elemento que emite una cantidad estable de radiación mientras se convierte lentamente en un elemento no radiactivo. (p. 167)

recessive (rih SE sihv): describes a trait that is covered over, or dominated, by another form of that trait and seems to disappear. (p. 130)

recesivo: describe un rasgo que está encubierto, o que es dominado, por otra forma del mismo rasgo y que parece no estar presente. (p. 130)

respiration: process by which producers and consumers release stored energy from food molecules. (p. 85)

respiración: proceso mediante el cual los organismos productores y consumidores liberan la energía almacenada en las moléculas de los alimentos. (p. 85)

ribosome: small cytoplasmic structure on which cells make their own proteins. (p. 44)

ribosoma: estructura citoplasmática pequeña en la que las células producen sus propias proteínas. (p. 44)

RNA: ribonucleic acid; a type of nucleic acid that carries codes for making proteins from the nucleus to the ribosomes. (p. 114)

ARN: ácido ribonucleico; tipo de ácido nucleico que transporta los códigos para la formación de proteínas del núcleo a los ribosomas. (p. 114)

S

scientific methods: procedures used to solve problems and answer questions that can include stating the problem, gathering information, forming a hypothesis, testing the hypothesis with an experiment, analyzing data, and drawing conclusions. (p. 9)

métodos científicos: procedimientos utilizados para solucionar problemas y responder a preguntas; puede incluir el establecimiento de un problema, recopilación de información, formulación de una hipótesis, comprobación de la hipótesis con un experimento, análisis de la información y presentación de conclusiones. (p. 9)

sedimentary rock: a type of rock, such as limestone, that is most likely to contain fossils and is formed when layers of sand, silt, clay, or mud are cemented and compacted together or when minerals are deposited from a solution. (p. 166)

roca sedimentaria: tipo de roca, como la piedra caliza, con alta probabilidad de contener fósiles y que se forma cuando las capas de arena, sedimento, arcilla o lodo son cementadas y compactadas o cuando los minerales de una solución son depositados. (p. 166)

sex-linked gene: an allele inherited on a sex chromosome and that can cause human genetic disorders such as color blindness and hemophilia. (p. 141)

gen ligado al sexo: un alelo heredado en un cromosoma sexual y que puede causar desórdenes genéticos humanos como daltonismo y hemofilia. (p. 141)

sexual reproduction: a type of reproduction in which two sex cells, usually an egg and a sperm, join to form a zygote, which will develop into a new organism with a unique identity. (p. 106)

reproducción sexual: tipo de reproducción en la que dos células sexuales, generalmente un óvulo y un espermatozoide, se unen para formar un zigoto, el cual se desarrollará para formar un nuevo organismo con identidad única. (p. 106)

species: group of organisms that share similar characteristics and can reproduce among themselves producing fertile offspring. (p. 156)

especie: grupo de organismos que comparten características similares entre sí y que pueden reproducirse entre ellos dando lugar a una descendencia fértil. (p. 156)

sperm: haploid sex cell formed in the male reproductive organs. (p. 106)

spontaneous generation: idea that living things come from nonliving things. (p. 21)

espermatozoides: células sexuales haploides que se forman en los órganos reproductores masculinos. (p. 106)

generación espontánea: idea que sostiene que los seres vivos proceden de seres inertes. (p. 21)

T

theory: explanation of things or events based on scientific knowledge resulting from many observations and experiments. (p. 12)

tissue: group of similar cells that work together to do one job. (p. 47)

teoría: explicación de cosas o eventos basándose en el conocimiento científico resultante de muchas observaciones y experimentos. (p. 12)

tejido: grupo de células similares que trabajan conjuntamente para hacer una tarea. (p. 47)

V

variable: something in an experiment that can change. (p. 11)

variation: inherited trait that makes an individual different from other members of the same species and results from a mutation in the organism's genes. (p. 160)

vestigial (veh STIH jee ul) structure: structure, such as the human appendix, that doesn't seem to have a function and may once have functioned in the body of an ancestor. (p. 170)

virus: a strand of hereditary material surrounded by a protein coating. (p. 54)

variable: condición que puede cambiar en un experimento. (p. 11)

variación: rasgo heredado que hace que un individuo sea diferente a otros miembros de su misma especie como resultado de una mutación de sus genes. (p. 160)

estructura vestigial: estructura, como el apéndice humano, que no parece tener alguna función pero que pudo haber funcionado en el cuerpo de un antepasado. (p. 170)

virus: cadena de material hereditario rodeada por una membrana proteica. (p. 54)

Z

zygote: new diploid cell formed when a sperm fertilizes an egg; will divide by mitosis and develop into a new organism. (p. 106)

zigoto: célula diploide nueva formada cuando un espermatozoide fertiliza a un óvulo; se dividirá por mitosis y se desarrollará para formar un nuevo organismo. (p. 106)

Italic numbers = illustration/photo **Bold numbers = vocabulary term**
lab = indicates a page on which the entry is used in a lab
act = indicates a page on which the entry is used in an activity

A

B

C

Index

Index

Index

M

N

Index

Magnification Key: Magnifications listed are the magnifications at which images were originally photographed.
LM–Light Microscope
SEM–Scanning Electron Microscope
TEM–Transmission Electron Microscope

Acknowledgments: Glencoe would like to acknowledge the artists and agencies who participated in illustrating this program: Absolute Science Illustration; Andrew Evansen; Argosy; Articulate Graphics; Craig Attebery represented by Frank & Jeff Lavaty; CHK America; John Edwards and Associates; Gagliano Graphics; Pedro Julio Gonzalez represented by Melissa Turk & The Artist Network; Robert Hynes represented by Mendola Ltd.; Morgan Cain & Associates; JTH Illustration; Laurie O'Keefe; Matthew Pippin represented by Beranbaum Artist's Representative; Precision Graphics; Publisher's Art; Rolin Graphics, Inc.; Wendy Smith represented by Melissa Turk & The Artist Network; Kevin Torline represented by Berendsen and Associates, Inc.; WILDlife ART; Phil Wilson represented by Cliff Knecht Artist Representative; Zoo Botanica.

Photo Credits

Cover Andrew Syred/Science Photo Library/Photo Researchers; **i ii** Andrew Syred/Science Photo Library/Photo Researchers; **iv** (bkgd)John Evans, (inset)Andrew Syred/Science Photo Library/Photo Researchers; **v** (t)PhotoDisc, (b)John Evans; **vi** (l)John Evans, (r)Geoff Butler; **vii** (l)John Evans, (r)PhotoDisc; **viii** PhotoDisc; **ix** Aaron Haupt Photography; **x** Dave B. Fleetham/Tom Stack & Assoc.; **1** Glencoe photo; **2–3** Bob Jacobson/International Stock; **3** Richard Hutchings/PhotoEdit; **4** courtesy of ABI PRISM; **5** (t)Dominic Oldershaw, (c)Dr. John Carpten, (b)Aaron Haupt; **6–7** A. Witte/C. Mahaney/Getty Images; **8** Kjell B. Sandved/Visuals Unlimited; **10 11** Mark Burnett; **13** Tek Image/Science Photo Library/Photo Researchers; **14 15** Mark Burnett; **16** (t)Michael Abbey/Science Source/Photo Researchers, (bl)Aaron Haupt, (br)Michael Delannoy/Visuals Unlimited; **17** Mark Burnett; **18** (tcr)A. Glauberman/Photo Researchers, (tr)Mark Burnett, (bl bcl br)Runk/Schoenberger from Grant Heilman, (others)Dwight Kuhn; **19** (t)Bill Beaty/Animals Animals, (bl)Tom & Therisa Stack/Tom Stack & Assoc., (br)Michael Fogden/Earth Scenes; **20** Aaron Haupt; **21** Geoff Butler; **24** (t)Arthur C. Smith III From Grant Heilman, (bl)Hal Beral/Visuals Unlimited, (br)Larry L. Miller/Photo Researchers; **25** Doug Perrine/Innerspace Visions; **26** (l)Brandon D. Cole, (r)Gregory Ochocki/Photo Researchers; **27** (l)Zig Leszczynski/Animals Animals, (r)R. Andrew Odum/Peter Arnold, Inc.; **28** Alvin E. Staffan; **29** Geoff Butler; **30** (t)Jan Hinsch/Science Photo Library/Photo Researchers, (b)Mark Burnett; **31** Mark Burnett; **32** Marc Von Roosmalen/AP; **33** (l)Mark Burnet, (r)Will & Deni McIntyre/Photo Researchers; **34** Hal Beral/Visuals Unlimited; **35** Jeff Greenberg/Rainbow; **36** Dwight Kuhn; **37** Dave Spier/Visuals Unlimited; **38–39** Nancy Kedersha/Science Photo Library/Photo Researchers; **41** David M. Phillips/Visuals Unlimited; **42** (t)Don Fawcett/Photo Researchers, (b)M. Schliwa/Visuals Unlimited; **44** (t)George B. Chapman/Visuals Unlimited, (b)P. Motta & T. Naguro/Science Photo Library/Photo Researchers; **45** (t)Don Fawcett/Photo Researchers, (b)Biophoto Associates/Photo Researchers; **49** (l)Biophoto Associates/Photo Researchers, (r)Matt Meadows; **50–51** (bkgd)David M. Phillips/Visuals Unlimited; **50** (cw from top)Kathy Talaro/Visuals Unlimited, Michael Abbey/Visuals Unlimited, Michael Gabridge/Visuals Unlimited, David M. Phillips/Visuals Unlimited, David M. Phillips/Visuals Unlimited, courtesy Nikon Instruments Inc.; **51** (tl)Michael Abbey/Visuals Unlimited, (tr)Bob Krist/CORBIS, (cl)courtesy Olympus Corporation, (cr)James W. Evarts, (bl)Karl Aufderheide/Visuals Unlimited, (br)Lawrence Migdale/Stock Boston/PictureQuest; **54** (l)Richard J. Green/Photo Researchers, (c)Dr. J.F.J.M. van der Heuvel, (r)Gelderblom/Eye of Science/Photo Researchers; **57** Pam Wilson/Texas Dept. of Health; **58 59** Matt Meadows; **60** (t)Quest/Science Photo Library/Photo (b)courtesy California University; **61** (l)Keith Porter/Photo Researchers, (r)NIBSC/Science Photo Library/Photo Researchers; **63** Biophoto Associates/Science Source/Photo Researchers; **64** P. Motta & T. Naguro/Science Photo Library/Photo Researchers; **66–67** Jane Grushow/Grant Heilman Photography; **69** Bob Daemmrich; **71** (t)Runk/Schoenberger from Grant Heilman, (b)Klaus Guldbrandsen/Science Photo Library/Photo Researchers; **76** (l)John Fowler, (r)Richard Hamilton Smith/CORBIS; **77** KS Studios; **78** Aaron Haupt; **79** Visuals Unlimited; **80** Biophoto Associates/Science Source/Photo Researchers; **82** Matt Meadows; **84** Craig Lovell/CORBIS; **85** John Fowler; **86** David M. Phillips/Visuals Unlimited; **87** (l)Grant Heilman Photography, (r)Bios (Klein/Hubert)/Peter Arnold; **88** (t)Runk/Schoenberger from Grant Heilman, (b)Matt Meadows; **89** Matt Meadows; **90** Lappa/Marquart; **91** CNRI/Science Photo Library/Photo Researchers; **92** Biophoto Associates/Science Source/Photo Researchers; **96–97** Zig Leszcynski/Animals Animals; **98** (l)Dave B. Fleetham/Tom Stack & Assoc., (r)Cabisco/Visuals Unlimited; **100** Cabisco/Visuals Unlimited; **101** (tl)Michael Abbey/Visuals Unlimited, (others)John D. Cunningham/Visuals Unlimited; **102** (l)Matt Meadows, (r)Nigel Cattlin/Photo Researchers; **103** (l)Barry L. Runk from Grant Heilman, (r)Runk/Schoenberger from Grant Heilman; **104** (l)Walker England/Photo Researchers, (r)Tom Stack & Assoc.; **105** Runk/Schoenberger from Grant Heilman; **106** Dr. Dennis Kunkel/PhotoTake NYC; **107** (tl)Gerald & Buff Corsi/Visuals Unlimited, (bl)Susan McCartney/Photo Researchers, (r)Fred Bruenner/Peter Arnold, Inc.; **109** (l)John D. Cunningham/Visuals Unlimited, (c)Jen & Des Bartlett/Bruce Coleman, Inc., (r)Breck P. Kent; **110** (tl)Artville, (tr)Tim Fehr, (c)Bob Daemmrich/Stock Boston/PictureQuest, (bl)Troy Mary Parlee/Index Stock/PictureQuest, (br)Jeffery Myers/Southern Stock/PictureQuest; **116** Stewart Cohen/Stone/Getty Images; **118** (t)Tom McHugh/Photo Researchers, (b)file photo; **119** Monica Dalmasso/Stone/Getty Images; **120** (t)Philip Lee Harvey/Stone, (b)Lester V. Bergman/CORBIS; **122** Walker England/Photo Researchers; **124** Barry L. Runk from Grant Heilman; **125** Cabisco/Visuals Unlimited; **126–127** Ashley Cooper/CORBIS; **127** Geoff Butler; **128** Stewart Cohen/Stone/Getty Images; **131** (bkgd)Jane Grushow from Grant Heilman, (others)Special Collections, National Agriculture Library; **132** Barry L. Runk From Grant Heilman; **134** Richard Hutchings/Photo Researchers; **136** (l)Grant Heilman Photography, (r)Gemma Giannini from Grant Heilman; **137** Raymond Gehman/CORBIS; **138** Dan McCoy from Rainbow; **139** (l)Phil Roach/Ipol, Inc., (r)CNRI/Science Photo Library/Photo Researchers; **140** Gopal Murti/PhotoTake, NYC; **141** Tim Davis/Photo Researchers; **142** (t)Renee Stockdale/Animals Animals, (b)Alan & Sandy

Credits

Carey/Photo Researchers; **145** Tom Meyers/Photo Researchers; **146** (t)Runk/Schoenberger from Grant Heilman, (b)Mark Burnett; **147** Richard Hutchings; **148** KS Studios; **153** CNRI/Science Photo Library/Photo Researchers; **154–155** B.G. Thomson/Photo Researchers; **157** Barbera Cushman/DRK Photo; **158** (l c)Tui De Roy/Minden Pictures, (r)Tim Davis/Photo Researchers; **160** (l)Gregory G. Dimijian, M.D./Photo Researchers, (r)Patti Murray/Animals Animals; **161** (l)George McCarthy/CORBIS, (r)Arthur Morris/Visuals Unlimited; **162** (l)Joe McDonald/Animals Animals, (c)Tom McHugh/Photo Researchers, (r)Tim Davis/Photo Researchers; **163** James Richardson/Visuals Unlimited; **164** Frans Lanting/Minden Pictures; **165** (l)Dominique Braud/Earth Scenes, (c)Carr Clifton/Minden Pictures, (r)John Cancalosi/DRK Photo; **166** (t)Larry Ulrich/DRK Photo, (cl)Ken Lucas/Visuals Unlimited, (cr)John Cancalosi/Peter Arnold, Inc., (bl)John Cancalosi/DRK Photo, (br)Sinclair Stammers/Science Photo Library/Photo Researchers; **167** John Kieffer/Peter Arnold, Inc.; **170** (l)Kees Van Den Berg/Photo Researchers, (r)Doug Martin; **171** Peter Veit/DRK Photo; **172** (l)Mark E. Gibson, (r)Gerard Lacz/Animals Animals; **173** (l)Michael Dick/Animals Animals, (r)Carolyn A. McKeone/Photo Researchers; **174** (l)John Reader/Science Photo Library/Photo Researchers, (r)Archivo Iconografico, S.A./CORBIS; **175** Francois Ducasse/Rapho/Photo Researchers; **176–177** Aaron Haupt; **176** Kenneth W. Fink/Photo Researchers; **178** (t)Oliver Meckes/E.O.S/Gelderblom/Photo Researchers, (b)Lara Jo Regan/Saba; **179** (l)Koster-Survival/Animals Animals, (r)Stephen J. Keasemann/DRK Photo; **181** Tom Tietz/Stone/Getty Images; **182** Gregory G. Dimijian, M.D./Photo Researchers; **183** Patti Murray/Animals Animals; **184** PhotoDisc; **186** Tom Pantages; **190** Michell D. Bridwell/PhotoEdit, Inc.; **191** (t)Mark Burnett, (b)Dominic Oldershaw; **192** StudiOhio; **193** Timothy Fuller; **194** Aaron Haupt; **196** KS Studios; **197** Matt Meadows; **198 199** Aaron Haupt; **200** Sinclair Stammers/Science Photo Library/Photo Researchers; **201** Amanita Pictures; **202** Bob Daemmrich; **204** Davis Barber/PhotoEdit, Inc.; **222** Matt Meadows; **223** (l)Dr. Richard Kessel, (c)NIBSC/Science Photo Library/Photo Researchers, (r)David John/Visuals Unlimited; **224** (t)Runk/Schoenberger from Grant Heilman, (bl)Andrew Syred/Science Photo Library/Photo Researchers, (br)Rich Brommer; **225** (tr)G.R. Roberts, (l)Ralph Reinhold/Earth Scenes, (br)Scott Johnson/Animals Animals; **226** Martin Harvey/DRK Photo.